PLUGIN
Publishing company

Architecture Material Sourcebook

Copyright by Plugin Publishing Company

All rights reserved. No part of this book may be reproduced in any form without written permission of the copyright owners. All images in this book have been reproduced with the knowledge and prior consent of the artists concerned and no responsibility is accepted by producer publisher or printer for any infringement of copyright or otherwise, Arising from the contents of the publication.

- Publication: Plugin Publishing Company
- Publisher: Sukam, Im
- Writer: Sukjwa, Park
- Printing: Abcds Co., Ltd.
- Address: 502 / 9, Hugok-ro, Ilsan 3(Sam)-dong, Ilsanseo-gu, Goyang-si, Gyeonggi-do, South Korea.
- Tel(Fax): +82-31-907-1772 (+82-31-907-1773)
- E-mail: 2twodesign@naver.com

₩97,000WON

Introduction

In this book the trend of concrete development at present and future has been espounded with a far-sighted strategy, that is reasonably utilizing the resource, protecting the natural environment, which human existance relies on, and realizing the sustainable development.

The word 'architecture' contains numerous ideas, philosophical thought and comprehensive meaning. In the contemporary civilization, people encounter meaning and reality of architecture, birth and rebirth of diverse and complicated phenomenon, efforts and concept of subject of architecture and other diverse architectural beauty and conveniences. And the word also has other tiny elements. It includes room, garden, stair, roof, window and other internal and external elements. Thus, mountain, lake, forest, grass, trees, animals, stones and water are included. It is also used as art which includes crafts, sculpture and photo. Therefore, architecture itself has so diverse and huge meaning. Reality of architecture is experienced through our senses.

Then, how do we accept and understand architecture? Architecture includes what we experience, feel, listen to, and smell, and also includes geometric result of volume, various architectural materials forming surface, building decoration, color, texture, connection and transparent.

Architecture Material Sourcebook

Therefore, it may mean everything of building. Second issue is on who we are? What we act, experience and continue is architecture. It is also to see other's doing. As mentioned already, architecture is related to almost all processes in our life.

Architecture can be done in any space or place. It is formed in a related space with other space, adjacent exterior space or corresponding space. And, the space has a topological relation instead of geometric relation. And, people create continual experiences like moving or entering through architecture.

No matter how expensive house you buy and how expensive furniture you put, it does not make your life beautiful or your place better. Proper time and history are required to obtain what we want. And, it must be produced and nurtured by ourselves. Architecture is to create the most important and fundamental stage among these elements. That is also desperately necessary to keep internal stability as an individual or a healthy member of a society.

To contemporary people, it is important for experts who design and develop concept of architecture as a space to make a design of visual scenery in front of them and also environments for their eyes. Seaside, where we often go, sandy beach and far away port light are visual outputs that are embodied through architecture and the stage of life we expected. Accordingly, architecture will continue as long as we live in this world. As always, architecture is something we create at all times and we experience harmony with various new materials. And, almost all life events, activities and outputs are embodied in the stage of architecture. Architecture contains all we produce and creates and maintains culture.

Therefore, I am quite sure that the understanding of new created architectures such as material, technique and spatial formative completion of all architectures in this book could contribute to the perfection of architecture to be born in the future.

Sukjwa, Park

Architectural material of architectures in this book are divided and indicated in the following pictogram as architectural materials are usually overlapped in their usage.

Architecture **Material** Sourcebook

C	Material Type - **Concrete**
B	Material Type - **Brick**
G	Material Type - **Glass**
M	Material Type - **Steel & Metal**
W	Material Type - **Wood**
S	Material Type - **Stone**
P	Material Type - **Plastic**

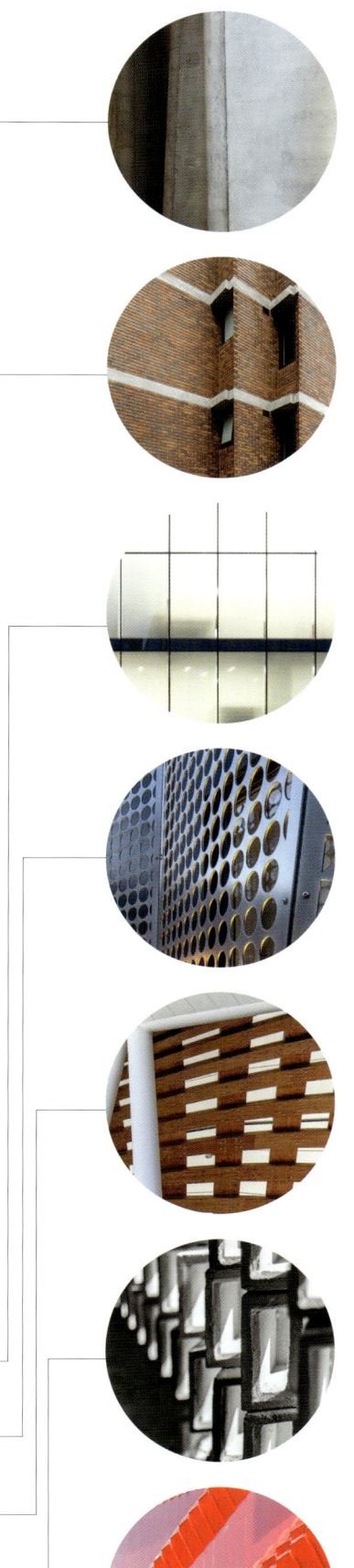

Architecture Material Sourcebook

table of contents

008~019 **Regional Music and Dance Conservatory** Boris Bouchet Architectes
020~031 **B&E Goulandris Museum of Contemporary Art** Vikelas Architects
032~039 **Pharmacy on Dzirciema Street** Substance Architects
040~049 **Benjamin Carrion House Renovation** Bernardo Bustamante Arquitectos
050~057 **The Reborn House** Alhumaidhi Architects
058~065 **The Odori Hotel** Alhumaidhi Architects
066~075 **Tortosa Law Courts** Camps Felip Arquitecturia
076~087 **Viewpoint House** Jim Caumeron Design
088~099 **FTE Office Building** Plan Architect
100~113 **PG&E Embarcadero & Potrero Building** Stanley Saitowitz | Natoma Architects
114~123 **Greenhouse Orchid Punta del Este** Mateo Nunes Da Rosa
124~135 **Rural Hotel in an Olive Grove** GANA Arquitectura
136~143 **Agency Giboire Morbihan Offices** a/LTA Architects
144~153 **The Centr'Al Building** B-architecten
154~163 **The Framed House** Crest Architecture
164~173 **Travertine walls enclose Casa ZTG in Guadalajara** 1540 Arquitectura
174~181 **Renovation of No.1 Sinopec Gas Station** TJAD Original
182~193 **Kaolin Court Housing** Stolon Studio Ltd. + Baca Architects

194~211	**The Sao Felix da Marinha House** Raulino Silva	294~305	**Shenzhen Pingshan Art Museum** Vector Architects
212~223	**The House No.6** Sara Kalantary + Reza Sayadiyan	306~319	**Es Pou House in Formentera** Marià Castelló
224~231	**The Casa da Mole** Marchetti Bonetti	320~327	**The Building of flats Sucre 812** Ana Smud + Alberto Smud
232~245	**The Antoine de Ruffi School** TAUTEM Architecture + bmc2 architectes	328~339	**Dong Phuong Y Dao Medical Center** Landmak Architecture
246~253	**The 55 Southbank Boulevard** Bates Smart	340~353	**The Biosphere** Chain10 Architecture & Interior Design
254~263	**The White Canvas House** ACA Architects	354~367	**Sala Samui Chaweng Beach Resort, Phase 02** onion Architecture
264~275	**SMOORE Liutang Industrial Park Shenzhen** CM Design	368~379	**Grand Palais Cinema** Antonio Virga Architecte
276~285	**Runxuan Textile Office Building** Masanori Design Studio	380~383	**Cima Morelos Complex** Minuz Workshop + Miguel Montor Architecture
286~293	**Landscape House** FORM	Kouichi Kimura Architects		

Regional Music and Dance Conservatory

Architects: Boris Bouchet Architectes, Studio 1984
Location: Le Pradet, France
Area: 953 m²
Photographs: Benoit Alazard

Manufacturers: Carrières de Provence
Engineering: Calder Ingénierie
Main Material: Estaillade stone

The Conservatory of Music is at the heart of a much broader public project. It serves as a catalyst for a refined urban lifestyle, breathing new life into the entire public pedestrian space. The use of massive stone in the construction is a striking environmental performance. With little transformation at a nearby source, it makes for a low carbon footprint. The site proposed by the Pradet City Council is the plot of the former Jean Jaures school, comfortably nested in the historic downtown neighbourhood. This policy choice offers an alternative to the automobile-dominated lifestyle that would have otherwise dictated a location at the entrance of the city with abundant parking. The challenge of installing equipment of this scale within the residential fabric of the heart of a seaside village is a major consideration in this project.

How then is architecture capable of bearing the cultural and social graft between the users hailing from the entire French metropolitan territory and the inhabitants concerned about the transformation of their environment? The first choice is the superposition of the programme over 3 levels. Space thus freed makes it possible to keep the big trees and create a central void, conceived as a third amenity between the media centre and the conservatory, nurturing a breath of fresh air in the dense and private fabric of the neighbourhood. In its relationship with the site, the conservatory offers the first response to this contradictory demand that consists of creating a facility that is both a bold symbol of public renewal and a discrete element of the historic residential fabric.

Material Type: **Stone**
(Estaillade stone)

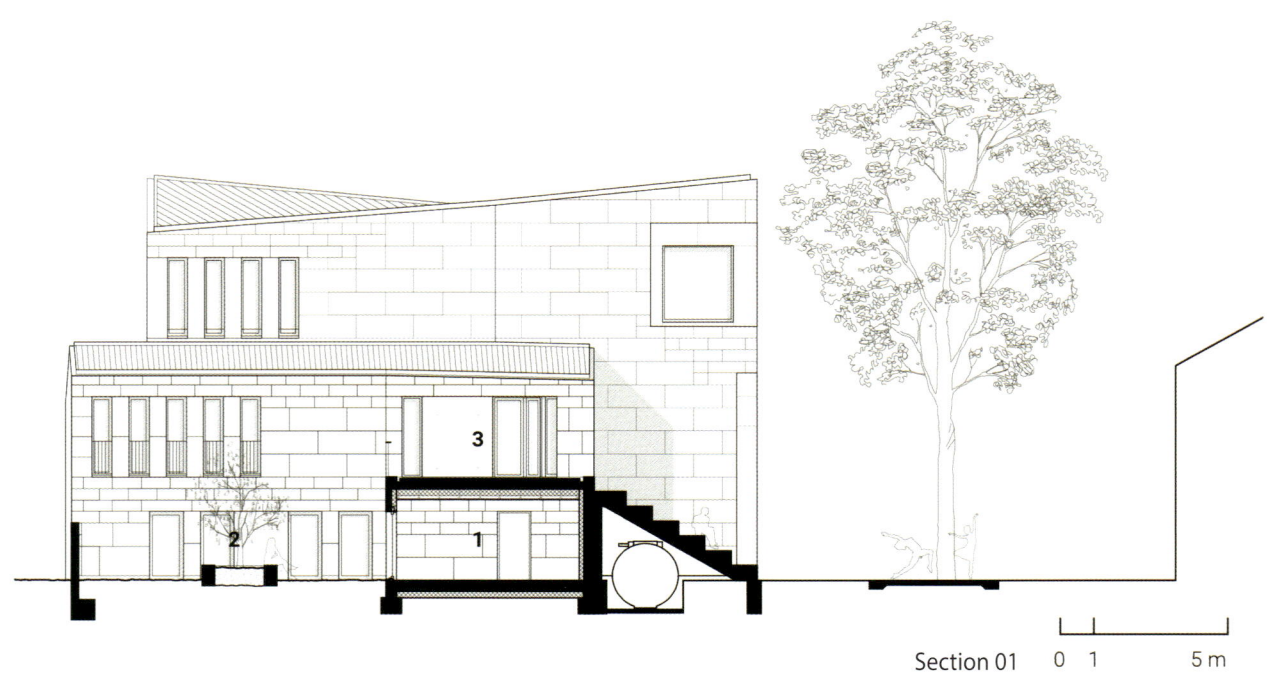

Section 01   0 1   5 m

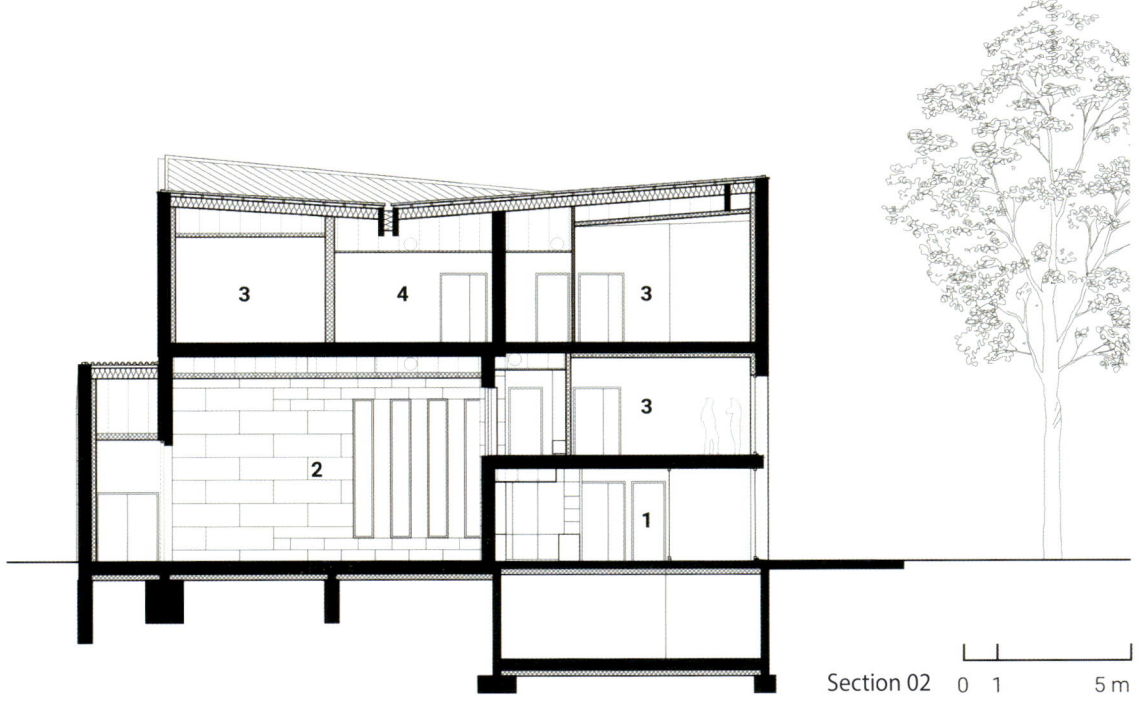

Section 02 0 1 5 m

Material Type: **Concrete**
(Concrete Hardening – Hard-Cem)

Concrete Hardening – Hard-Cem

To help maximize the durability of a design and have it remain sustainable, Hard-Cem provides construction professionals with an award-winning integral hardener.

Added to a concrete mix directly, this admixture uses proprietary technology to increase the hardness of the concrete paste and reduce fine and coarse aggregate exposure. That in turn increases the concrete's ability to resist degradation and wear from erosion and abrasion, making it capable of lasting twice as long as regular concrete in harsh environments. It does all this without the need for time-consuming surface applications like liquid hardeners or health hazardous applications like dry shake hardeners.

It's a hardener free of volatile organic compounds that can be applied to any type of concrete slab, whether it's vertical, horizontal, or entirely unique. So design teams and builders can have the optimal durability and sustainability their project needs.

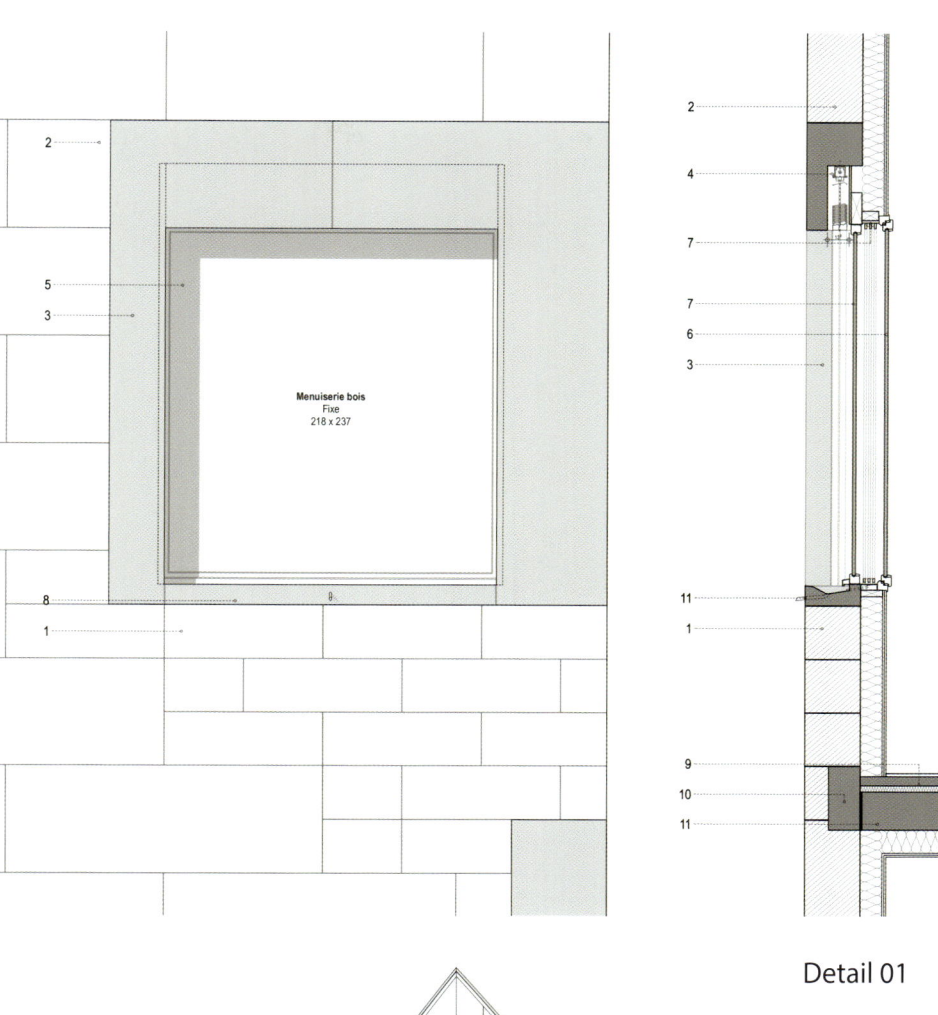

Detail 01

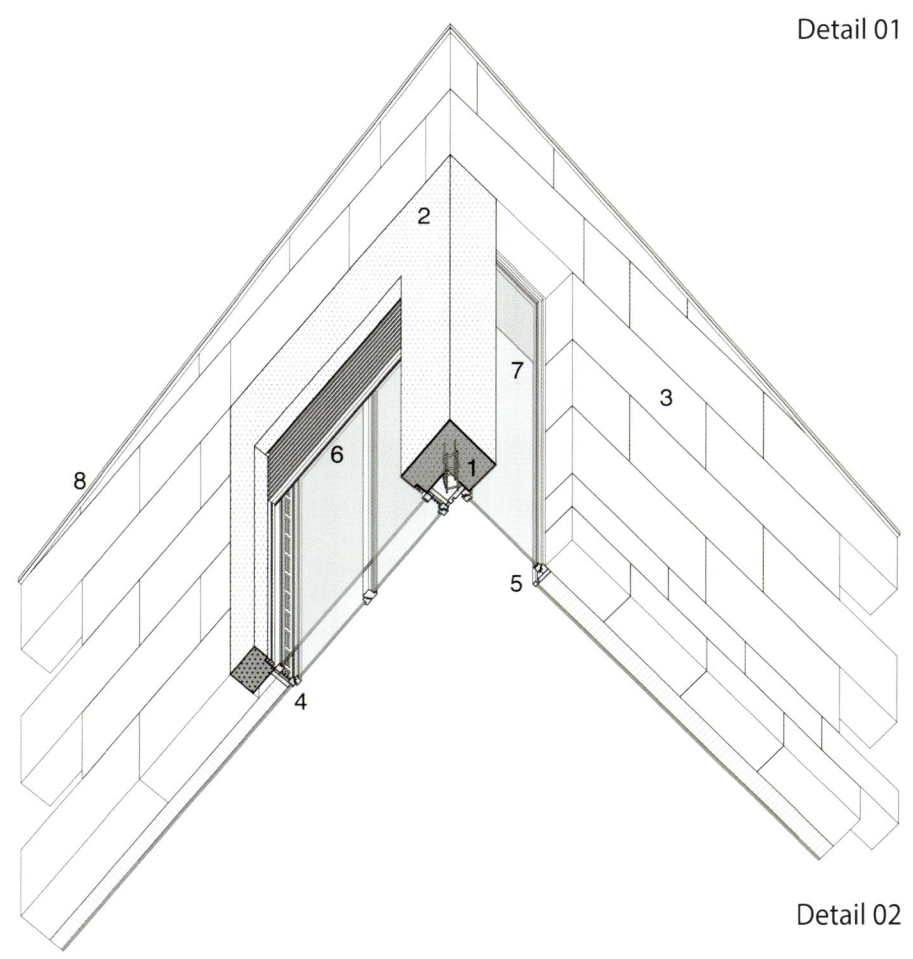

Detail 02

013

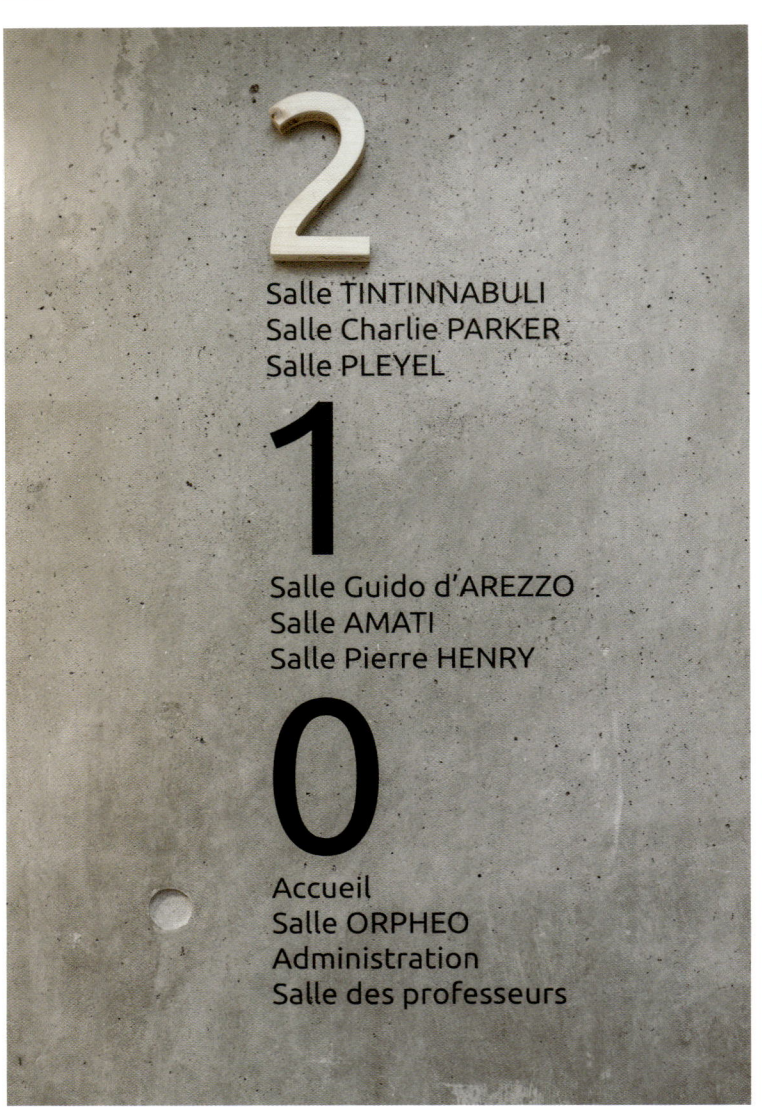

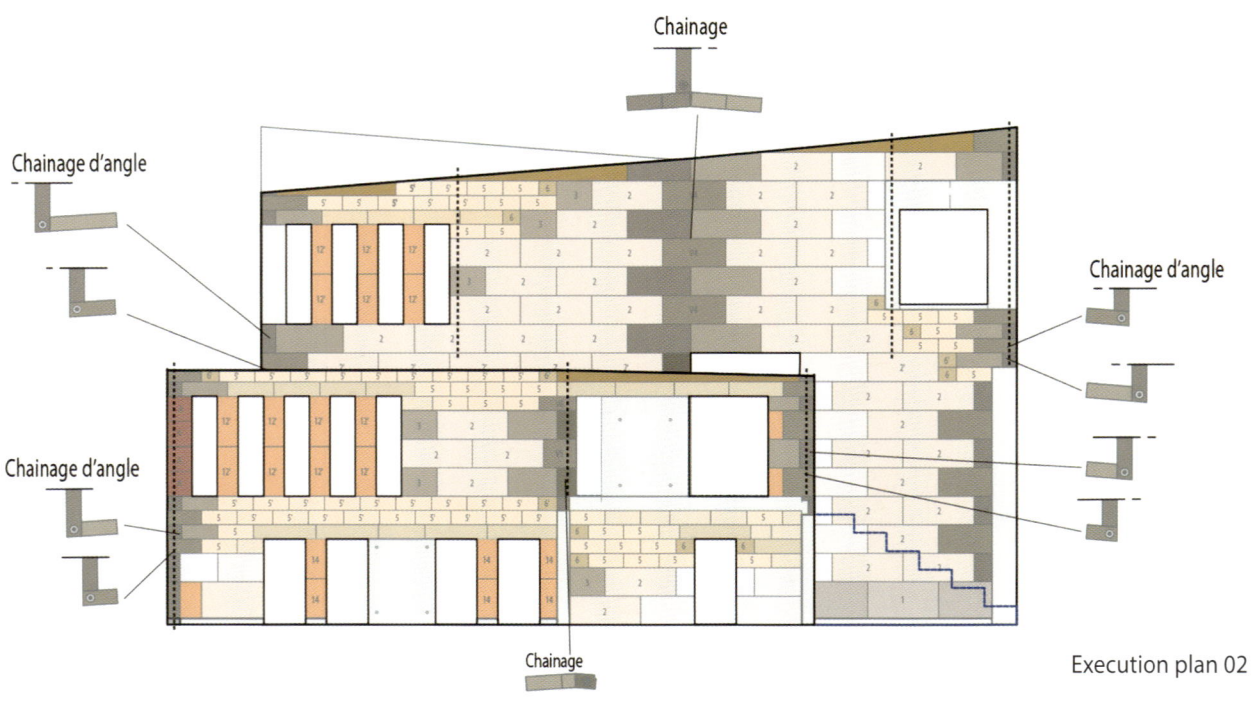

Execution plan 02

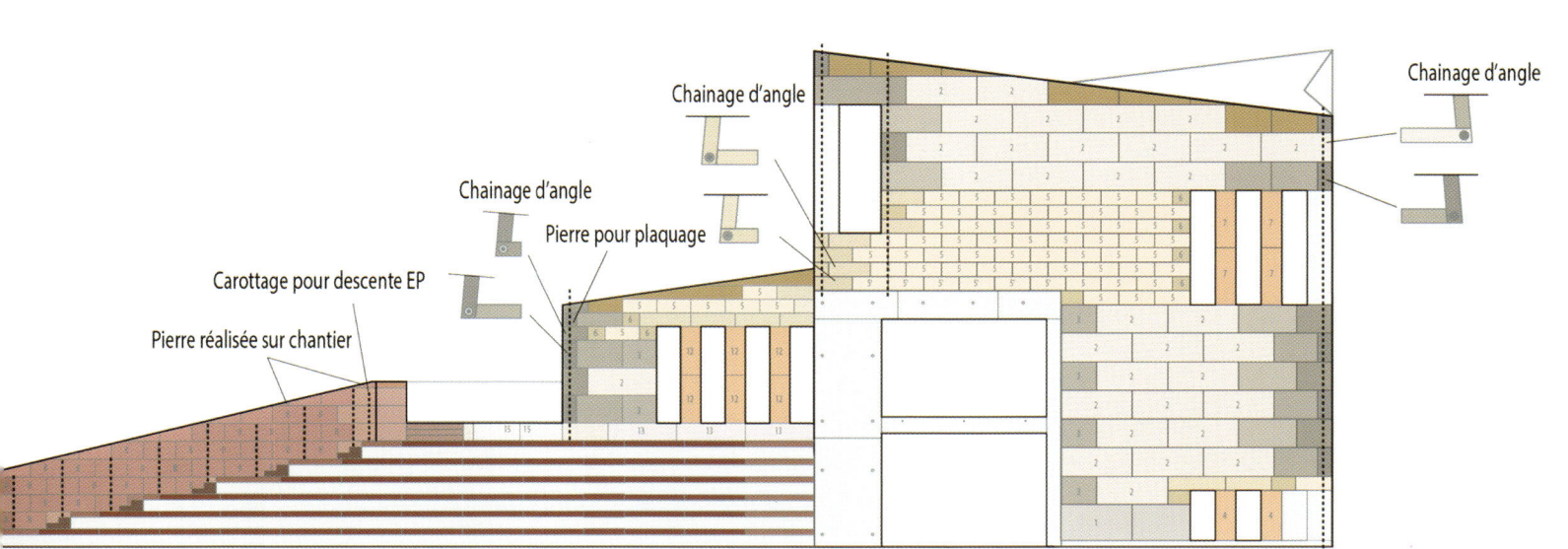

Execution plan 01

The project is based on a double reading. From the hillsides or the city centre, it is perceived as an architecture that has been present for centuries. Its indented volume, sparsely pierced, associates it at the domestic scale with the large houses nearby, built with the same materials, clear mineral facades, and tiled roofs. Meanwhile, in the tenuous space in the courtyard, the discovery of the unusual scale of the borings and the double-height front portico evoke a more monumental image of the facility and affirm its public nature.

The conservatory is built in Estaillade stone cut in the quarries of Oppède located just a few kilometres from Pradet. This technical choice was first guided by discretion in a historic neighbourhood. The conservatory faces the Church of Saint-Raymond Nonnat, which was also built during the 19th century in the same stone. It is this prolific production and proximity with the operators working with massive stone in the quarries in Provence that explain the undying economic relevance of stone.

This is a symbolic and technical response to the needs for acoustic isolation arising from the conservatory activities with regard to its neighbours. Of course, it is possible to create acoustic protections with synthetic materials or concrete, but there is something vital about enveloping the music in 35 cm of solid stones. The general form of the butterfly plan comprising non-orthogonal walls also determines the acoustic quality of the music rooms.

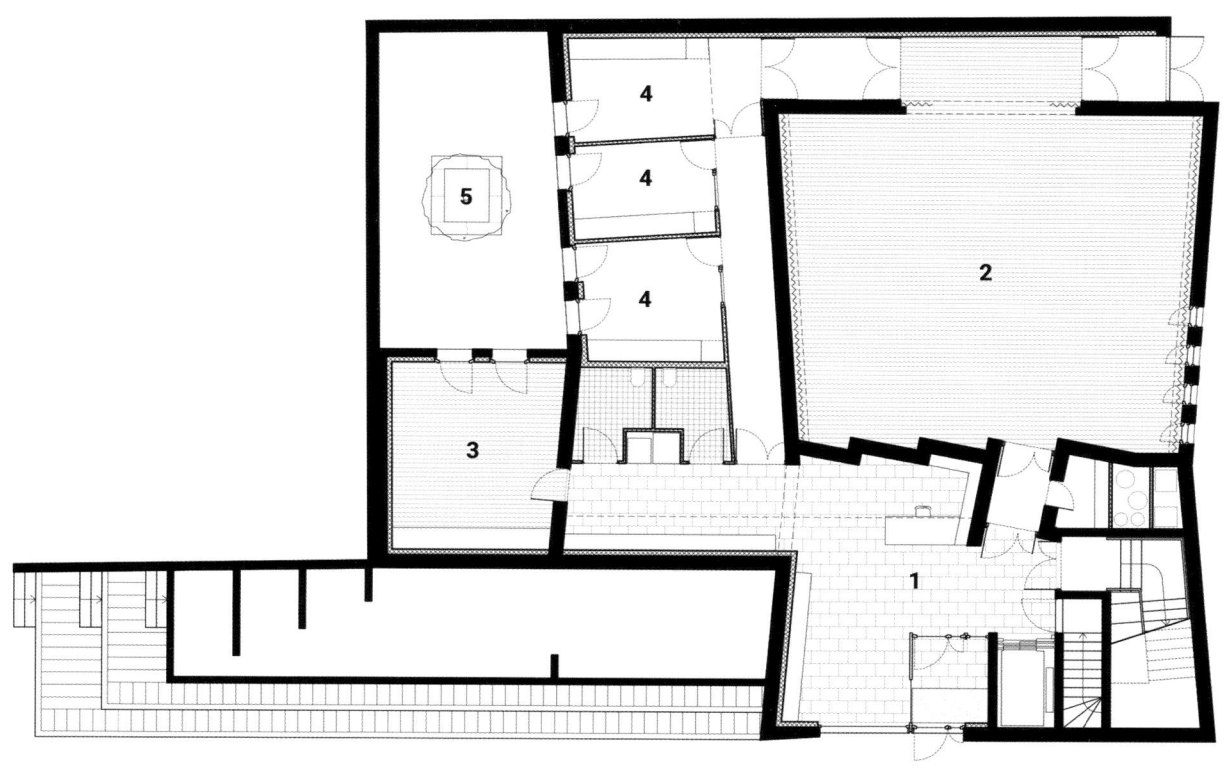

Plan-Ground Floor 0 1 5 m

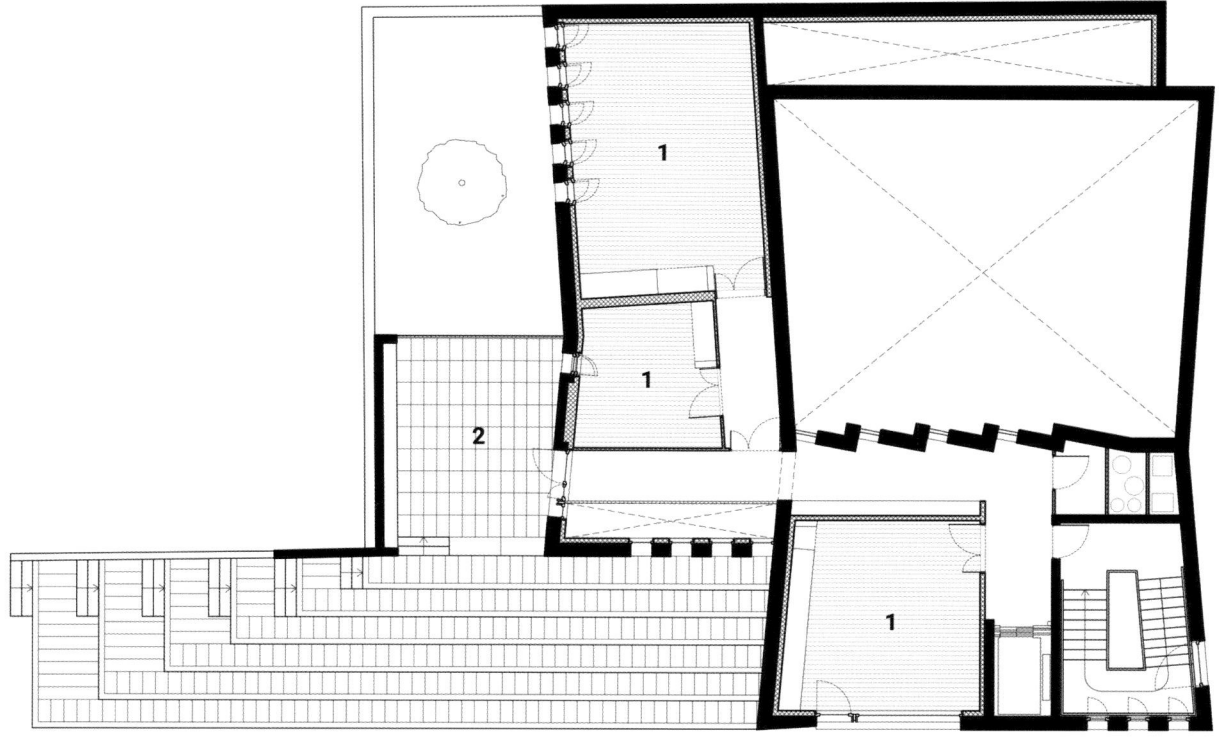

Plan-1st Floor 0 1 5 m

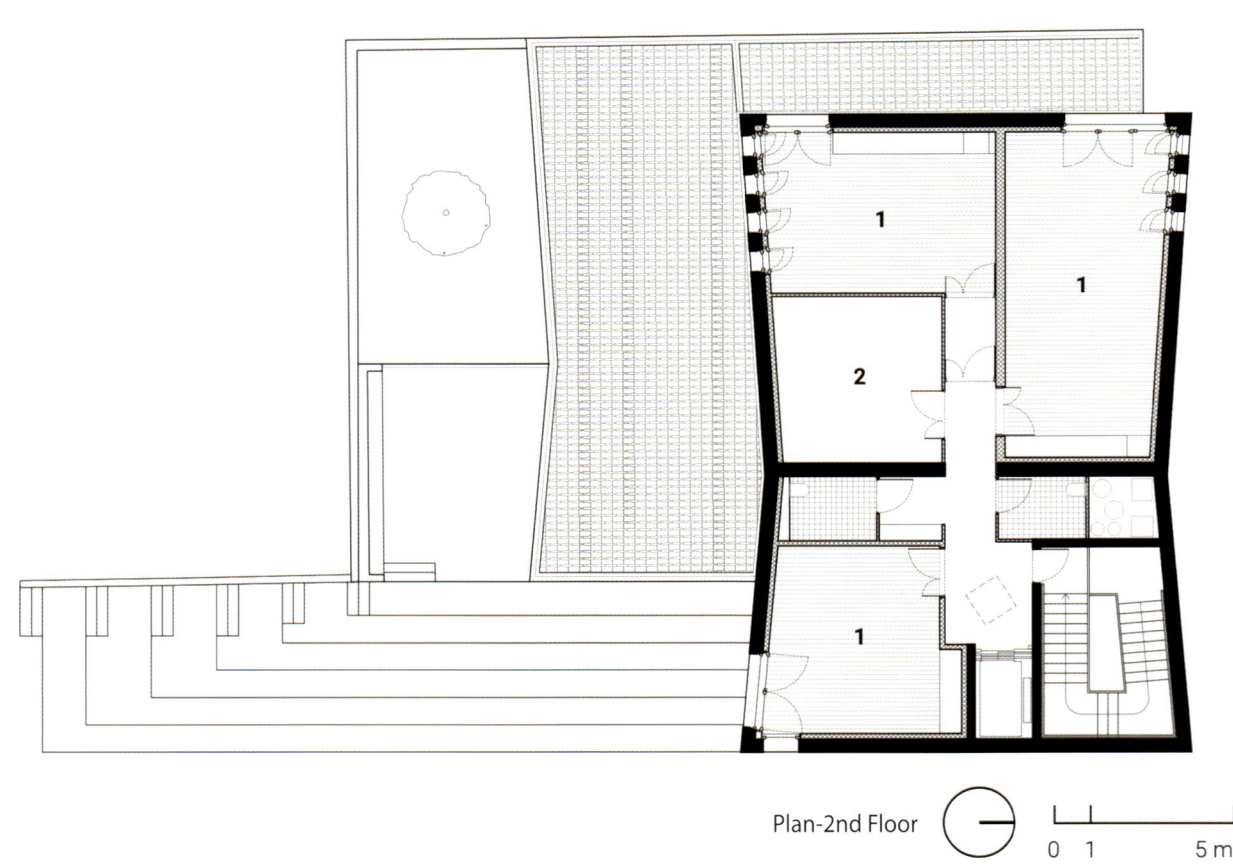

Plan-2nd Floor 0 1 5 m

Material Type: **Wood**
(StructureCraft wood)

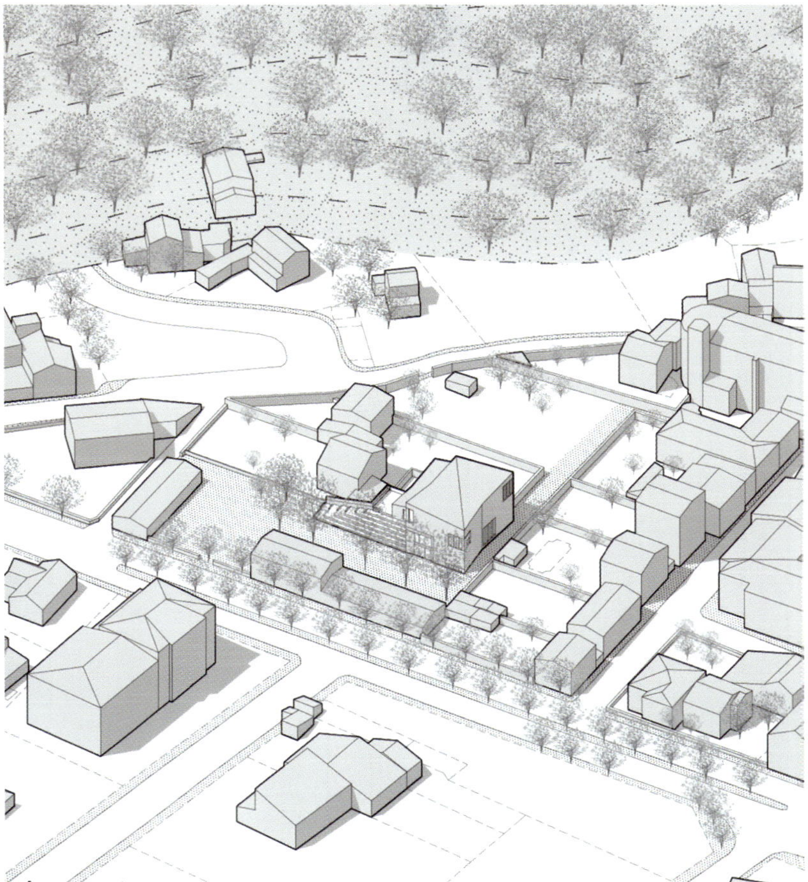

Axonometry

019

B&E Goulandris Museum of Contemporary Art

Architects: Vikelas Architects
Location: Athina Museum, Greece
Area: 7250 m²
Photographs: Christophoros Doulgeris

Head of Design and Supervision: Alexis Vikelas
Preliminary Design: Renata Metheniti
Detail Design: Katerina Giokari
Main Material: Edessa ; ocher limestone panels

In this project a systematic attempt has been made, on the one hand, to restore and enhance an existing building in the neoclassical style, and on the other, to elaborate a contemporary addition, which together had to respond to a consolidated program of functions with demands and contents specials. The new museum is located on a corner site, in a densely built environment of Athens, close to the historic center. It faces the square of the Church of San Spiridón, which was remodeled by the B&E Goulandris Foundation, as a donation to the municipality of Athens.

The museum seeks, due to its presence and unique content, to serve as a pole of attraction of international importance and scope. In terms of function, it combines the character of an exhibition and educational place, including library spaces, conference room, video room and children's workshop, while a cafe-restaurant is placed near the entrance on the mezzanine, accessing the sheltered outdoor patio. The first and second floors are used for works by foreign artists, while the third and fourth floors house works from the Greek collection. The gallery for the temporary exhibition is located on the first level of the basement.

Concept

The premises are functionally distributed around a central staircase, into which diffuse natural light is introduced through the glass-enclosed elevators. From the third floor up, the spaces are arranged on two axes, converging at the core point of vertical circulation, which serves as the predominant landmark, with elaborate hidden lighting details and a skylight at the top. The project had many peculiarities. The first was the position of the site. The narrow sidewalks of the Eratosthenous Str., The streetcar line and the sloping sidewalk, minimized the adequate construction site, while there was a need for a very deep excavation 20 to 27 meters deep. The requirement of the Ministry of Culture to preserve even the back of the walls of the houses of the existing old town, made the excavation even more difficult. A spatial metal framework was structured in phases, in order to support the retaining walls and receive horizontal loads from the excavation slope. The gradual construction of the basement, allows the previous metal frame to be disassembled step by step.

Elevation-A

ROOF
FIFTH FLOOR
FOURTH FLOOR
THIRD FLOOR
SECOND FLOOR
FIRST FLOOR
MEZZANINE
GROUND FLOOR

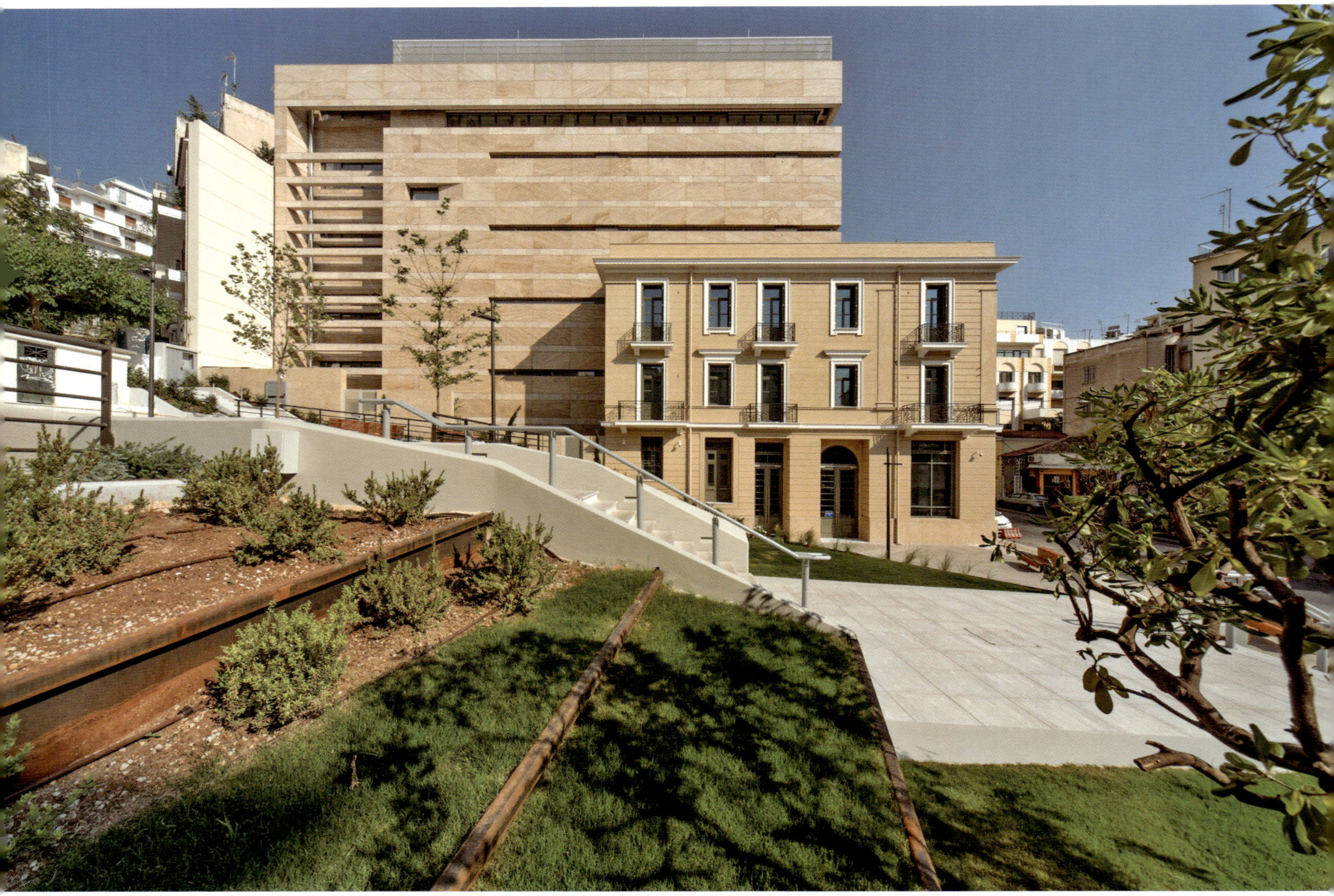

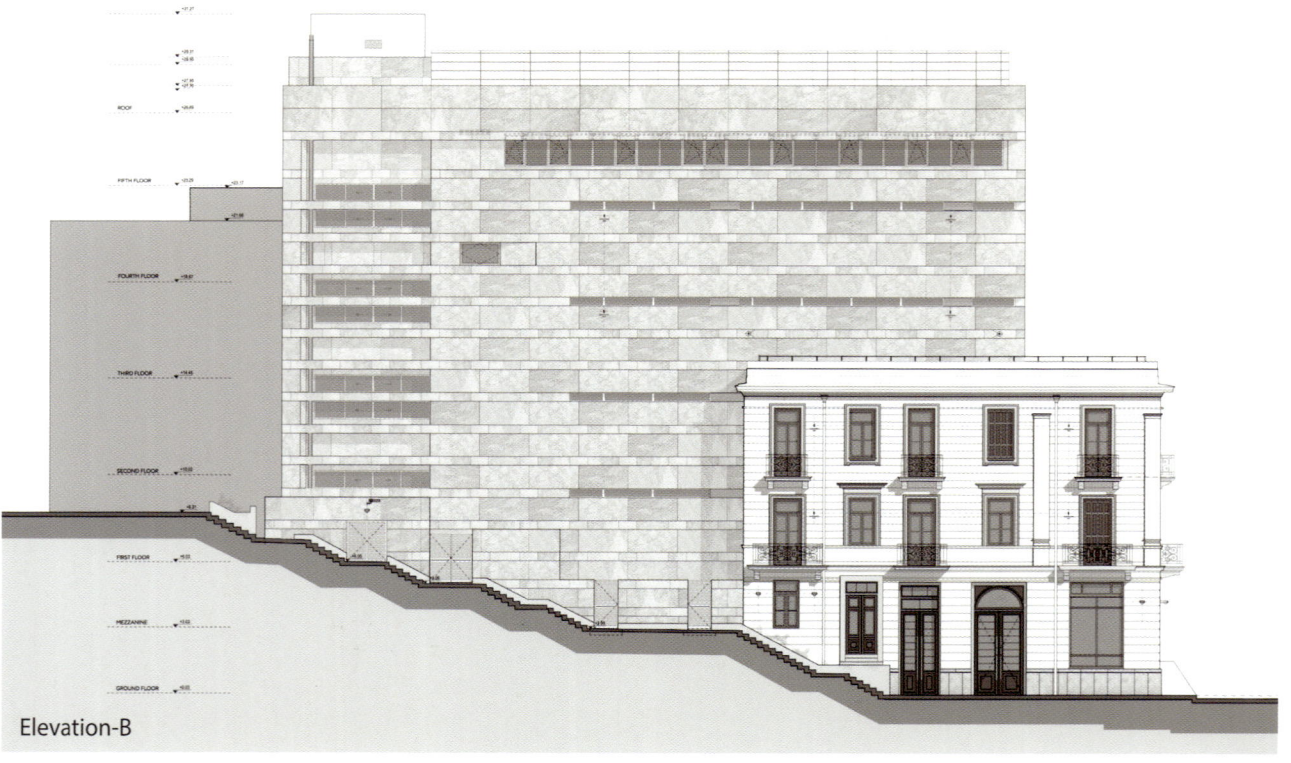

Elevation-B

S Material Type: **Stone Cladding System**
(Edessa ocher ; limestone panels)

Material Type: **Glass**
(Clerestory windows with structural glazing aluminum system)

Materials

The elevation design approach involves an effort to deal with the high rise of the extension, as well as visual abstraction within an urban environment that is aesthetically quite overloaded. The building remains in the foreground, while the new addition is recessed, thus creating a distinguishable volume with a contemporary morphological vocabulary.

The new cladding is covered with Edessa ocher limestone panels, in alternating horizontal zones of two different heights and finishes. The large size of the panel was an initial choice, serving a minimalist architectural style, dialogue with the preserved shell and highlighting the public character of the building. In some places, the panels give way to linear openings - clerestory windows, which let in natural light into the exhibition areas, or become fine rays, which indicate the presence of a circulation core behind them.

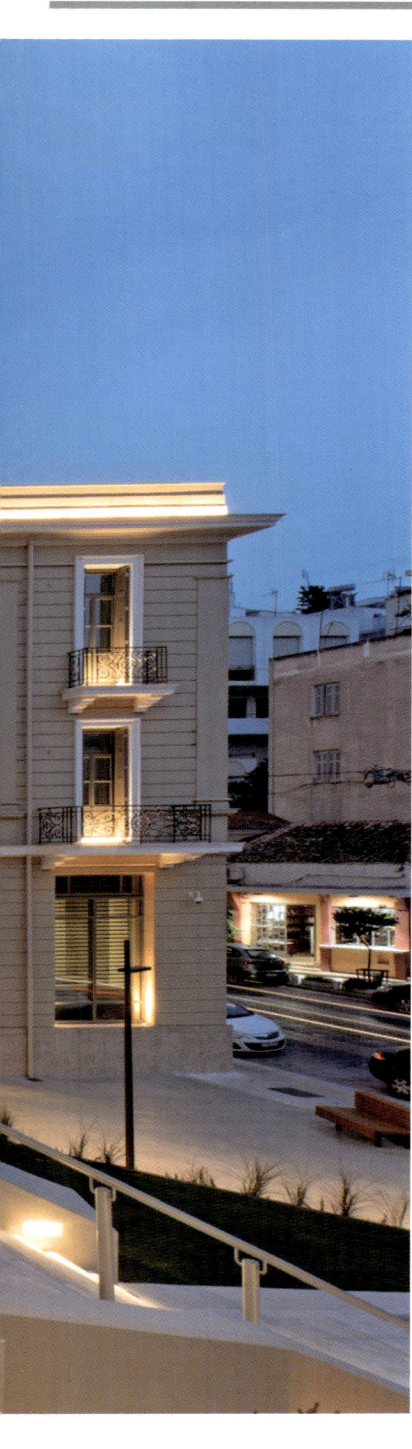

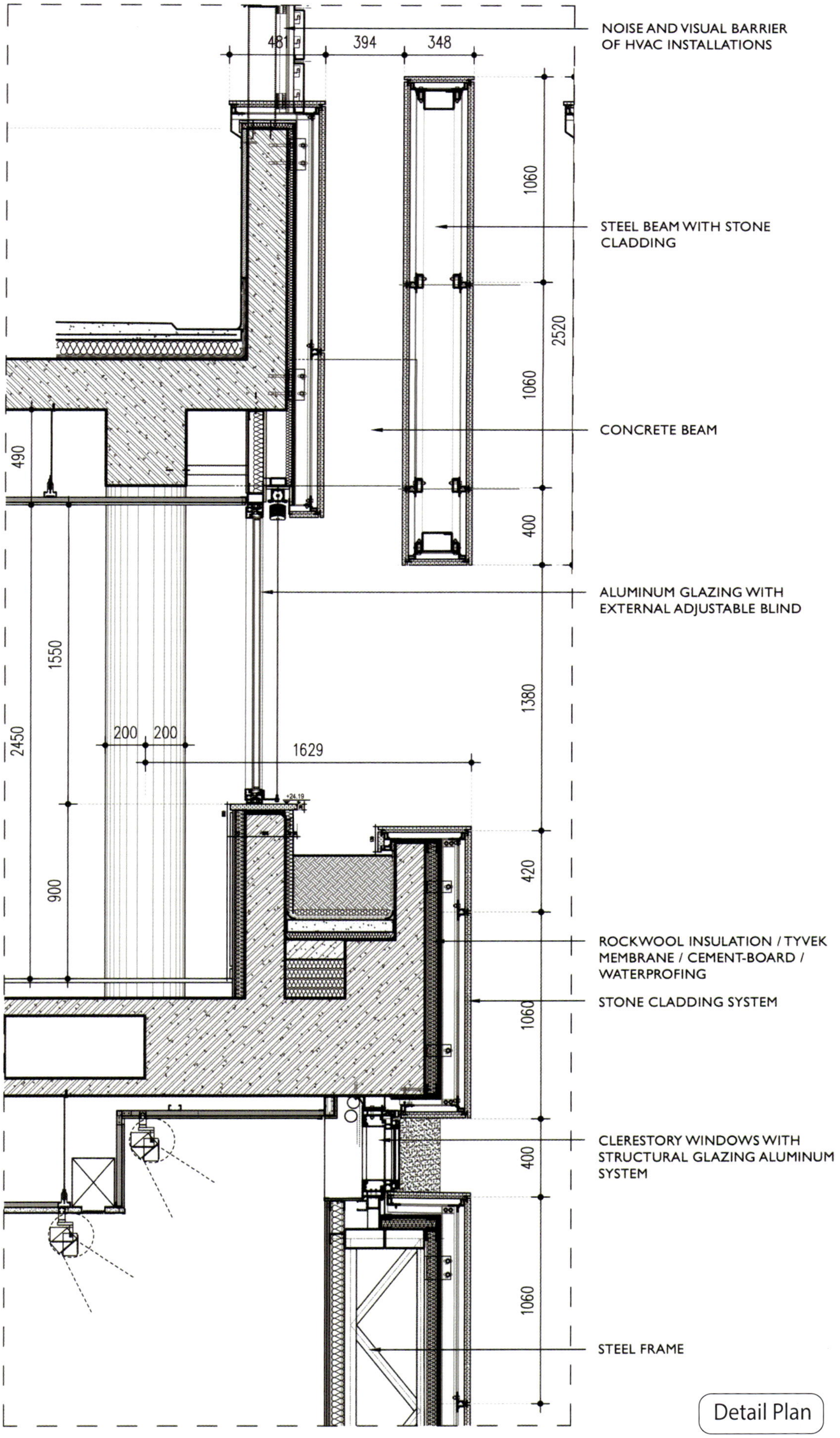

NOISE AND VISUAL BARRIER OF HVAC INSTALLATIONS
STEEL BEAM WITH STONE CLADDING
CONCRETE BEAM
ALUMINUM GLAZING WITH EXTERNAL ADJUSTABLE BLIND
ROCKWOOL INSULATION / TYVEK MEMBRANE / CEMENT-BOARD / WATERPROFING
STONE CLADDING SYSTEM
CLERESTORY WINDOWS WITH STRUCTURAL GLAZING ALUMINUM SYSTEM
STEEL FRAME
481
394
348
1060
2520
1060
400
490
1550
2450
200
200
1629
1380
420
900
1060
400
1060

Detail Plan

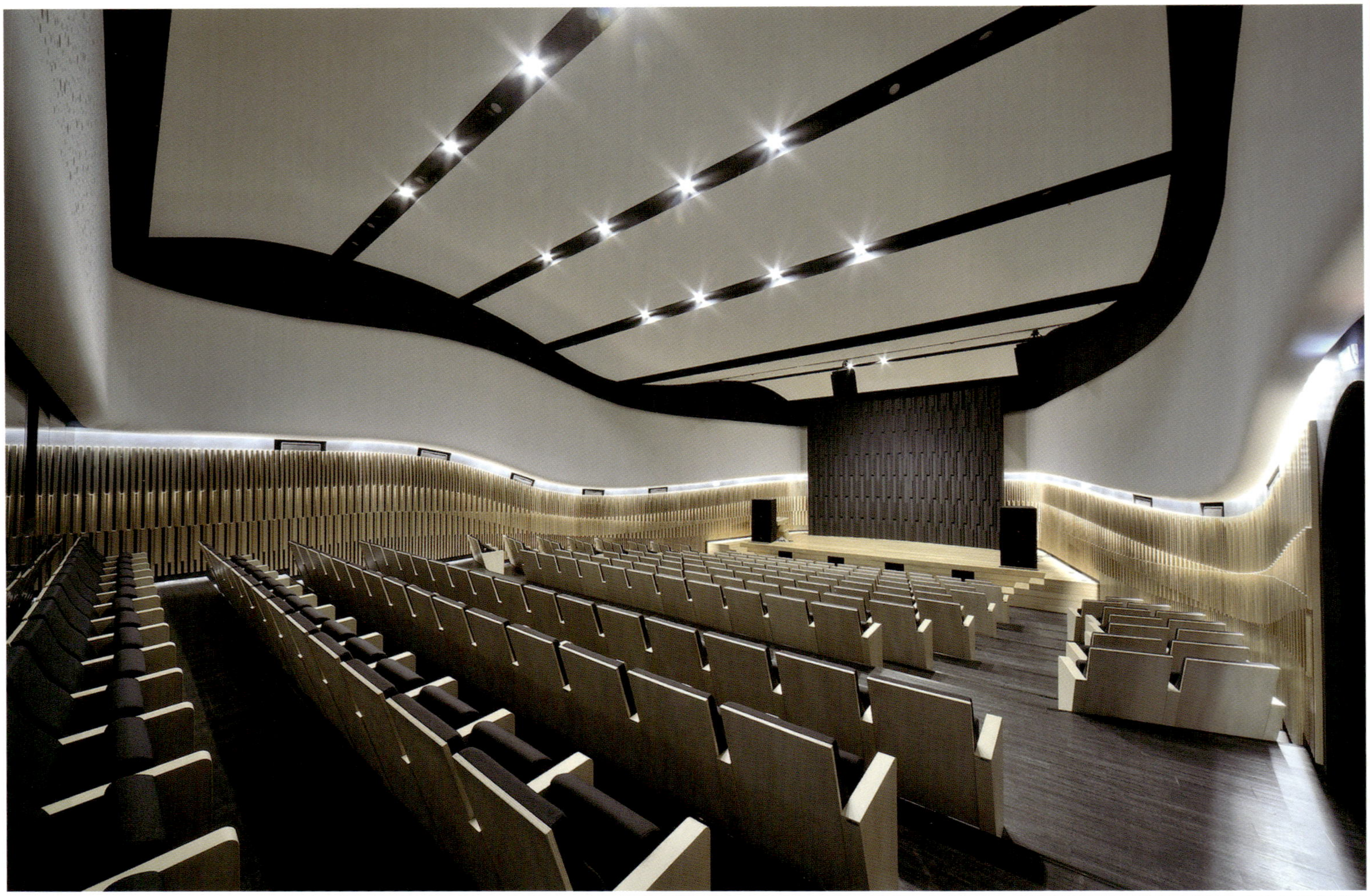

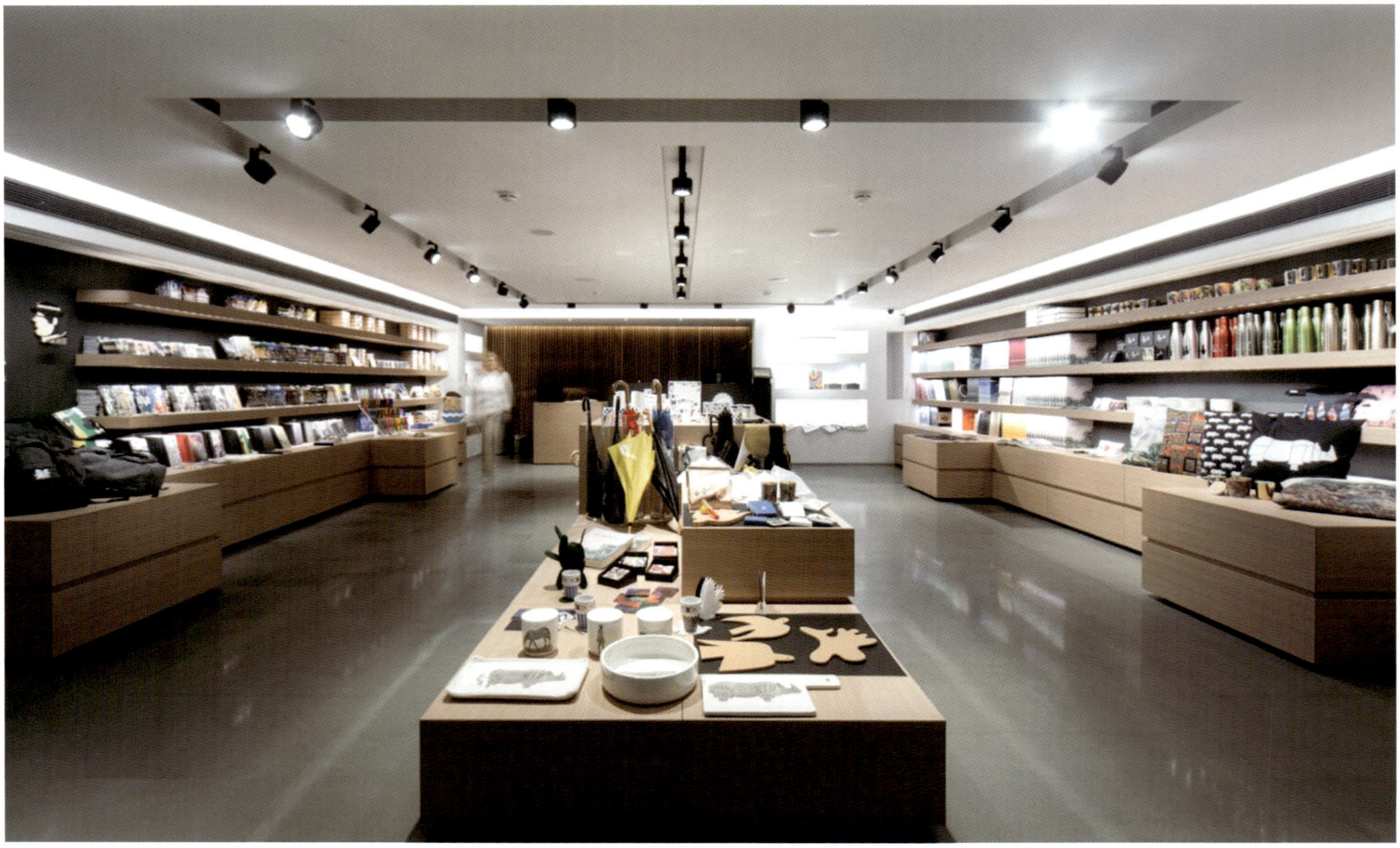

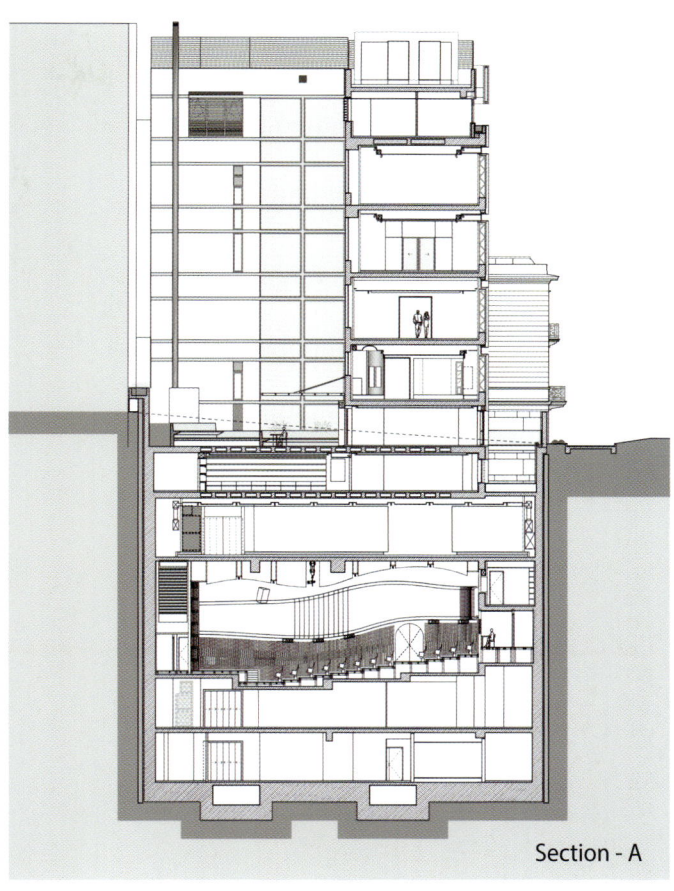

Section - A

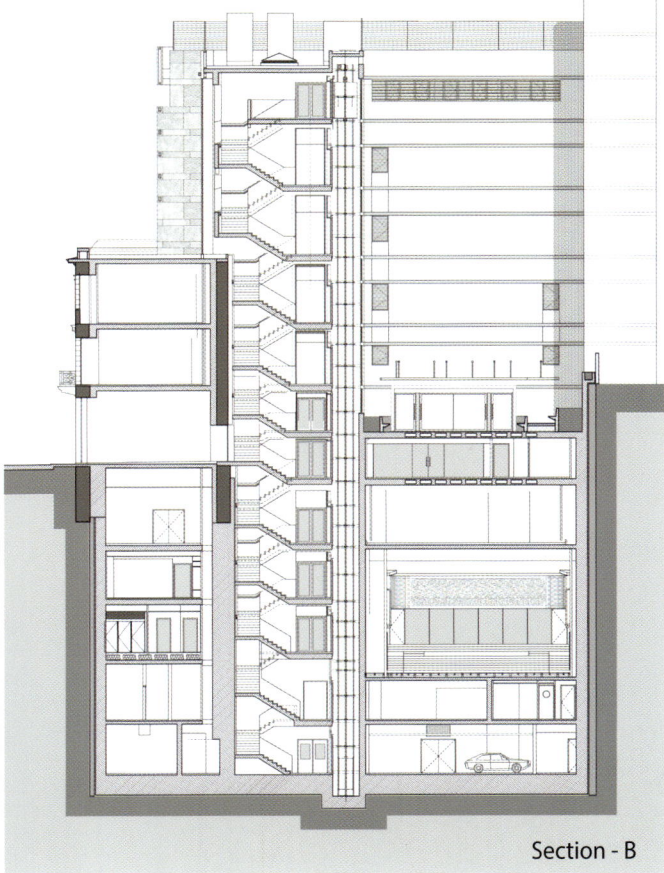

Section - B

027

low level plan
FOYER
MUSEUM SHOP
FOYER

mezzanine plan
CAFE
ATRIUM

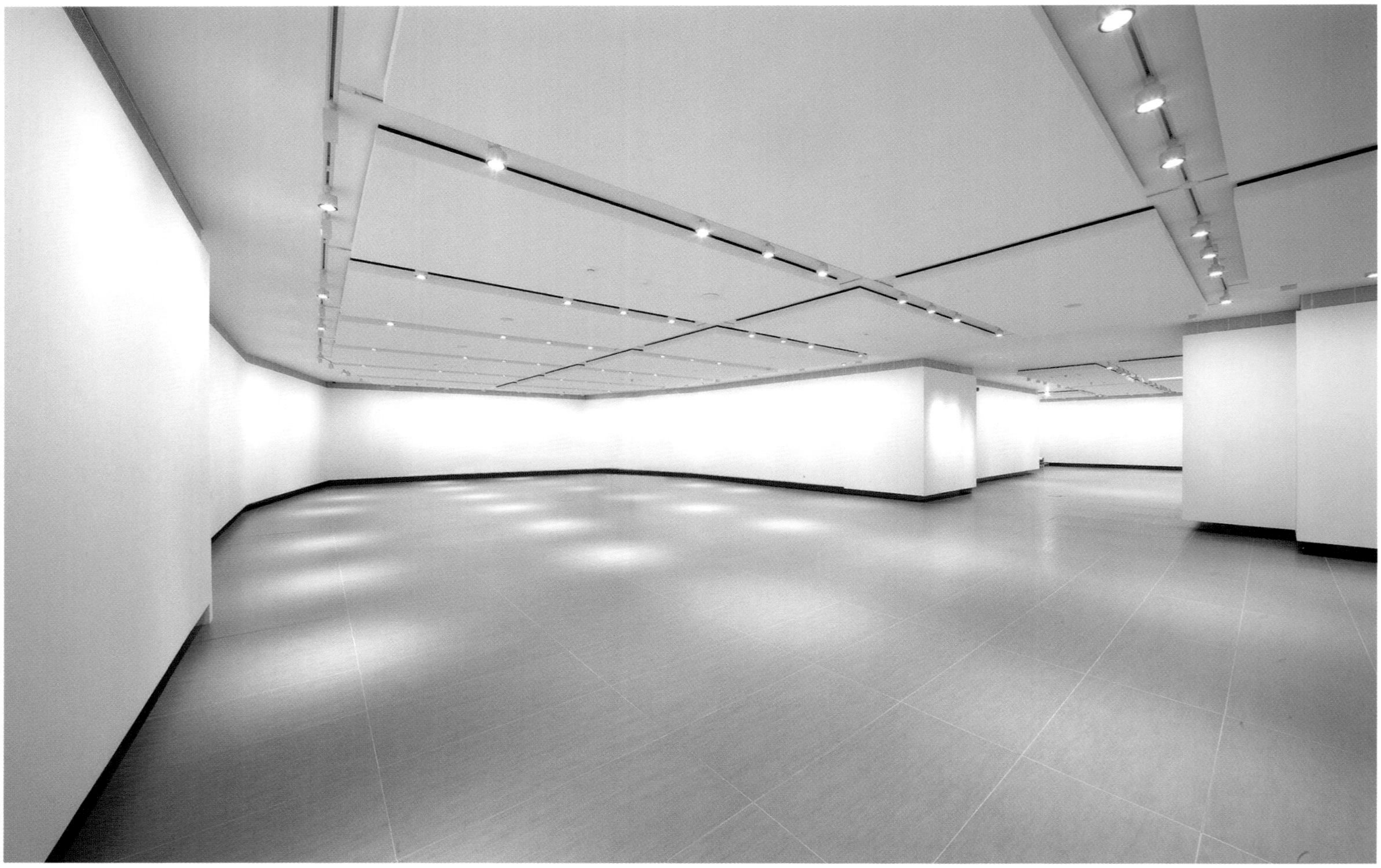

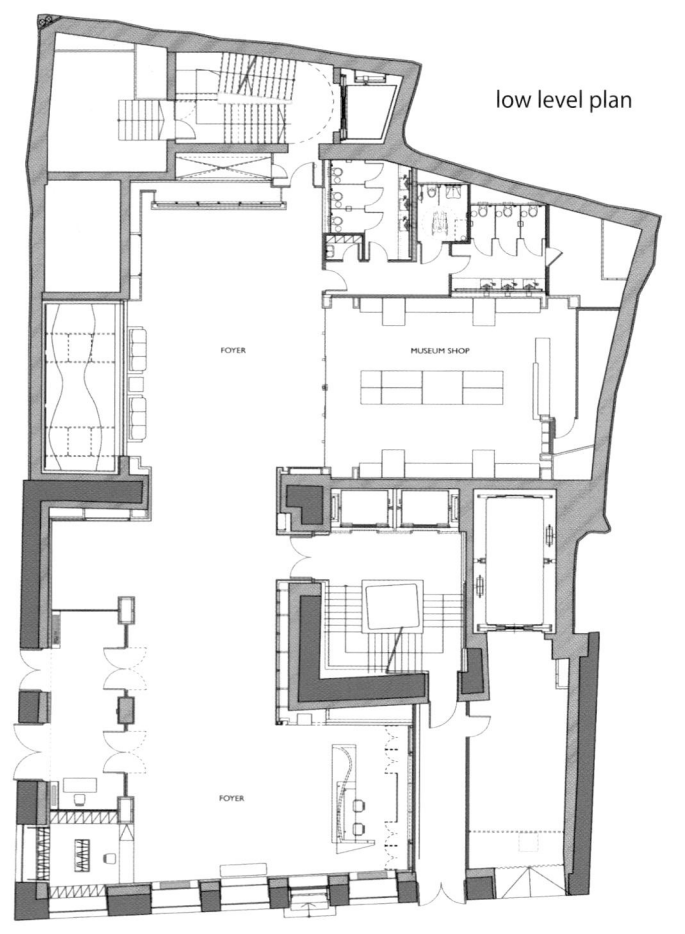

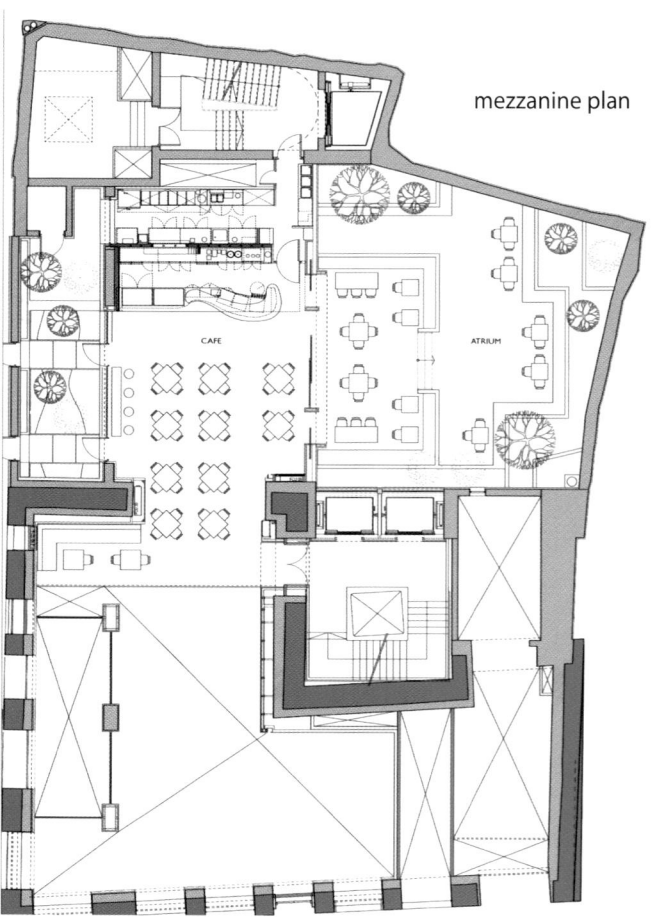

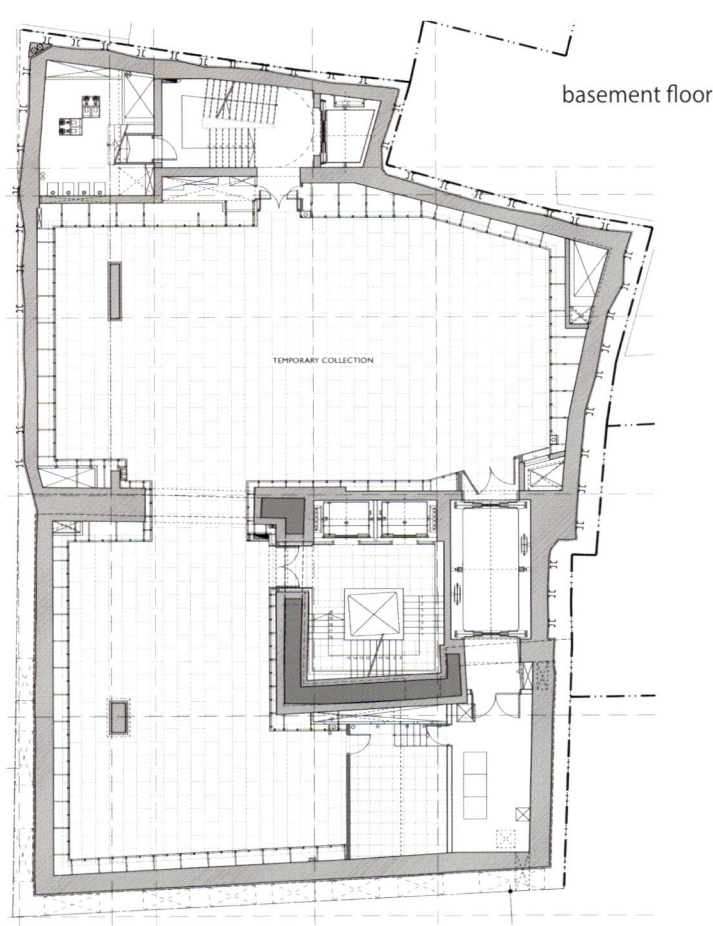

basement floor 1

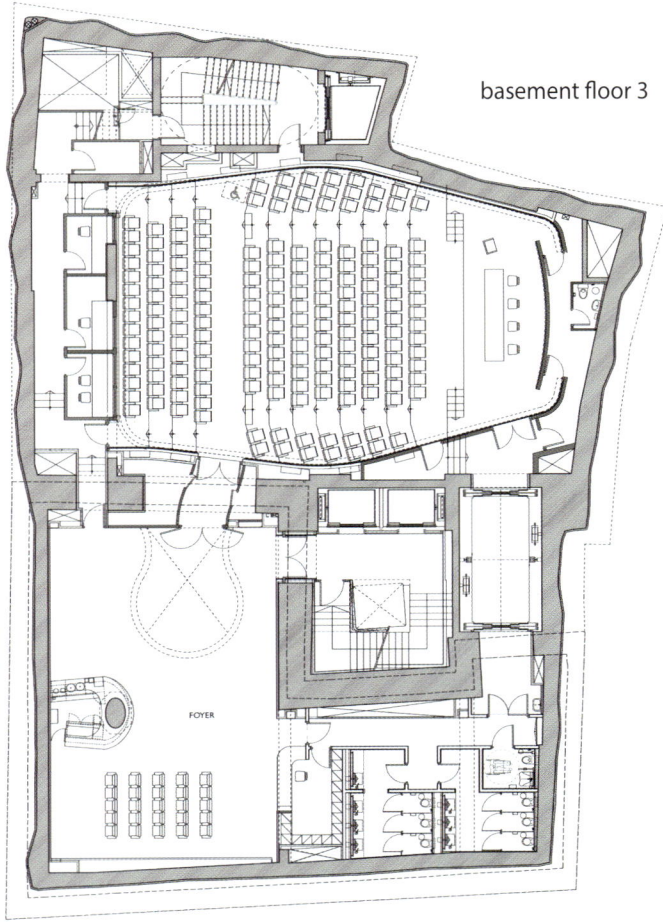

basement floor 3

The museum seeks, due to its presence and unique content, to serve as a pole of attraction of international importance and scope. In terms of function, it combines the character of an exhibition and educational place, including library spaces, conference room, video room and children's workshop, while a cafe-restaurant is placed near the entrance on the mezzanine, accessing the sheltered outdoor patio. The first and second floors are used for works by foreign artists, while the third and fourth floors house works from the Greek collection. The gallery for the temporary exhibition is located on the first level of the basement.

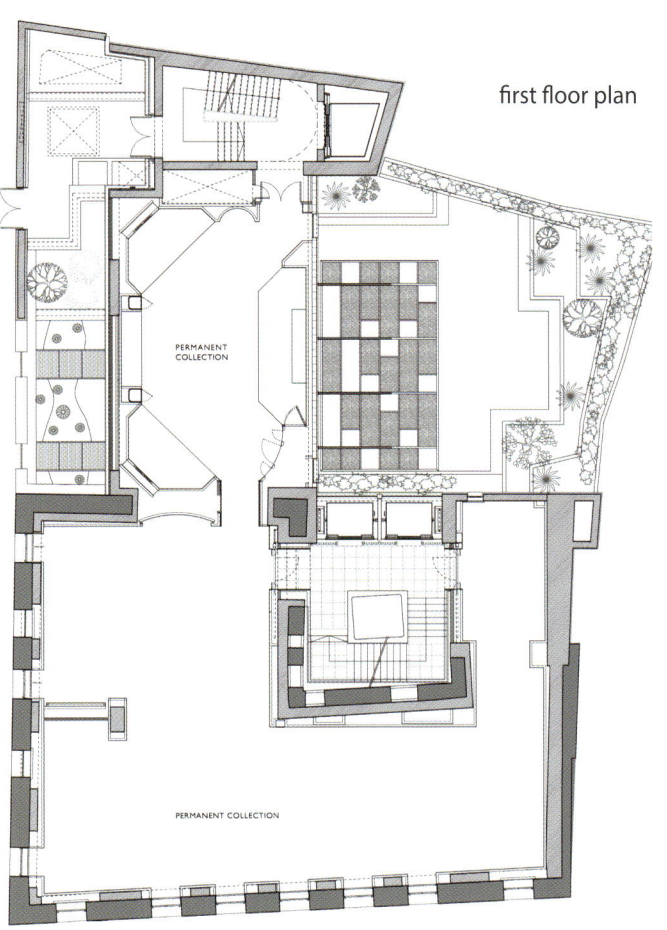

first floor plan

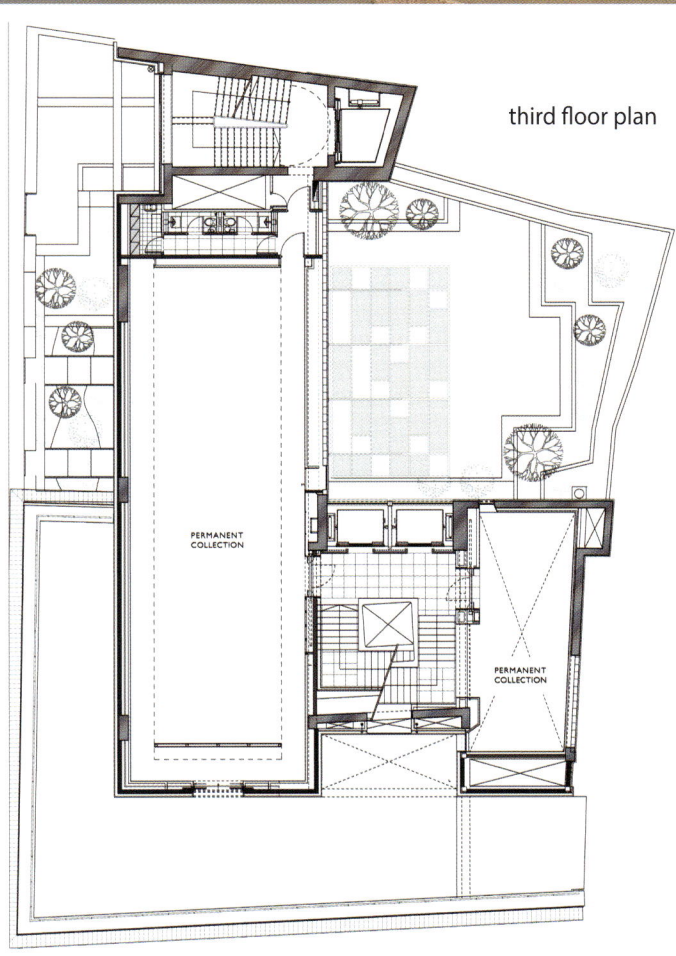

third floor plan

Pharmacy on Dzirciema Street

Architects: Substance
Location: Riga, Latvia
Area: 222 m²
Photographs: Substance

Lead Architects: Arnis Dimins
Lighting Designers: Gaismas Projektu Darbnīca
Main Material: Metal (perforated aluminium panels)

The new pharmacy on the corner of Jurmalas gatve and Dzirciema street is not just a building, but rather a peculiar hybrid between an environmental installation and a building. Its overall image is modest and conceptually assertive, but at the same time keenly different from the surrounding urban environment and attractively expressing the contents of the building. According to the location of the new pharmacy, the plot visually and functionally perceived as a small city square. The concept is based on the development of a public outdoor installation. Openwork sixteenth-corner metal construction basket two storey high is designed to be filled with the required function of the pharmacy.

In order to ensure the functionality of the newly created facility, the installation was designed with reference to the pharmacy, which would be easily recognisable from the potential buyers of drugs and other medicines. The building's facade installation frame uses the traditional pharmacy symbol – a cross as a module that would be left open or filled with perforated aluminium panels depending on the needs of the inner plans of the building. The idea of environmental installation arose not only due to the existing green square but also due to the local building authorities, who insisted that a traditional low-rise building from the urban planning view is not suitable for the corner of the block with five-storey building.

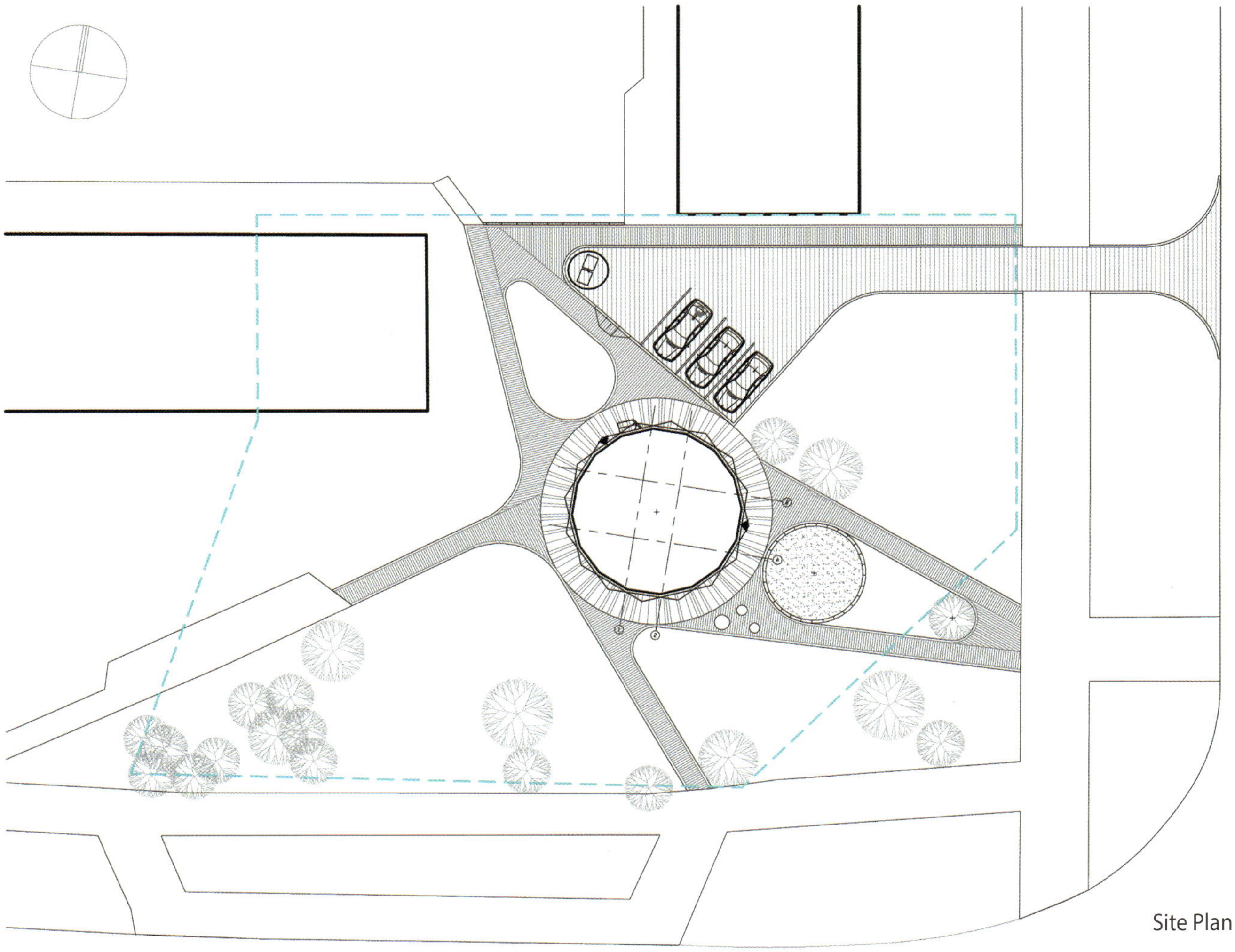

Site Plan

033

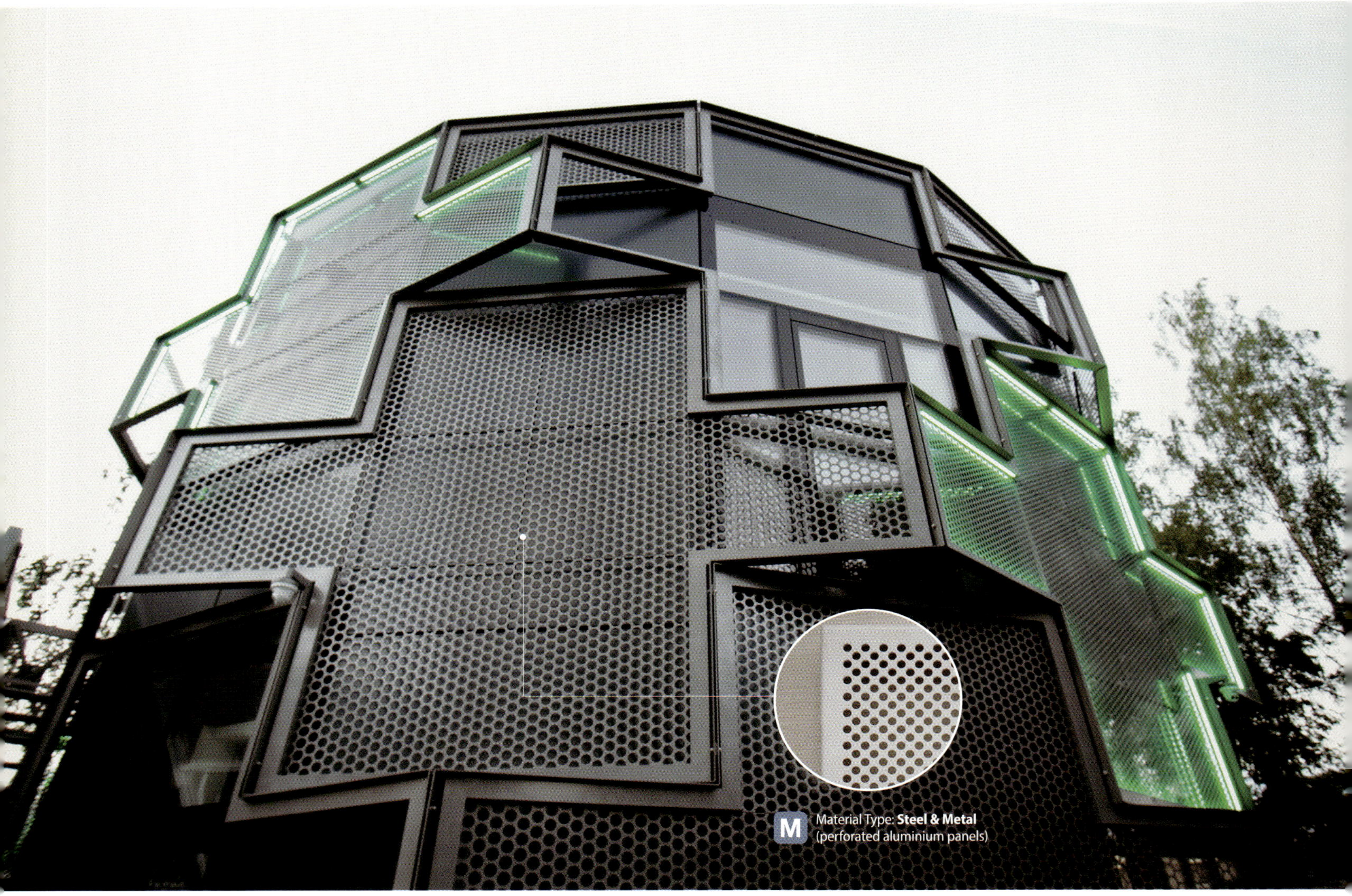

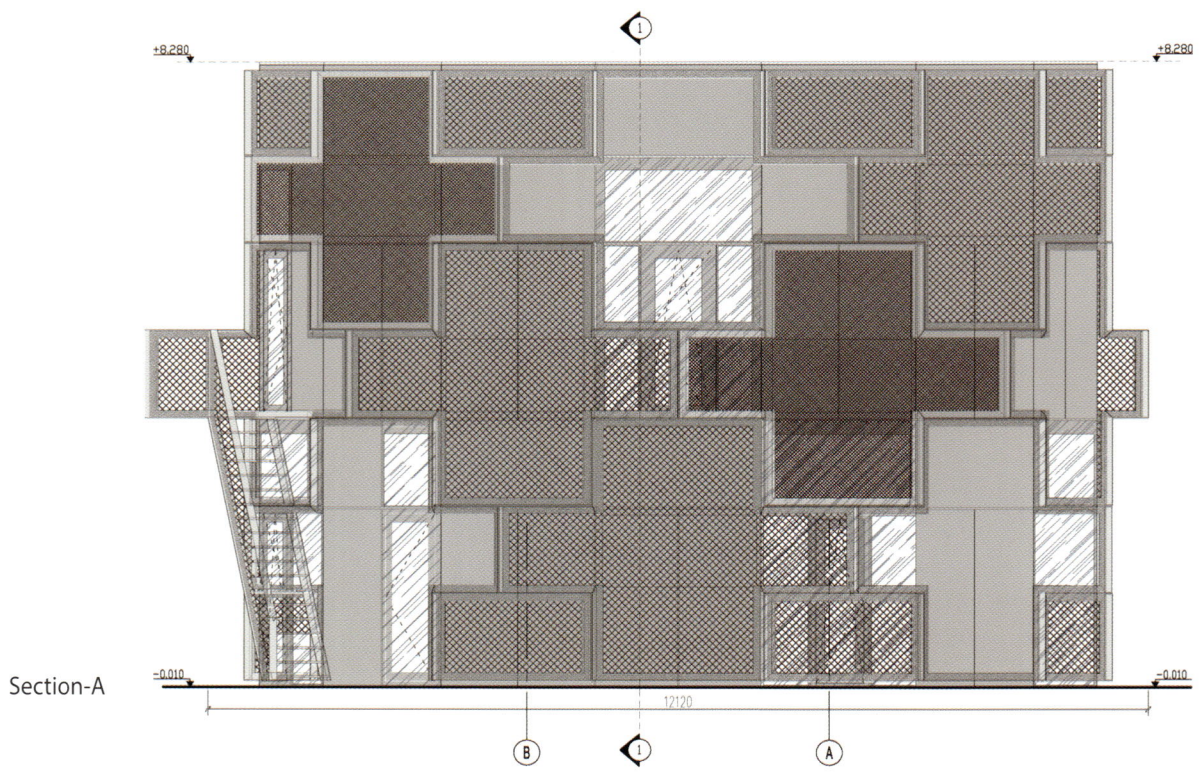

M
Material Type: Steel & Metal
(perforated aluminium panels)

+8.280
+8.280
1
-0.010
-0.010
12120
B
1
A

Section-A

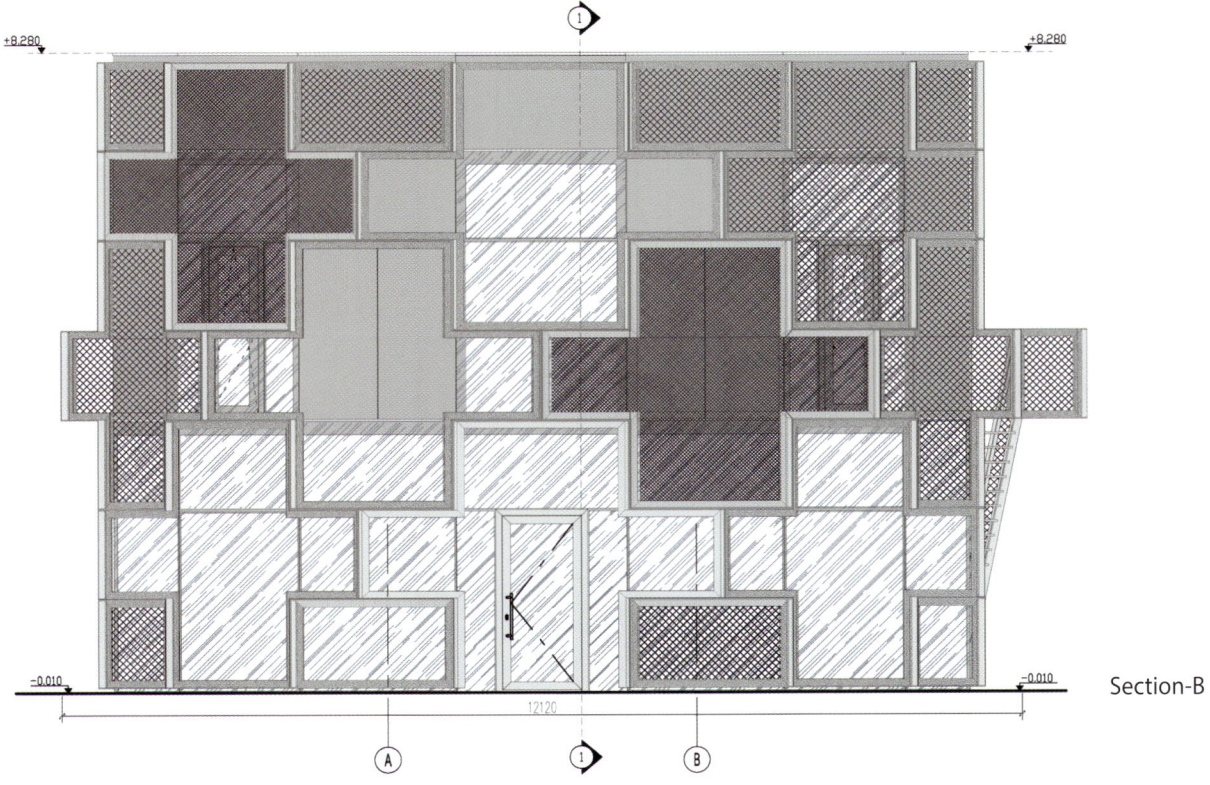

Section-B

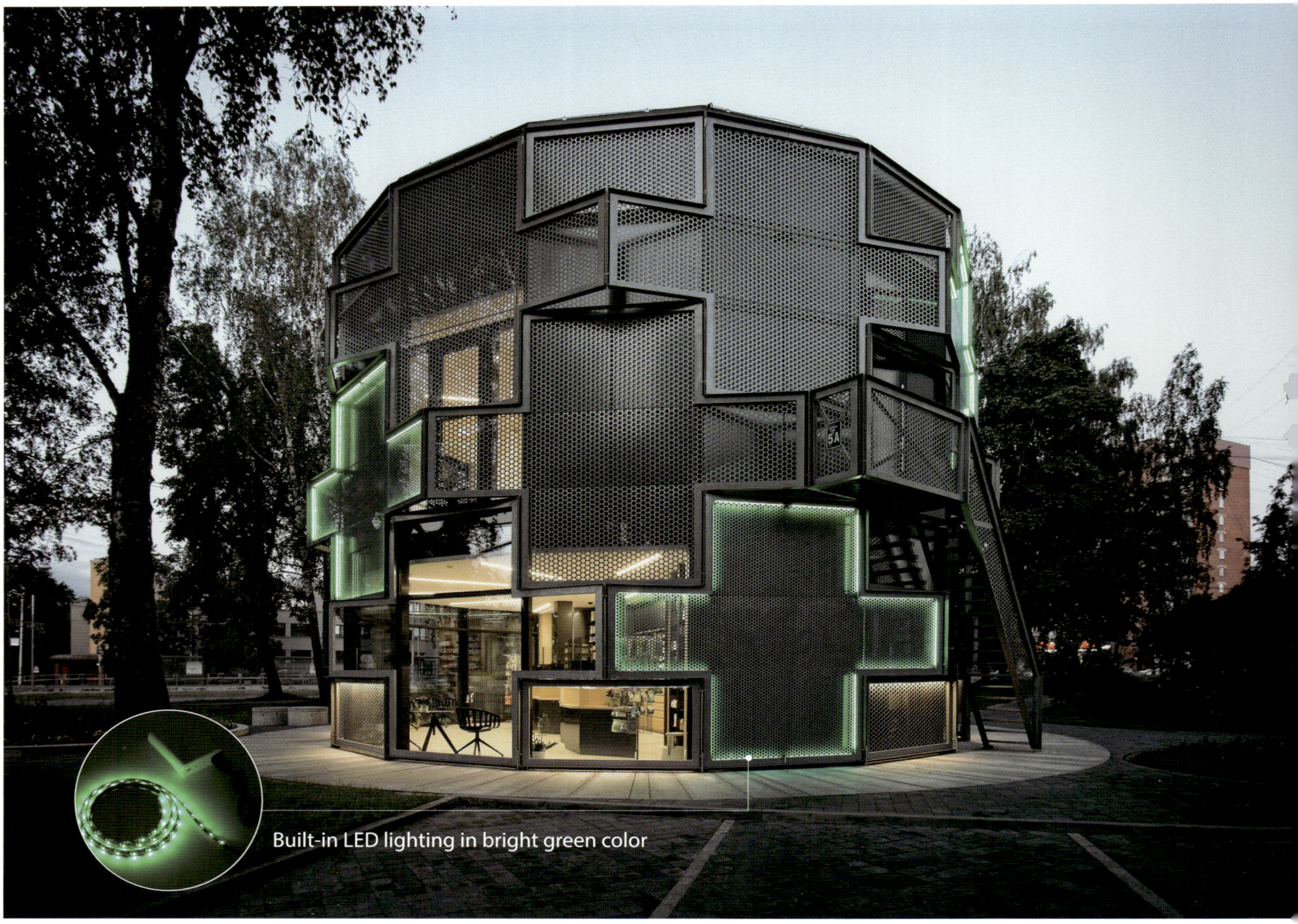

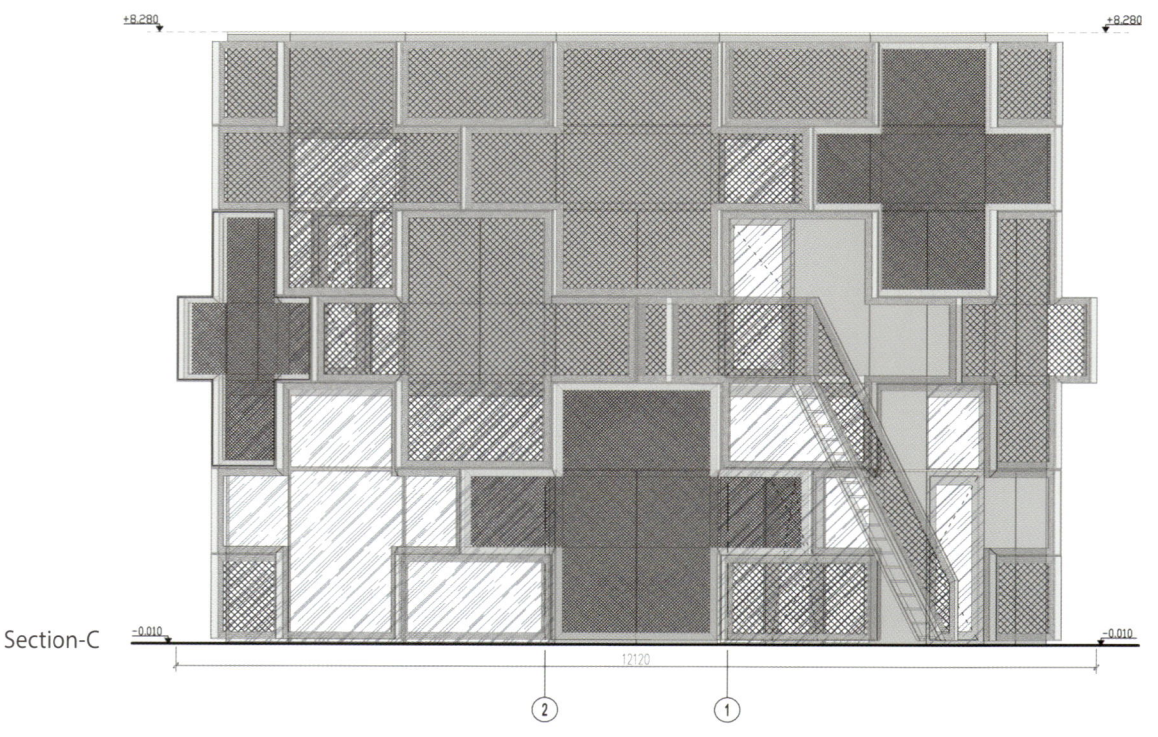

Built-in LED lighting in bright green color

+8.280
+8.280
-0.010
-0.010
12120
2
1
Section-C

Elevation-A

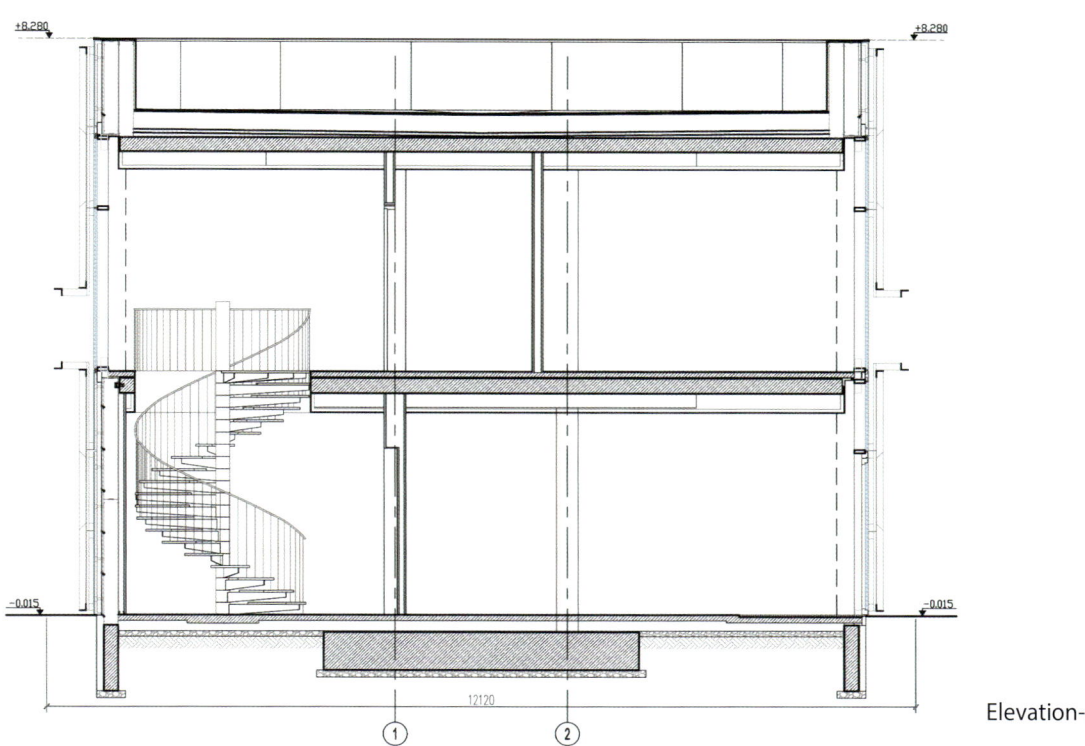

037

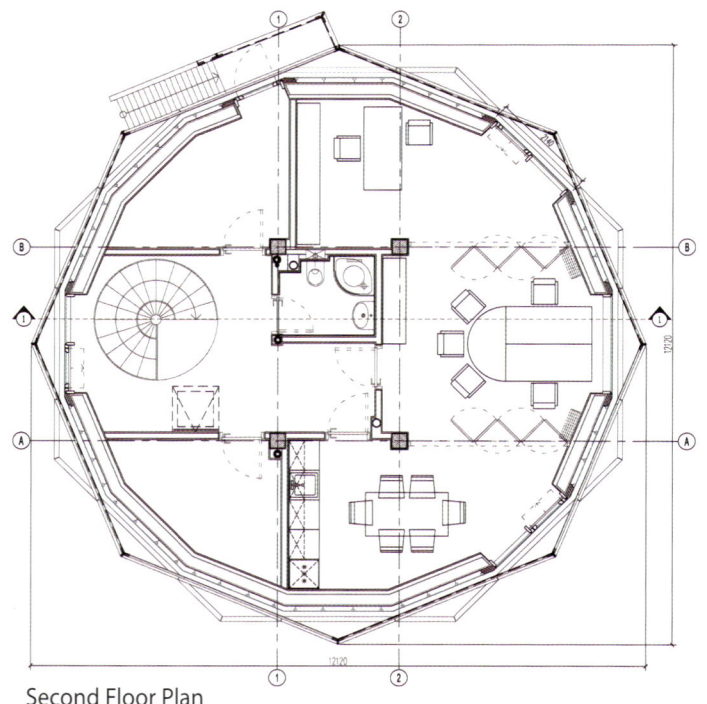

Second Floor Plan

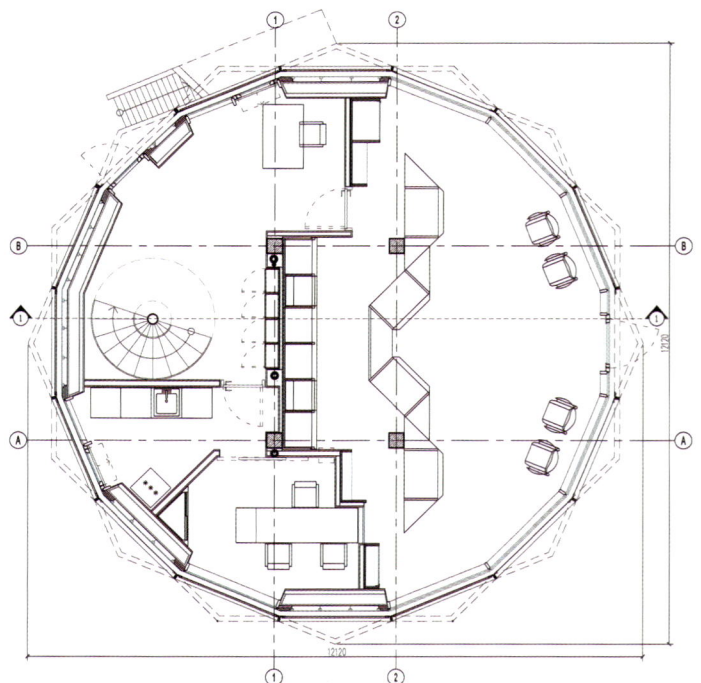

First Floor Plan

New installation is not only a design element of the parametric architecture but also some kind of the signboard of the pharmacy. During the periods when there is no intense sun lighting on the facade, individual crosses are illuminated with built-in LED lighting in bright green color. The building's bearing construction is based on four columns of reinforced concrete. The columns located in the middle of the building make the perimeter of the building fully subordinated to the parametric pattern of the facade installation.

In contrast to the visually saturated metal frame, the facade of the building made as simple as possible - with structural glazing and composite panels. Finally, the facade installation also significantly improves the energy performance of the new building and, despite the wide glazed openings in the base facade of the building, it is almost 0 consumer building, according to the actual energy efficiency calculation.

Benjamin Carrion House Renovation

Architects: Bernardo Bustamante Arquitectos
Location: Quito, Ecuador
Area: 10764 ft²
Photographs: Bicubik

Design Team: Bernardo Bustamante, Doménika Baquero
Engeneering: Iván Delgado
Main Material: Metal (Zinc steel slats Systems; Rheinzink)

The project initiates with the donation of the property from the family Benjamin Carrion, to the city. Providing the neighborhood with this equipment and an ancestral garden park, for the enjoyment of all. The result is the democratization of the private green, abundant in our South American cities. The new and rehabilitated infrastructure, a tribute from the family and the Municipality of Quito to the literary prestige of Benjamin Carrion, one of the most important writers and intellectuals of Ecuador. His great interest and contribution to science, literature and the artistic and cultural development of the country have made his name constantly resound when speaking about these subjects. The project offers a program focused on the cultural development and leisure of young people.

The re-functionalization of the Benjamin Carrion´s House is characterized by giving this property to the city, recovering an ancestral garden of Quito, that coexisted with the House of the Carrión´s Family since its origin, punctually in the neighborhood of Bellavista, and protects the modern architectural heritage. The proposal seeks to solve the new needs for the new use of the house to a Cultural Center, through interventions with transparent elements that relate correctly to the immediate environment, acting with a lot of respect for the existing heritage house. The house is located in a large lot of 2200m2, which has been almost intact over time, resisting the onslaught of real estate speculators for decades. It is located in one of the sectors with the highest capital gains in the city, so it has become a green oasis between large towers that occupy the entire surface of the plot.

For the design team, was a priority to maintain the garden, without touching the old trees, autochtonous of this Andean landscape.

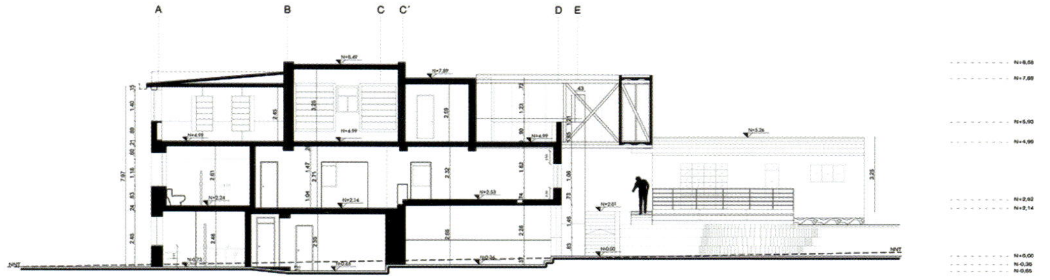

CORTE LONGITUDINAL B - B´

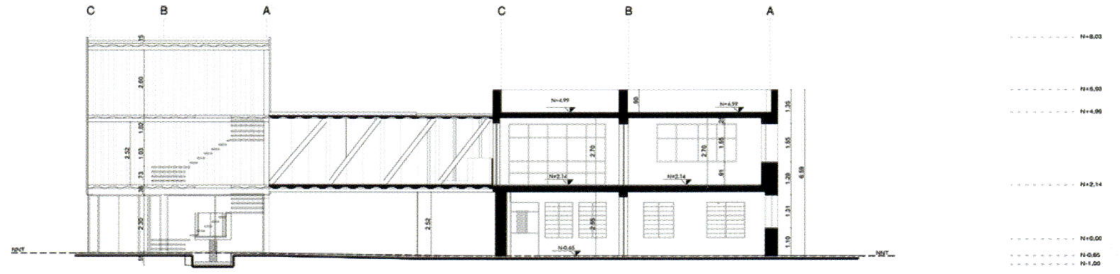

CORTE LONGITUDINAL C - C´

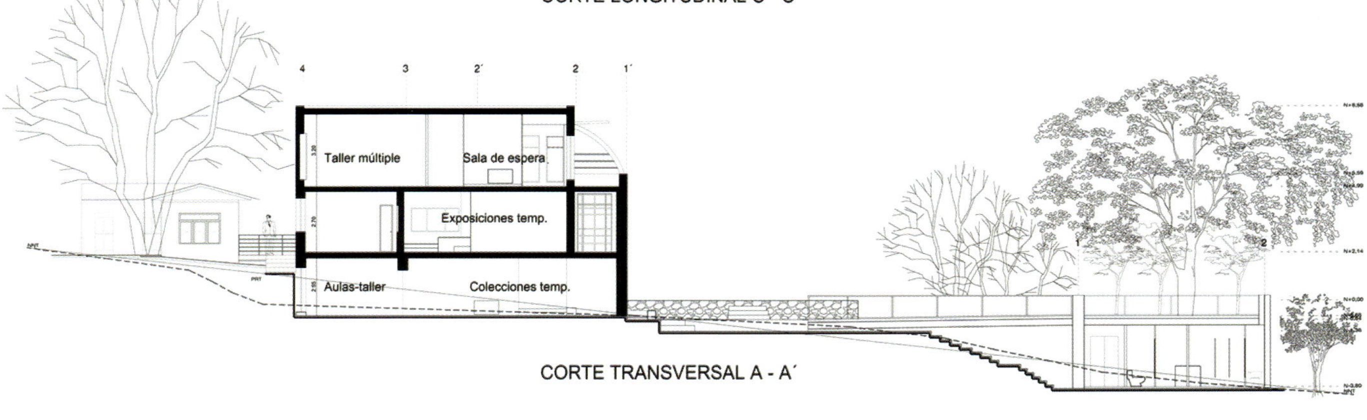

CORTE TRANSVERSAL A - A´

- Taller múltiple
- Sala de espera
- Exposiciones temp.
- Aulas-taller
- Colecciones temp.

FACHADA LATERAL DERECHA

FACHADA LATERAL IZQUIERDA

Material Type: Metal
(Zinc steel slats Systems; Rheinzink)

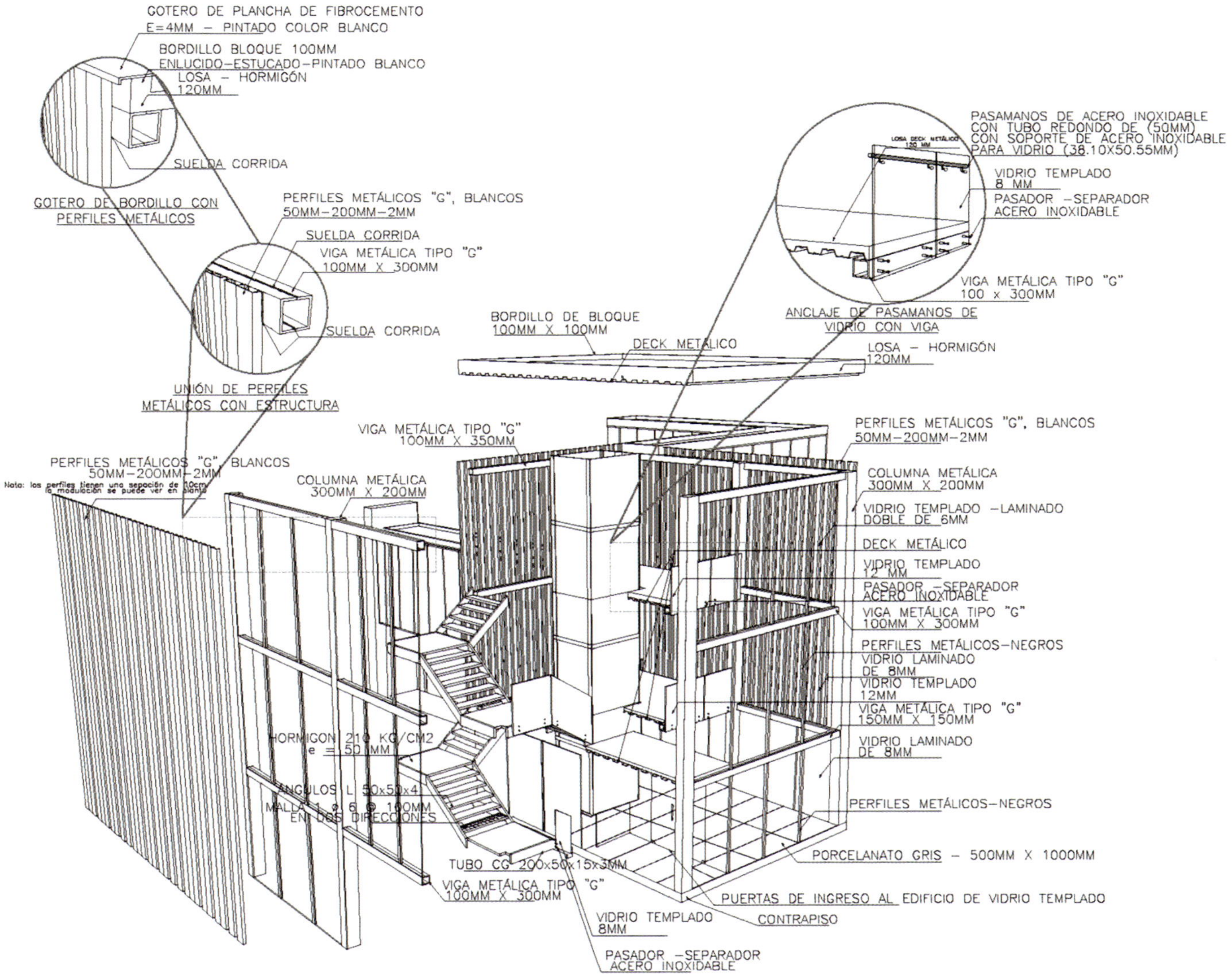

PERSPECTIVA EXPLOTADA DEL HALL DE INGRESO sin escala

Material Type: **Glass**
(Glass Curtain Wall System)

The intervention consists of two pieces required by the program: A new entrance hall that faces the access axis from the street, that almost without touching the existing house, connects it through two bridges in a very respectful way, solving the circulation system of the new program with only two incisions. The new hall is a translucent prism, nine meters high, covered by a second skin of steel slats, that nourish the natural light to the distributor, generating interesting effects in the interior during the sun translation.

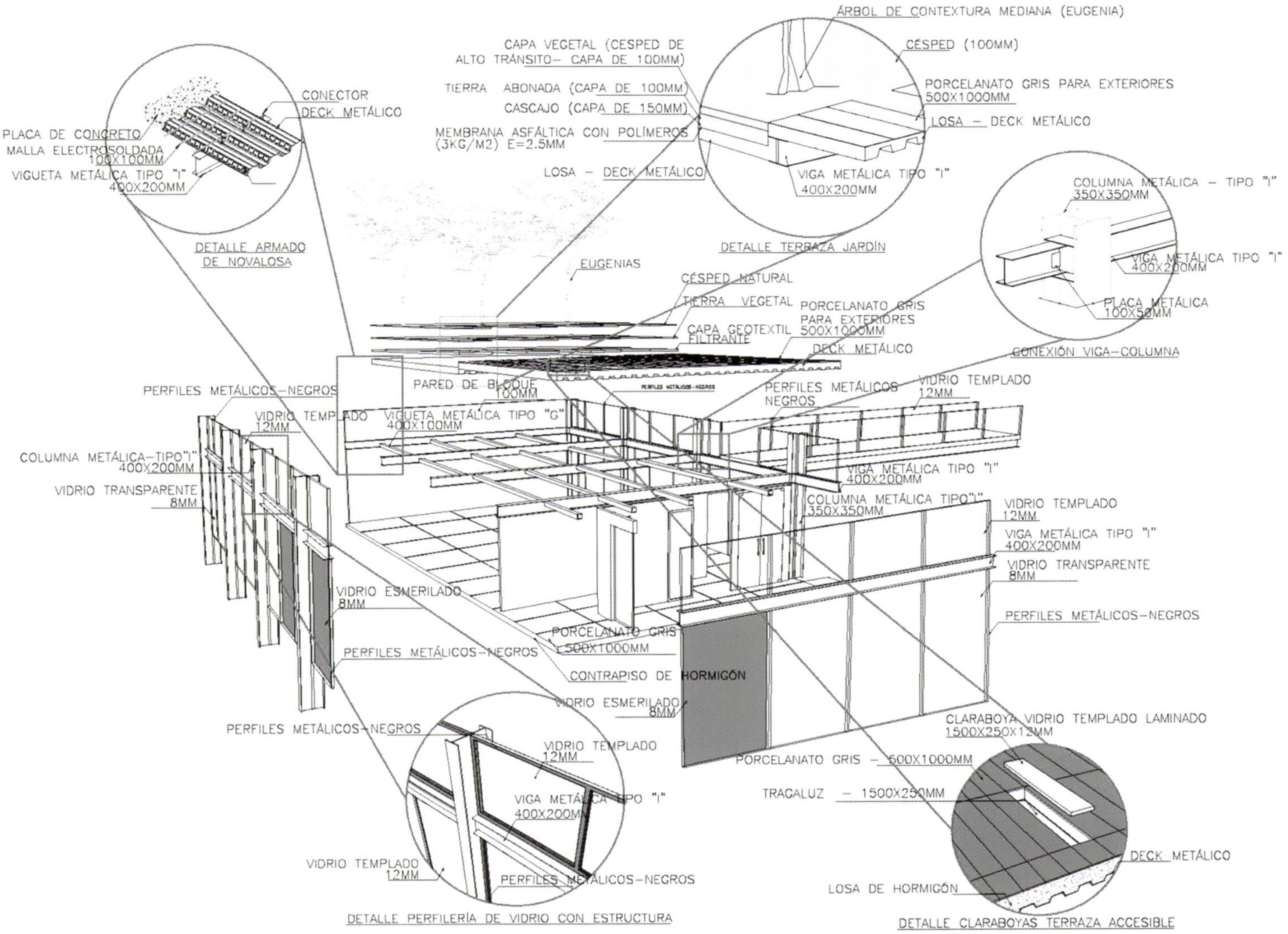

PERSPECTIVA EXPLOTADA SALON DE USO MULTIPLE sin escala

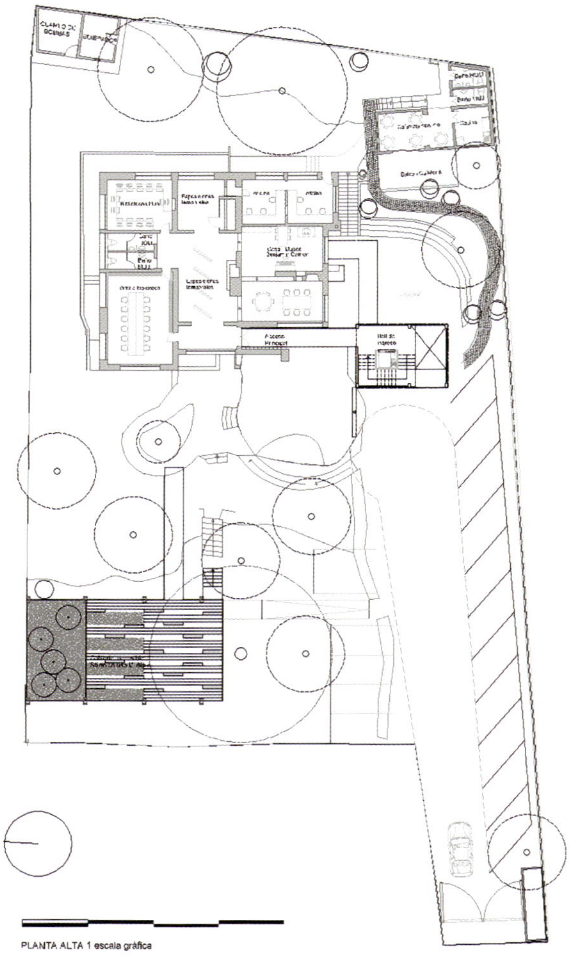

PLANTA ALTA 1 escala gráfica

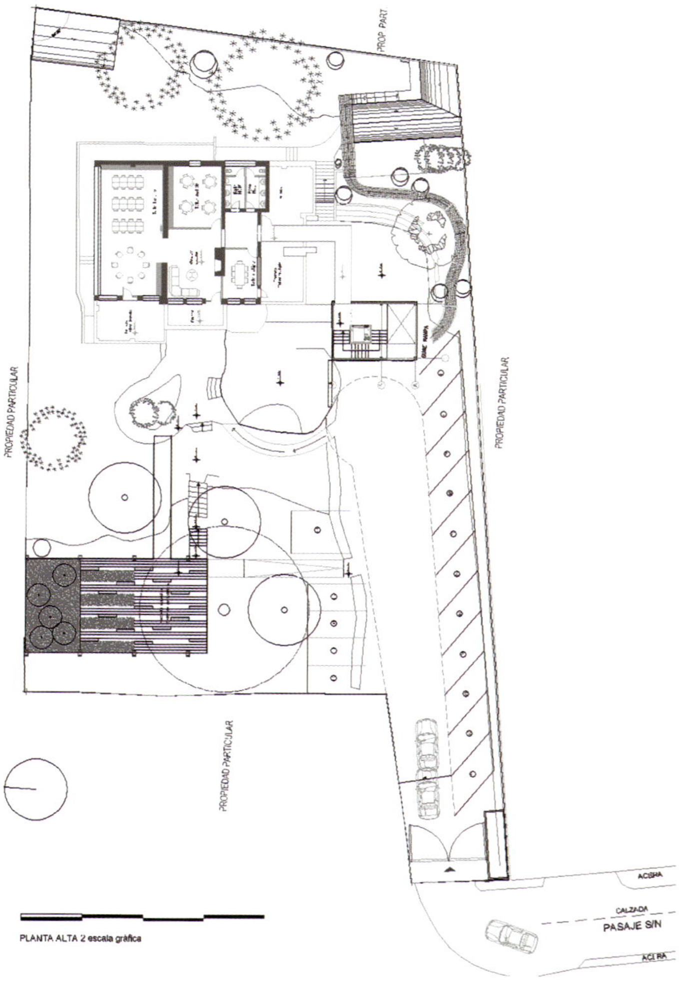

PLANTA ALTA 2 escala gráfica

SUBSUELO escala gráfica

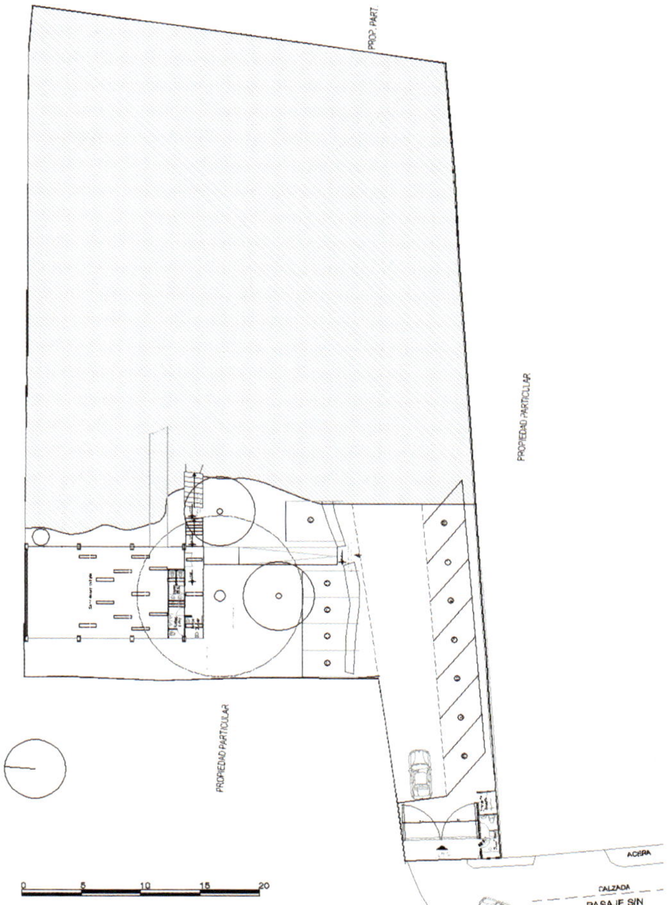

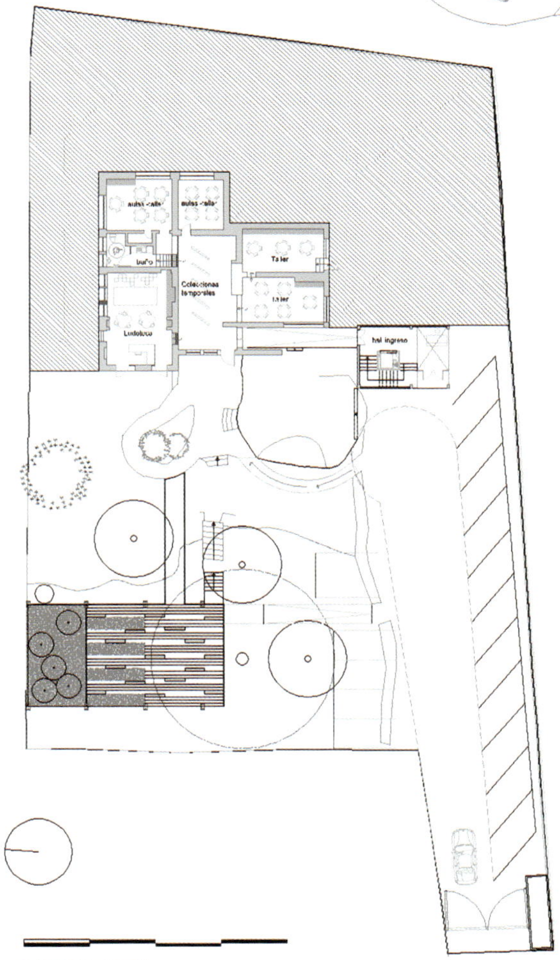

PLANTA BAJA escala gráfica

The second piece is a multipurpose hall, which for a requirement of the administration of Benjamin Carrion Foundation, needed to work autonomously to the House. The design looked for the only place in the garden where there were no trees, to place this piece, taking that garden to an upper stratum that connects with the patios of the house, through a bridge, allowing to approach the top of a large Guabo tree, that covers this green terrace with its shade. This volume is hidden in the natural topography of the lot, and its transparent skin allows a total connection with the gardens of the project. The original house was completely restored, maintaining its essence without changing its distribution, adapting the existing spaces to the new uses.

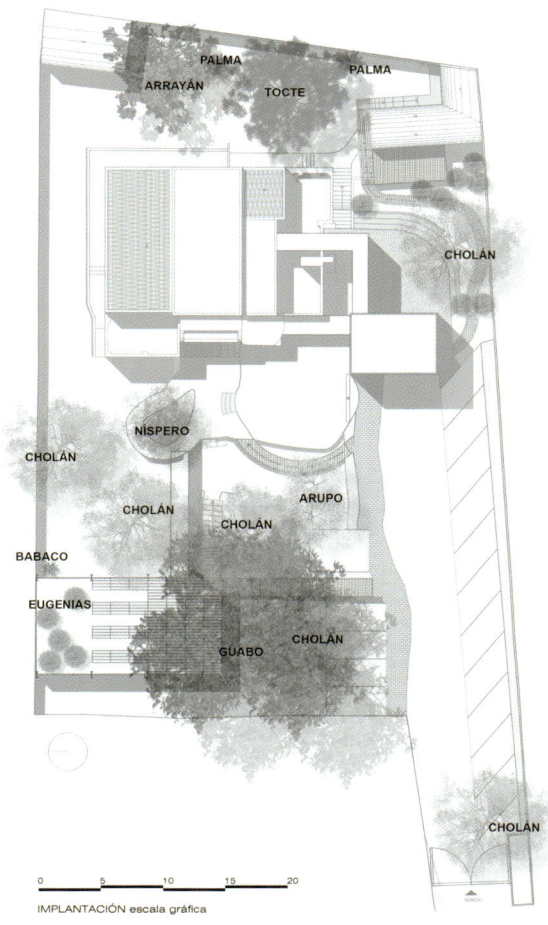

IMPLANTACIÓN escala gráfica

FACHADA FRONTAL

FACHADA POSTERIOR

The Reborn House

Architects: Alhumaidhi Architects
Location: Al-Bidea, Kuwait
Area: 1200 m²
Photographs: Gijo Paul George

Manufacturers: Franken-Schotter, Technal, Agape, MARAZZI, Poggenpohl, SALVATORI, Decormami, Jura
Main Material: Precast Concrete + Stone (Limestone Panels | Franken-Schotter)

Reborn House is a seaside property in Bidaa, Kuwait. With the western elevation facing a busy street, we opted to carve a series of openings out of the main mass. Cuts were made at acute angles and the windows were set within a slight tilt. By partially masking the windows, we were able to create shadows that minimize solar heat gain while allowing for diffused light to enter the house. On the other hand, the eastern elevation (facing the sea) benefits from deep set openings and balconies that maximize both views and shade. When the client first approached us, the initial foundation had already been laid out before they decided to change the design. The general spatial configuration from the previous design had to be retained for the organization of programs in the plan.

All main open spaces, such as dining, living and reception spaces, are situated to the East and the West while the services and corridors are along the Southern and Northern facades. Amenities, such as dressing rooms and bathrooms, receive light through the slit windows from these elevations that face the neighboring homes. This private villa includes a large interior atrium bringing in lots of natural light to the heart of the house, while the internal corridors connect the main functional spaces from both ends. The main atrium is lined with louvers that serve as form and function: shielding the view of movement along the staircase; and activating the triple high space with texture that creates a play of shadows throughout the day.

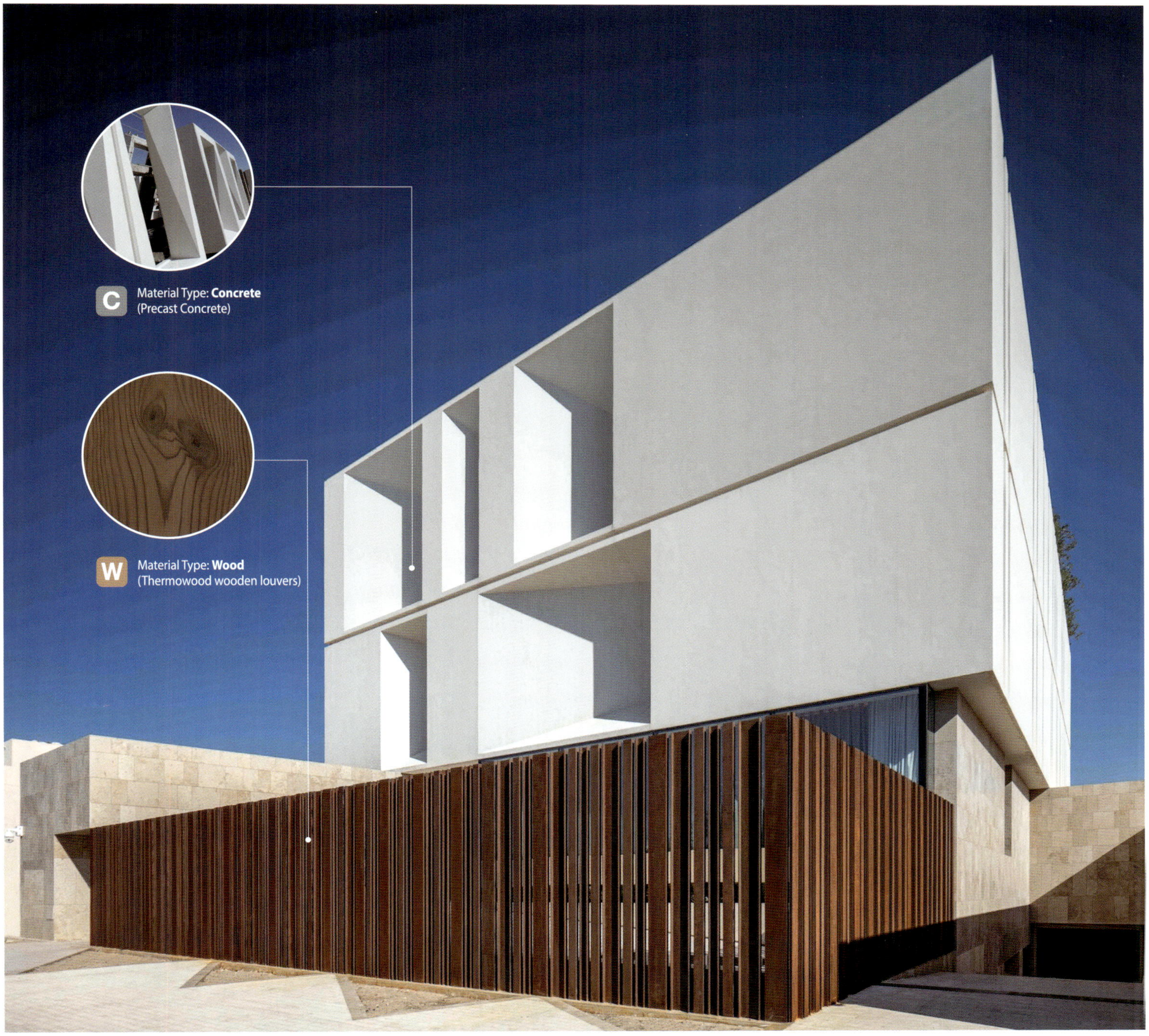

Material Type: Concrete
(Precast Concrete)

Material Type: Wood
(Thermowood wooden louvers)

The exterior volume consists of a white orthogonal mass sitting on a stone plinth that has been carved into for accessing the ground and basement levels. On the eastern façade, the carved mass acts as a sunken courtyard for the guest reception while the stone cladding continues to the interior spaces as flooring and wall cladding for the core wall. The stone wall, the white walls and the wooden louvers creates a playful contrast energizing the central atrium.

Material Type: **Stone**
(Limestone Panels | Franken-Schotter)

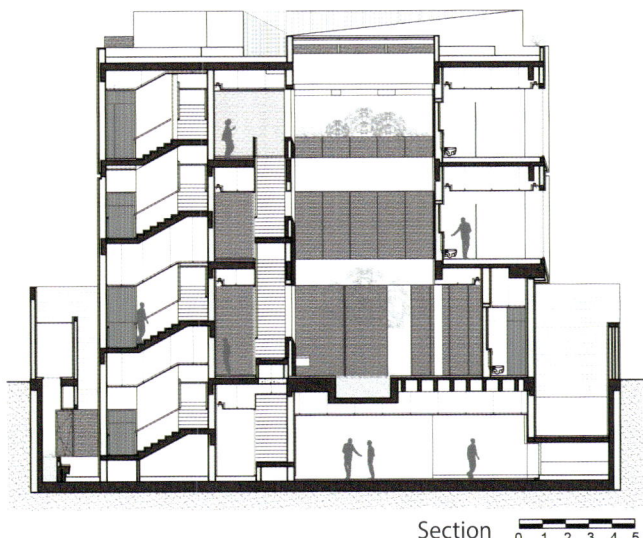

Section 0 1 2 3 4 5

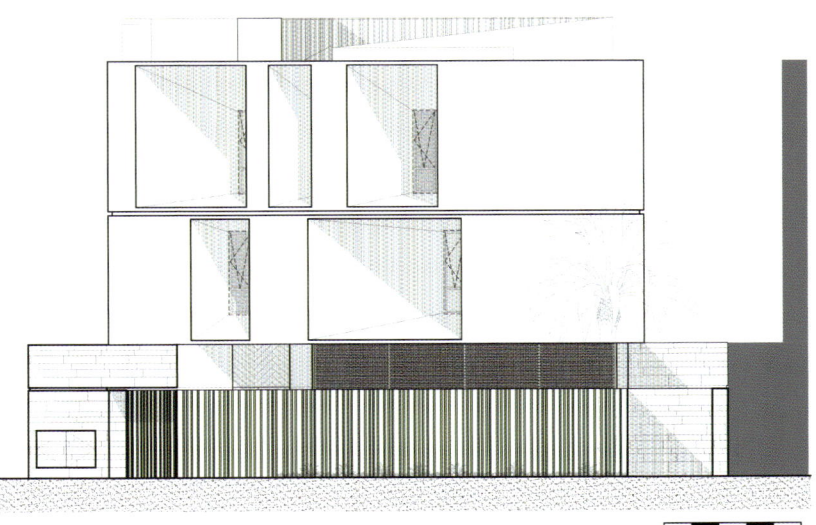

Elevation 0 1 2 3 4 5

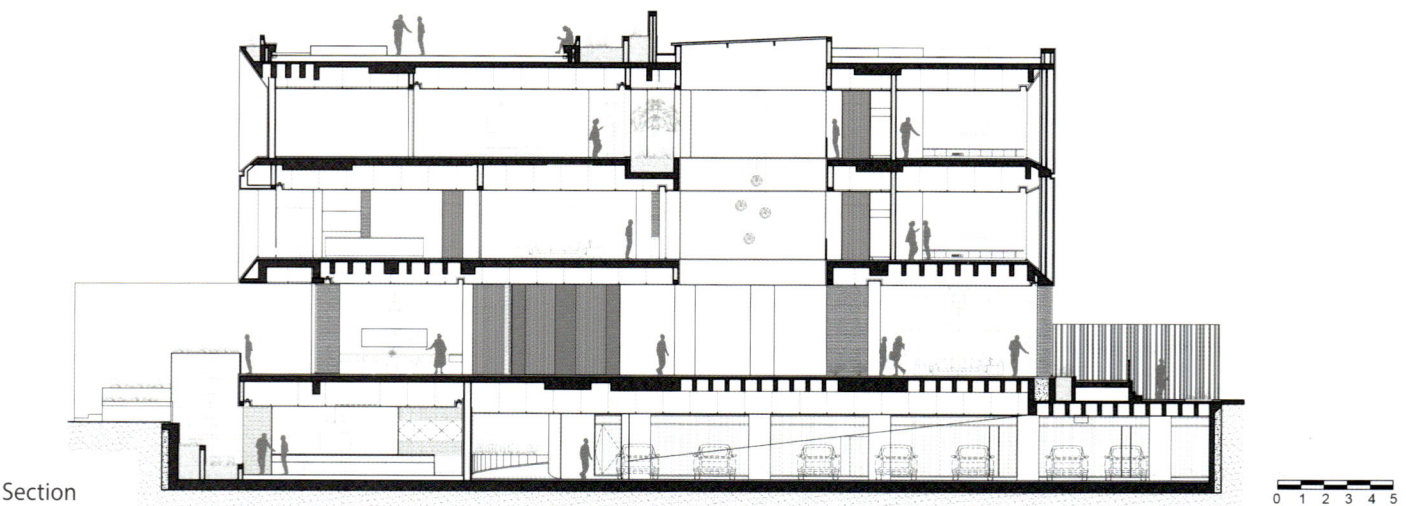
Section

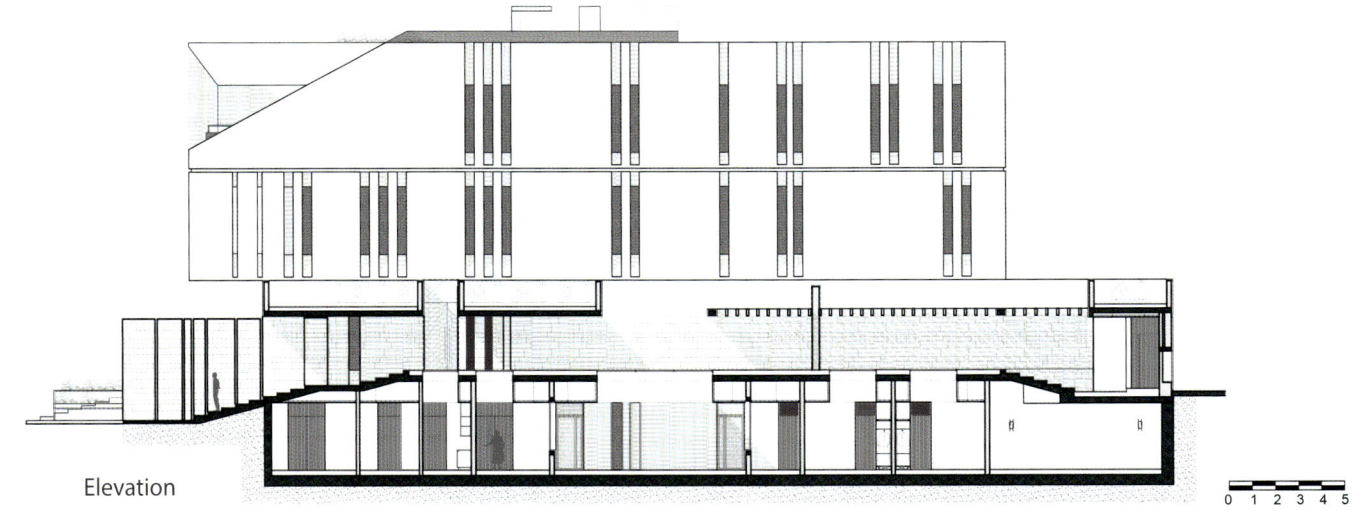

Elevation

0 1 2 3 4 5

055

Material Type: **Wood**
(Thermowood wooden louvers)

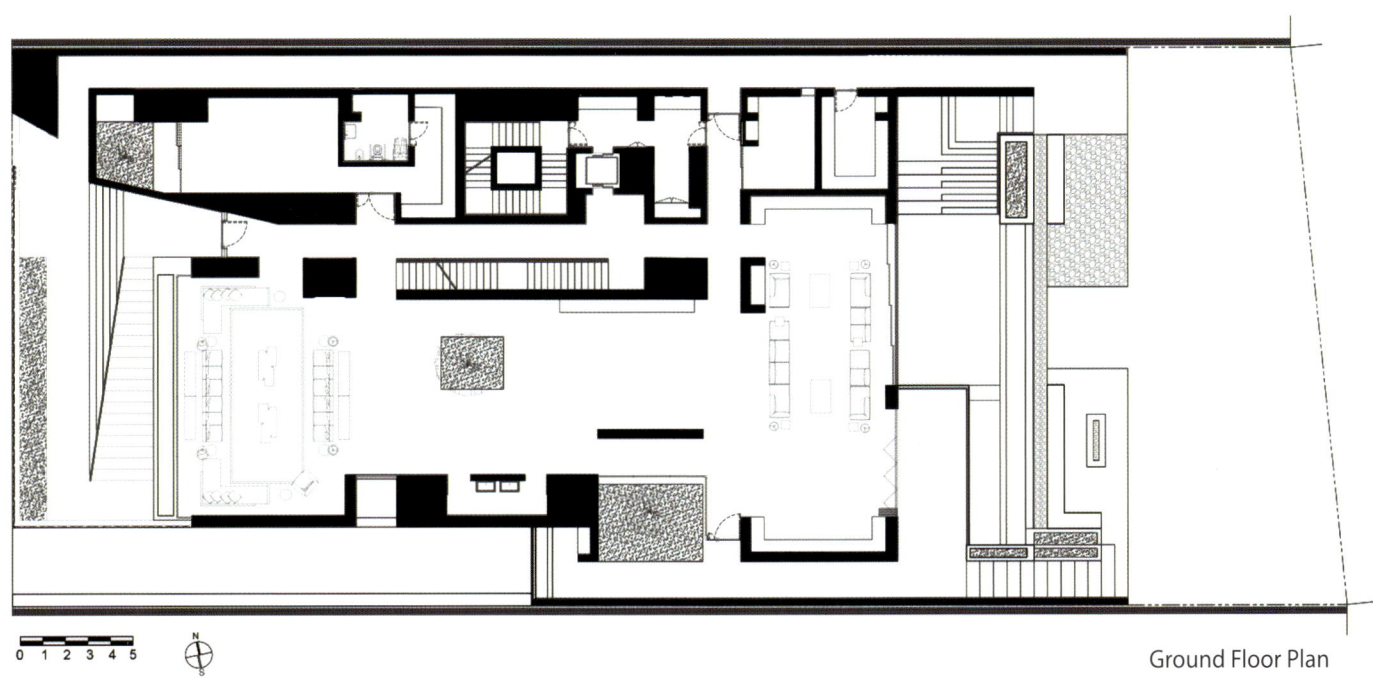

Ground Floor Plan

The Odori Hotel

Architects: Alhumaidhi Architects
Location: Kebayoran Lama, Indonesia
Area: 416 m²
Photographs: Mario Wibowo
Manufacturers: AutoDesk, GRAPHISOFT, Toto, Byoliving……
Architect In Charge: Rabani Kusuma Putra
Design Team: Gandrung, Aditya
Main Material: 10m thick cantilever concrete slab +paint that resembles CorTen (corrosion resistant) steel material

Odori hotel constructed in relatively narrow site with 6x18,8 meters area, which must have a very efficient space programs in designing, with 13 units of guest rooms, coffee shop which also functions as lobby and concierge, service area and as well roof garden on the top floor. Due to the narrow space to designed, this façade shall be facing towards to the road. This kind of Façade abstract is designed to make it look dominant among the neighborhood, which has variative function such as housing, boarding house, restaurant and retails. This abstractive façade formed in protruded windows. The shapes was deliberately created to be stand out and to represent a balcony.

However, the protruded window built with fixed glass by reason of boutique hotel requirement, which needed a high privacy and to eliminate the outside noises as well. This attractive façade effect with 10m thick cantilever concrete slab, also function as a bay window, which can be used for sitting area for occupants. Façade finish material used are quite vary. Namely wash paint that resembles CorTen (corrosion resistant) steel material, black metal plat stripes, and synthetic rattan which gives a natural impression. With massive shape of building and the used of varied materials, its shows a tectonic sculpture art in between the old buildings, the shape and its solid colors strengthen the characteristic, and make its more modern an unique.

C Material Type: **Concrete**
(10m thick cantilever concrete slab +paint that resembles CorTen (corrosion resistant) steel material)

The interior divided into 2 zone, suite and deluxe which connected by split level staircases. A strategy and solution for buildings with narrow lands. Among the room zones, there's a full void with skylight roof on top. it helps to distribute the skylight to the heart of the building and illuminate to the common area, café and as well the lobby, all day long, less waste energy to be pursued. One of the ideas why the roof garden made, is due to the lack of greenery on ground level, it can be enjoyed by the visitors as an common open area. Besides, this functionate as a dry garden, which consist of scattered white coral and lush shaded of Terminalia Catappa tree.

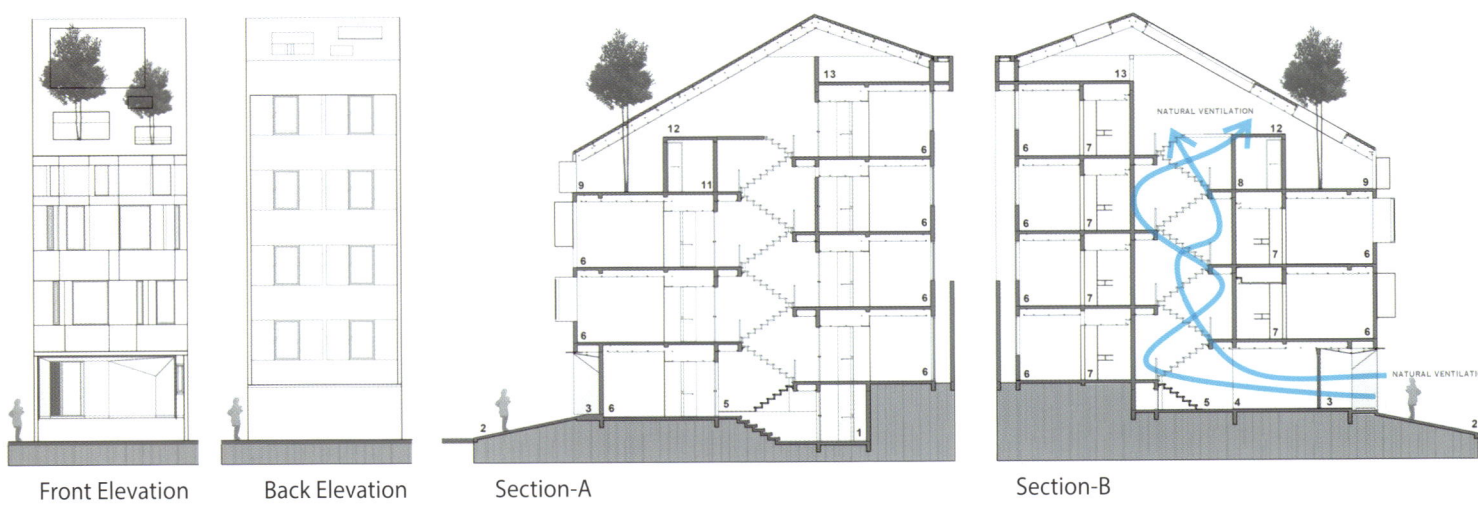

Front Elevation　　Back Elevation　　Section-A　　Section-B

 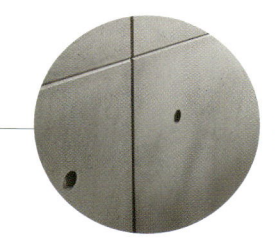

M Material Type: **Concrete**
(Black metal plat stripes)

C Material Type: **Concrete**
(Exposed concrete wall panel)

The building has 4 units of suite room and 9 units for deluxe room. Both has the same interior concepts, the difference is only at the floor area. The interior concept was made very compact regarding to the hotel occupants activities which regularly came for business travel. Bed is designed like very compact stage shape, which has a same function as Tatami (Traditional Japanese bed). This room also has a wood installation which has very poetic effects and also has clothes hanger functions. Bed and bathroom are divided by glass, to give spacious impression and ambiences feel so warmth with wood texture touch. White color stairs and railing was made from perforated metal material. Hence, the interior will seem more light and clean, the aim is to eliminate the stiffness impression for the building which has a very limited and narrow land.

W Material Type: **Concrete**
(Synthetic rattan)

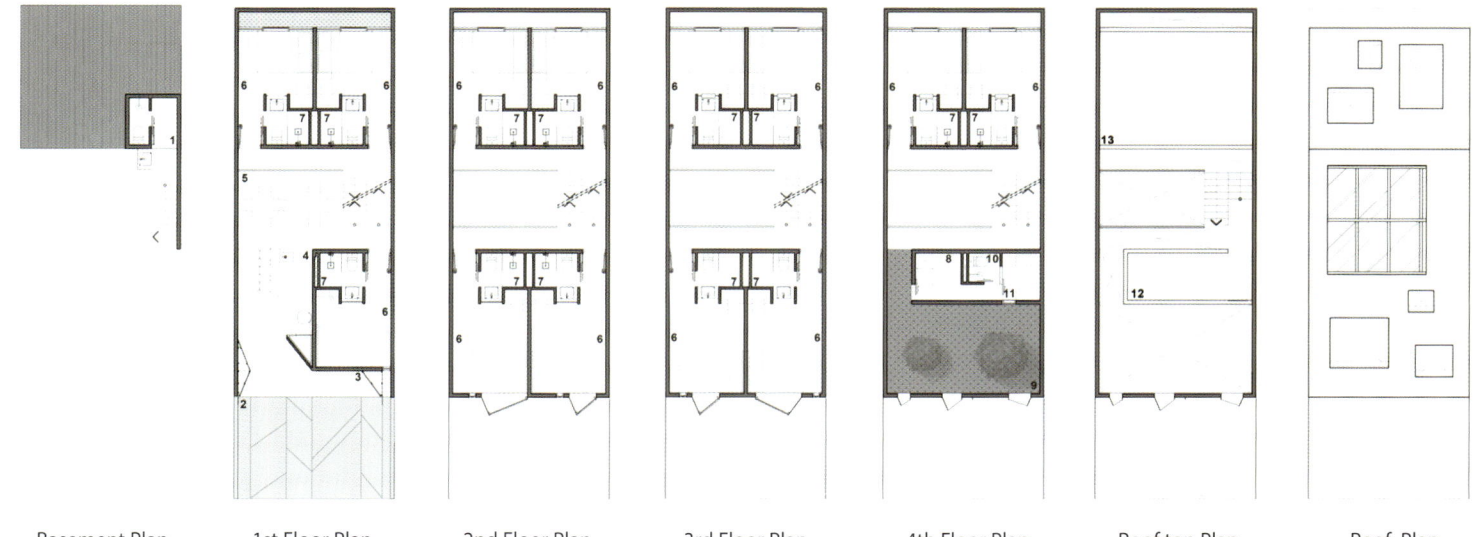

Basement Plan   1st Floor Plan   2nd Floor Plan   3rd Floor Plan   4th Floor Plan   Roof top Plan   Roof Plan

Tortosa Law Courts

Architects: Camps Felip Arquitecturia
Location: Tortosa, Spain
Area: 3467 m²
Photographs: Pedro Pegenaute, Josep Maria de Llobet
Manufacturers: AutoDesk, Erco, GEZE, COMLED, HERMO
Main Material: Concrete (corrugated prefab concrete pieces)

The city of Tortosa throughout its history is built from the overlapping of fortifications, walls, churches and palaces. Its historic center with narrow alleys and complex geometry, adapts on the hillside of the mountain. The new Law courts building responds effectively to the specific judicial program and its strict functional requirements and, to the topography and geometry of the site, located on one of the corners of la "Plaça dels Estudis".

The Law Courts are located in one of the streets that, from the Ebro river, go into and configure one of the main routes of the historic center and where gothic Palaces, the Cathedral Headquarters and the main buildings of its Renaissance past. All these constructions have in common the use of the stone of the area that configure a homogeneous chromatic environment with which the new architecture establishes continuity. An architecture that materializes and completes a fragment of a city that responds carefully to all the conditions of the place.

C Material Type: **Concrete**
(corrugated prefab concrete pieces + paint)

The project starts from a regular and ordered structure through a grid scheme modulated from the standard dimensions of the program's office unit. The program distinguishes the public access areas, which are solved with more open, permeable and continuous facades; and areas of restricted or private use, which are modulated and reated with the rhythm of the facade to allow possile future changes. The strict separation of public and private areas as well as prisoners or witnesses is also reflected in the paths and access cores. Public and private paths are carried out along the longitudinal facades. Placing the public and waiting areas on the north facade with views of the castle and the old town. Public and private are divided by a central area delimited by the customer service, which houses the restricted uses of offices and work.

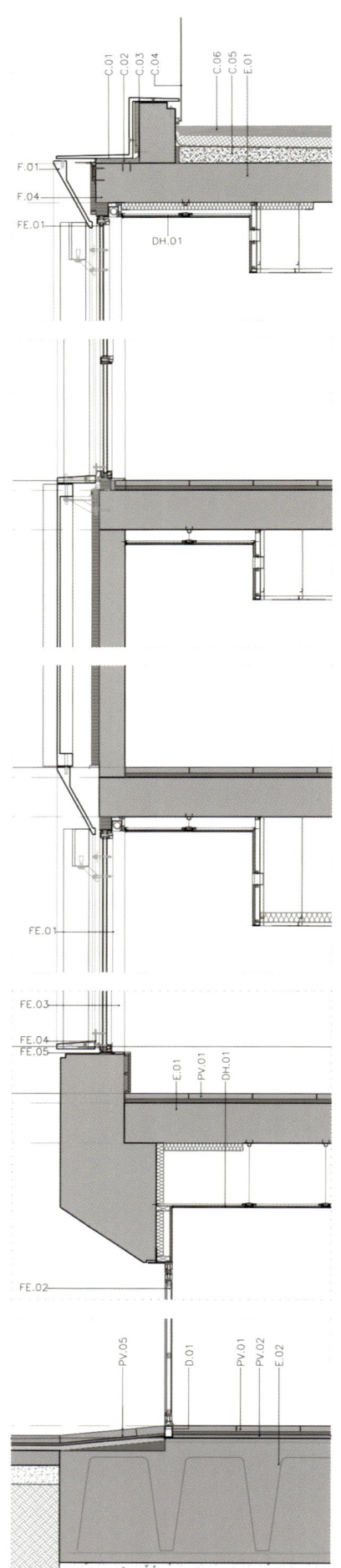

DETALLS FAÇANA

E.01 Forjado reticular de hormigón insitu e= 32 cm
E.02 Solera aligerada para elementos de tipo Caviti e=57+15cm
D.01 Pintura impermeabilizante
C.01 Làmina impermeable de refuerzo, tela asfàltica con media caña
C.02 Base de mortero
C.03 Acabado perimetral de cubierta GRC con fixación metàlica a murete perimetral
C.04 Barandilla de piezas tubulares de acero Ø=6cm cada 10cm i pasamano de pletina 5cm e=8mm
C.05 Formigó alleugerit de pendents e=8cm
C.06 Acabat de palet de riera
FE.01 Ventana con vidrio oscilobatente y vidrio fijo climalit (5+5)+12+(4+4) Stadip con carpinteria de aluminio tipo Unicity HI de Technal y vestigio de tubos de acero
FE.02 Puerta doble con revestimento metàlico con marco oculto con subestructura interior tubular reforzada y aislamiento de alta densidad, acabado de acero pintado color RAL Aluminio
FE.03 Remate interior de aluminio 3mm
FE.04 Platina de remate oberturas de aluminio 3mm
FE.05 Chapa de remate perimetral de las aberturas 4mm
PV.01 Pavimento de terrazo liso de grano pequeño de 50x50cm o formato rectangular sobre capa de mortero, color a definir per DF.
PV.02 Capa de aislamiento tipo corcho e=30mm con lámina de PVC con tapaporos de 96g/m2
PV.05 Pavimento exterior de losas de granito de 40x60x6 cm adherido con mortero cola, sobre base de regulación de mortero y solera de hormigón armat

DH.01 Falso techo de placas de hieso laminado de 13mm de grosor. Sistema fijo con entremado oculto con suspensión autoniveladora de barra roscada. Color negro

C Material Type: **Concrete**
(corrugated prefab concrete pieces + paint)

M Material Type: **Steel & Metal**
(Copper Surface - Classic Coated | TECU®)

TÍTOL DEL PROJECTE
CONSTRUCCIÓ EDIFICI JUDICIAL TORTOSA
A4 e: 1/50
OLGA FELIP ORDIS
JOSEP CAMPS POVILL

The detainees area is directly related to the private elevator that connects the cells to the main conference room, which has a protection element that preserves privacy between detainee and witness. There is a stairs and elevator public core related to the principal hall and in contact with the outside. Likewise, in relation with the private spaces there is a stairs and elevator core for restricted use. The volume of the building, responds to the alignements of the plot following the urban lines of the historic center: narrow streets with cloe-up views and foreshortening. A stony concrete plinth manufactured on site forms the first level of the ground floor and the first contact with the urban space and the topographic difference.

In contrast to the plinth, a second level with three floors is defined by an industrialized system of three modules of approximately 1000x3000 mm of prefab concrete with which it is possible to systematize the building while solving different construction situations: the corners, the openings in relief and the slat system...A play of light and shadows provides continuity to the entire facade. The internal structure is regular and ordered. On the other hand, the exterior volume responds to the irregularity of alignments and topography. Between the inside and outside, an interstitial space of variable thickness appears and allows to mediate between these two worlds, providing intimacy and permeability to workspaces and modulating natural light.

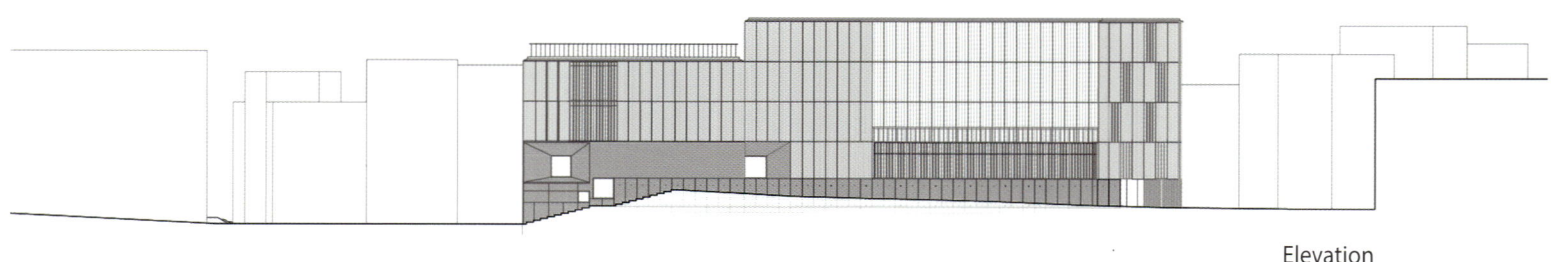

Elevation

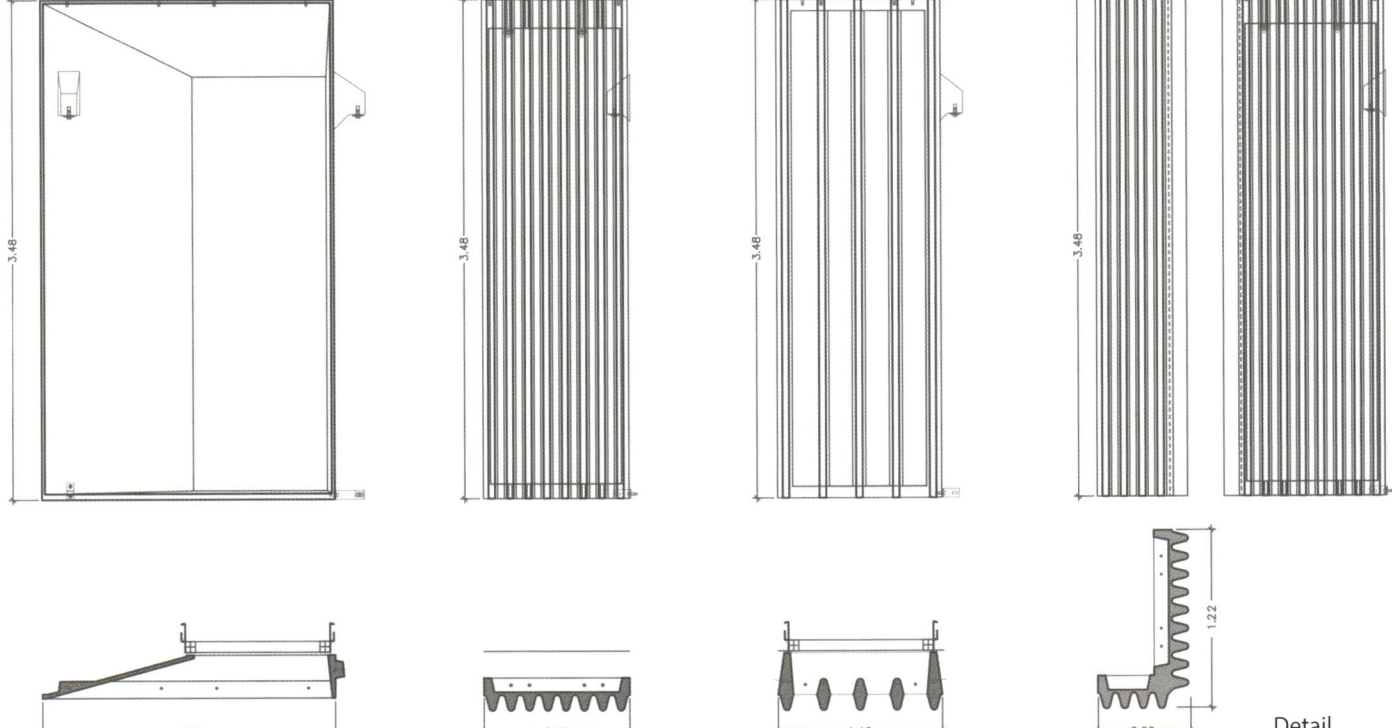

Detail

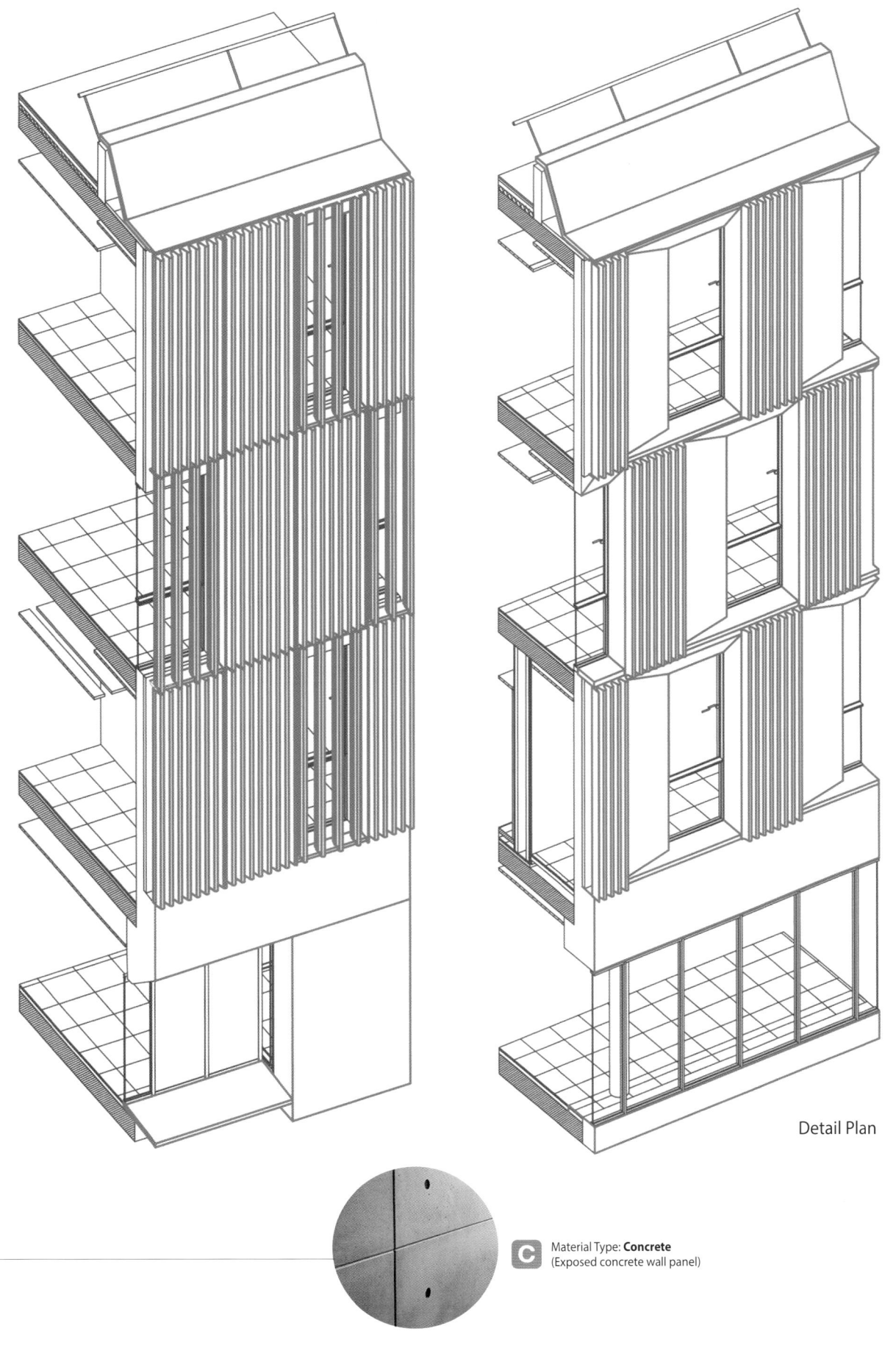

Detail Plan

C Material Type: **Concrete**
(Exposed concrete wall panel)

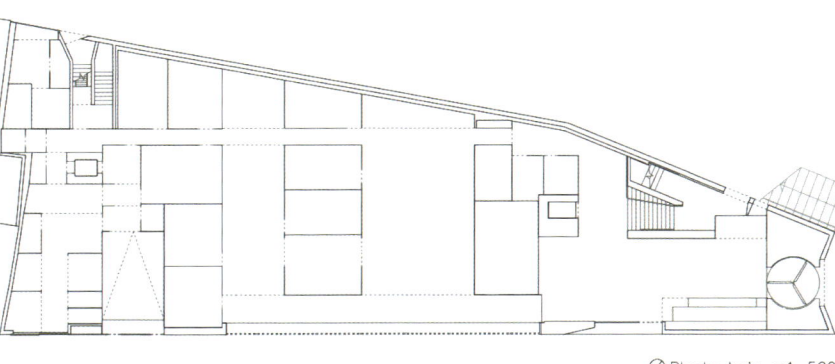
Planta baja e:1_500

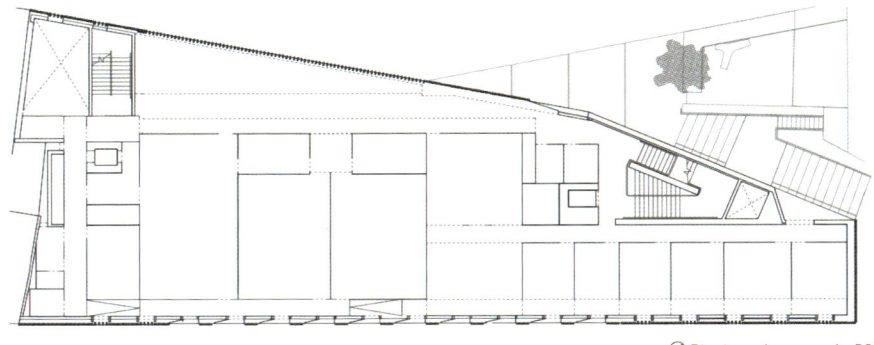
Planta primera e:1_500

Different gradients of thermal comfort and sun protection are identified in the interior spaces according their activity and orientation and good energy control is facilitated, reducing demand and energy consumption. Efficiency is optimized thanks to the use of natural lightning and use of low consumption elements.

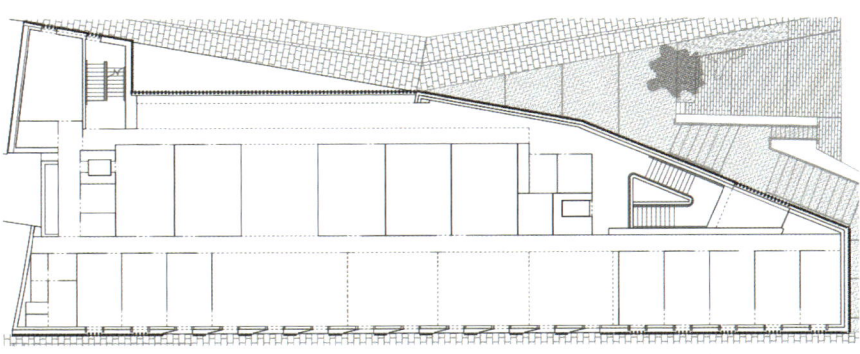
Planta segunda e:1_500

The facade of the new building uses the warm color palette of the environment to give texture and shape to the corrugated prefab concrete pieces that configure the building's envelope and trace the irregular shape of the place.
In this specific context and with a generic and at the same time rigorous and specific functional program, the new building for the Law Courts of Tortosa offers an attentive, respectful and contemporary look. A solid and at the same time permeable architecture that blends and merges with its surroundings without giving up its time.

075

Viewpoint House

Architects: Jim Caumeron Design
Location: Quezon City, Phillppines
Area: 400 m²
Photographs: Bien Alvarez

Architect In Charge: Jim Caumeron
Manufacturers: Blanco, Catalano, Hafele, ABK Re-work, Boysen, Ceasar Stone...
Main Material: Concrete (white concrete "hood")

This is a house with warm interior-core sheltered by a white concrete "hood" with trapezoidal niches that frame the outside views. It is called Viewpoint House designed by Jim Caumeron Design. Located in a dense subdivision in Quezon City, Philippines. the L-shaped lot called for an L-shaped plan to fit the hefty 400 square meter client's space requirements. The property's east and back side has no views because it is blocked by a fence-wall owned by neighbors. The west side on the other hand exposes the lot to the heat of the afternoon sun. This side also fronts the street access and a small community park that can pose an issue on privacy if the living areas were to face west.

Jim Caumeron proposed to highlight the view of the park by providing a huge picture window at the corner of the house. This serves as the "mother" window, raised above a standard human height to lessen the visibility of the living room from the street. When one is lounging in the living area, they can look up to the big window with the view of the crown of the tree. The entrance hallway and stairs were located on the west side with a ribbon window facing the street and the park. Above it is a wall of small punctured windows with a profile section-detail that is slanted to keep the rainwater out while still allowing natural air to seep through it. This wall blocks the sun's heat in the afternoon but at the same time allows the stair atrium to breathe.

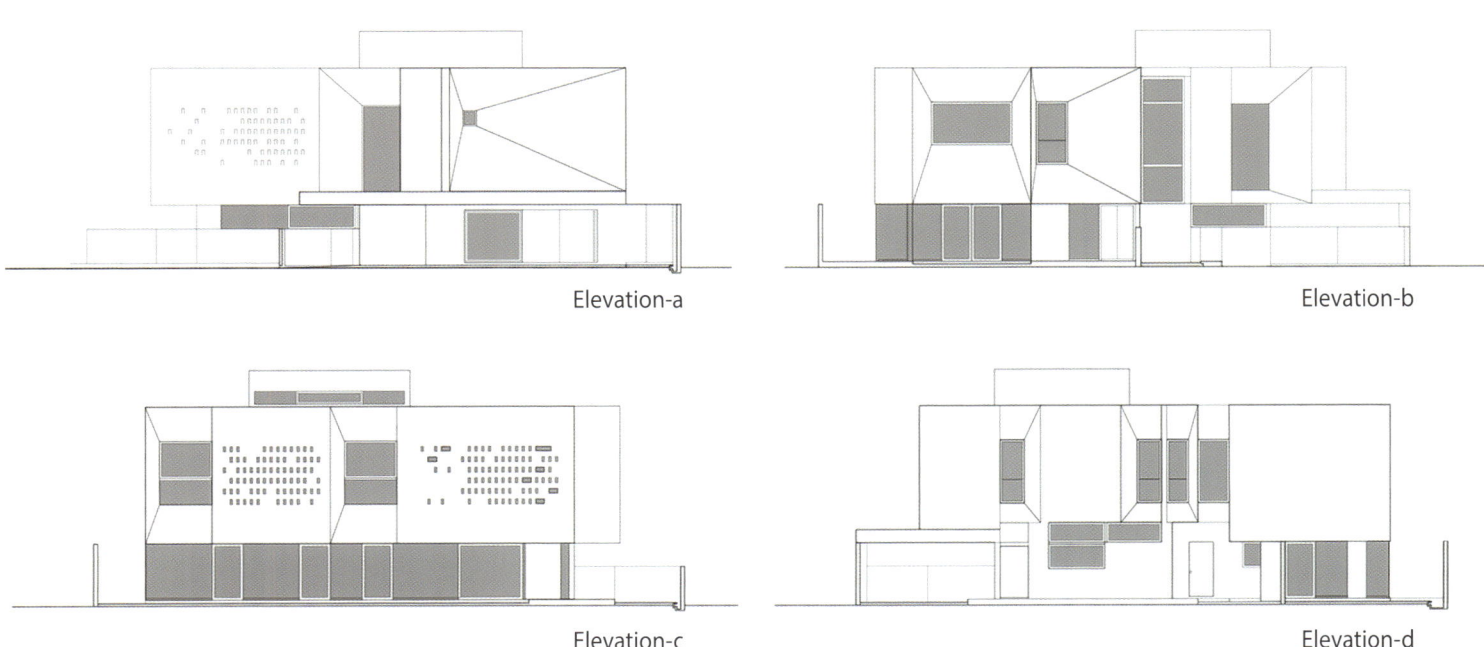

Elevation-a

Elevation-b

Elevation-c

Elevation-d

Material Type: Concrete
(white concrete "hood")

This is a house with warm interior-core sheltered by a white concrete "hood" with trapezoidal niches that frame the outside views.

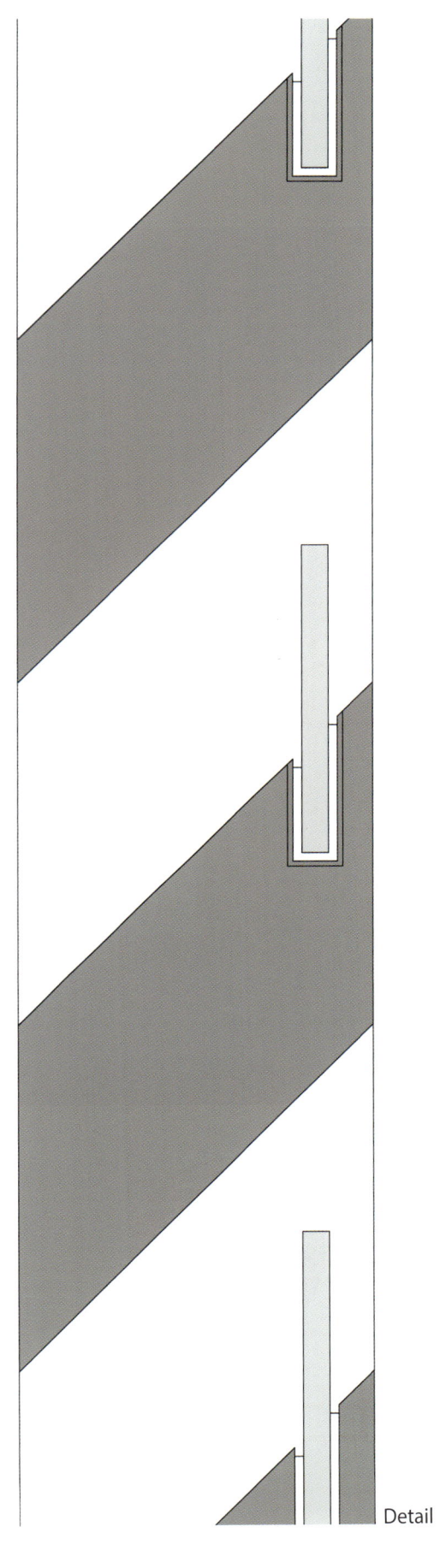

Detail

The setback was kept open as small private open areas of the dining room and the guests' room at the ground floor. The garage adjacent to the living areas is transformable into an extended party area, highlighted by a square frosted glass door from the living room. It is typical for Filipinos during fiestas, christening, house blessings etc. to welcome everyone in the community to feast. Normally, the garage area where a big table can be set up is the best space for it to keep the main living and dining areas private and indirectly exclusive to family and close friends.

At the second floor, two hallway windows are located on both ends for light and cross-ventilation. The hallway has a flank of doors that are recessed on the wall echoing the profile of the window viewports. Behind these recessed spaces are storage and closet spaces. One of these doors is the library with customized tables, lamps and bookshelves. The space was meant to be fully-air-conditioned but the architect insisted to make the picture window along the hallway to be openable.

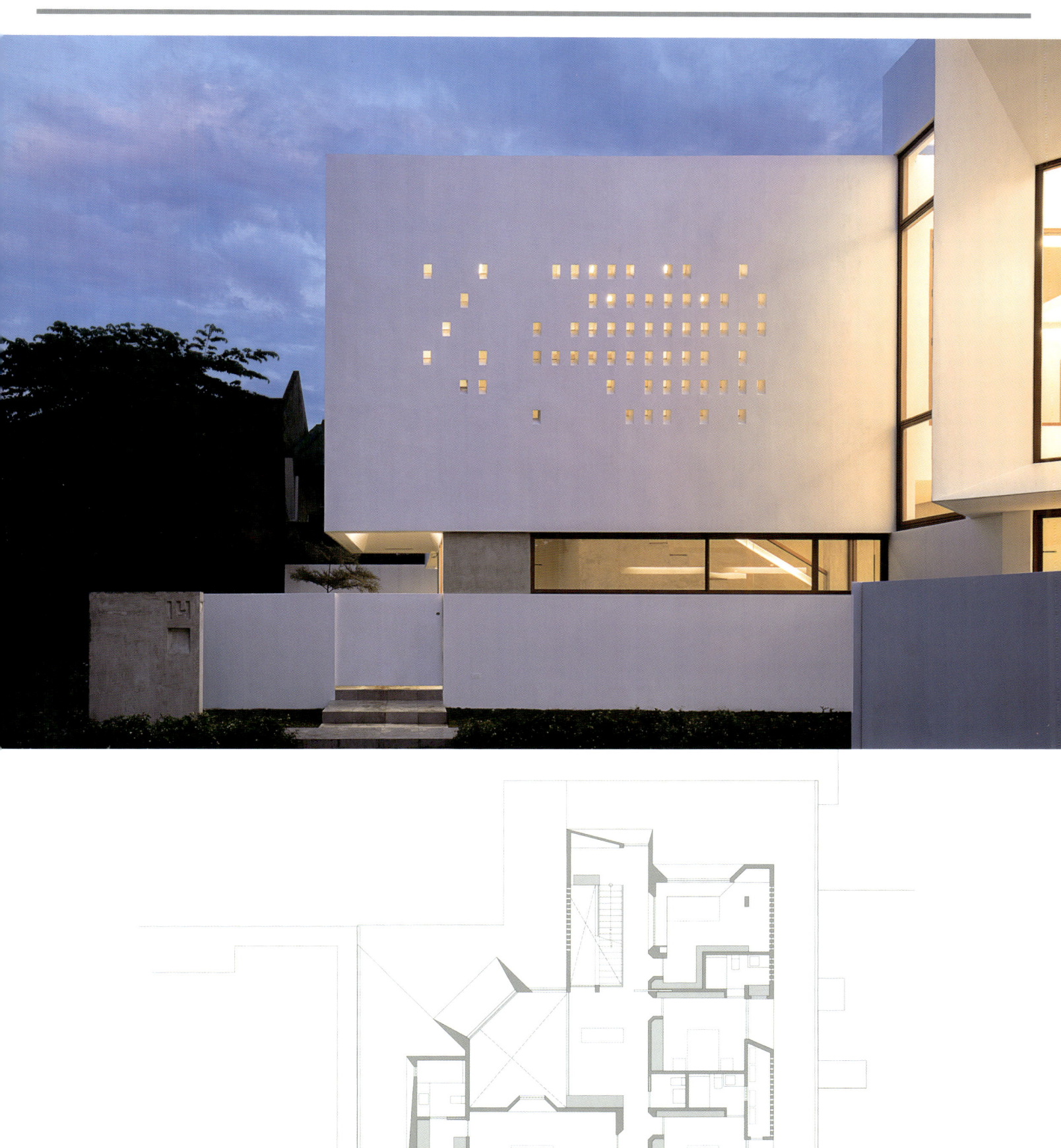

Floor Plan-A

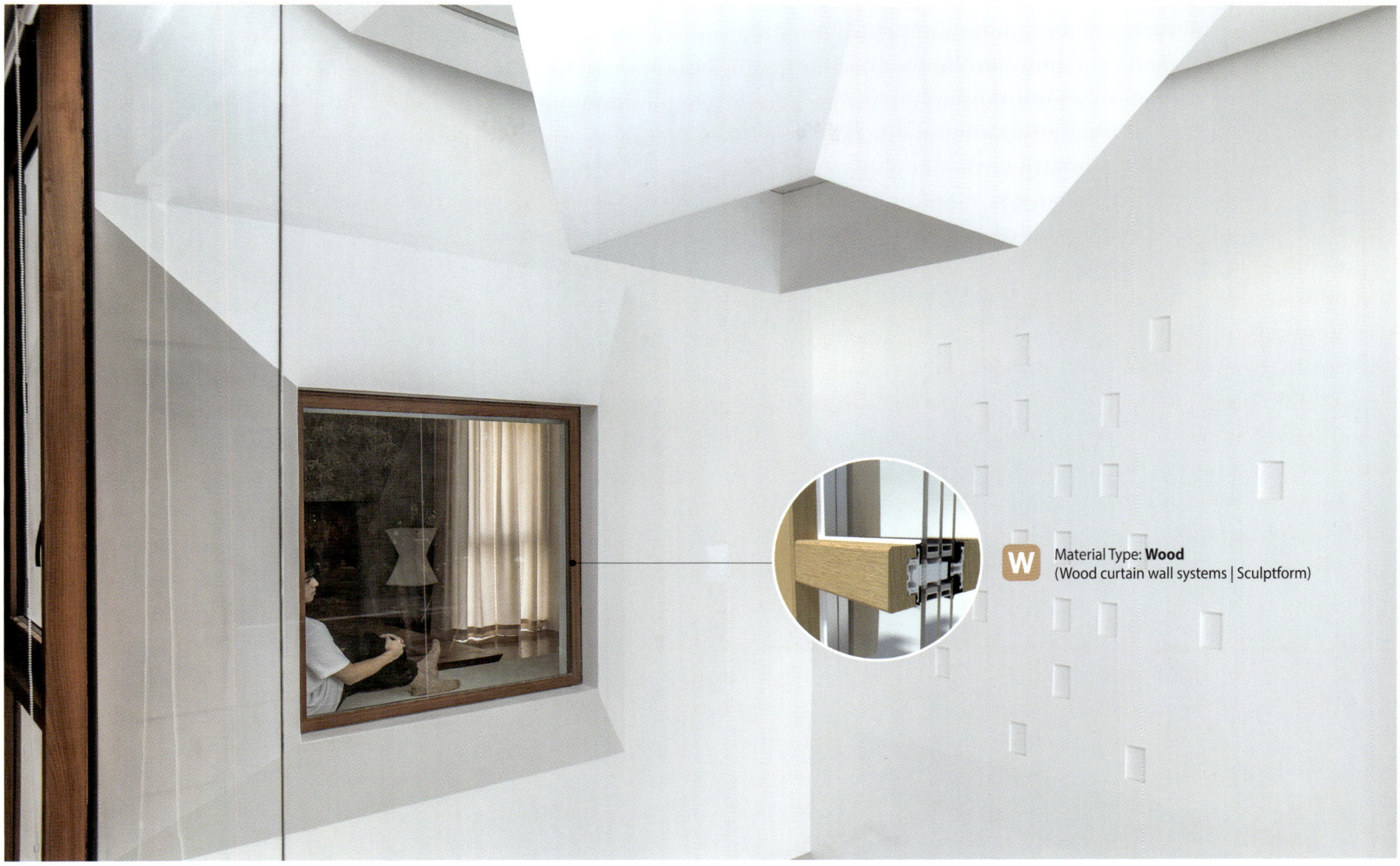

Material Type: Wood
(Wood curtain wall systems | Sculptform)

There are also small random windows on the opposite side of the library with few of them that are operable. The window frames were customized wooden casements with magnets for closing. The small windows reflect the punctured windows of the west façade and becomes a backdrop of the library interior when looking towards the hallway. A portion of the corridor is partly family room. When the air-condition is on, a pocket glass door can be pulled out from the wall making the hallway area part of the space.

Material Type: **Wood**
(Timber Click-on Battens | Sculptform)

Floor Plan-B

The owners' local hardwood collection was used for the second floor and wood finishes that dictated the color scheme of the house. The architect imagined the wood finishes as the warming and comforting element in the interior. The geometric visual language of the exterior can also be seen on the ceiling design and interior furnishings Jim Caumeron designed.

The "mother" window is also used as view portal by two interior windows – the family room and the masters' bedroom at the second floor. Instead of having a boring blank wall opposite the masters' bed, the architect provided a bay window seating that overlooks the living areas, family room and the view of the park. Same views can be enjoyed in the family room. This makes the tree at the corner of the park referenced as the viewpoint of the house.

FTE Office Building

Architects: Plan Architect
Location: Khet Suan Luang, Thailand
Area: 5644 m²
Photographs: Yama studio

Architect: Wara Jithpratuck, Naphasorn Kiatwinyoo, Sira Anugoolprayote
Manufacturers: American Standard, SCG, TOA, WDC, WILLY
Main Material: Panels cladding system(polymer concrete)
(Water Facade Panels in Uneatlantico | ULMA Architectural Solutions)

Fire Trade office is a 6-floors office building located in the suburban area of Bangkok, Thailand. The building is surrounded by a low-density residential area with plenty of open space. Fire Trade Engineering is the company that sells Fire-protection and life safety systems in the building, thus the client emphasis on efficiency and engineering aspect of the building. 1st floor of the building consisted of a parking area, mechanical room, and other service functions. The 2nd floor, comprised of a large semi-outdoor terrace that led to the reception area.

From the 3rd floor to the 6th floor, the typical plan of the office can be divided into 3 sections, the core, the semi-outdoor transition area, and the flexible open planning office. The core section of the building consisted of stairs, elevators, WC, and MEP room lined up along the western side of the building. The semi-outdoor section located between the core section and the office consisted of a common area with an enclosed pantry.

The office section consisted of a flexible open planning workspace. The small meeting rooms, photocopy room, and storage room are located at the center of the office floor for ease of access. The challenge of the project is how to create energy-efficient architecture while complying with the owner's Feng shui belief. To achieve that, the architect has to strategically use common design elements that meet both passive design principles and Feng shui requirements.

P **C** Material Type: **Panels (Concrete)**
(Polymer concrete for cladding system: graphite gray color)

Material: ULMA system is based on a system of primary and secondary profiles of extruded aluminum where the plate is held by a slit in the lower and upper face that allows the fixation to be completely hidden.

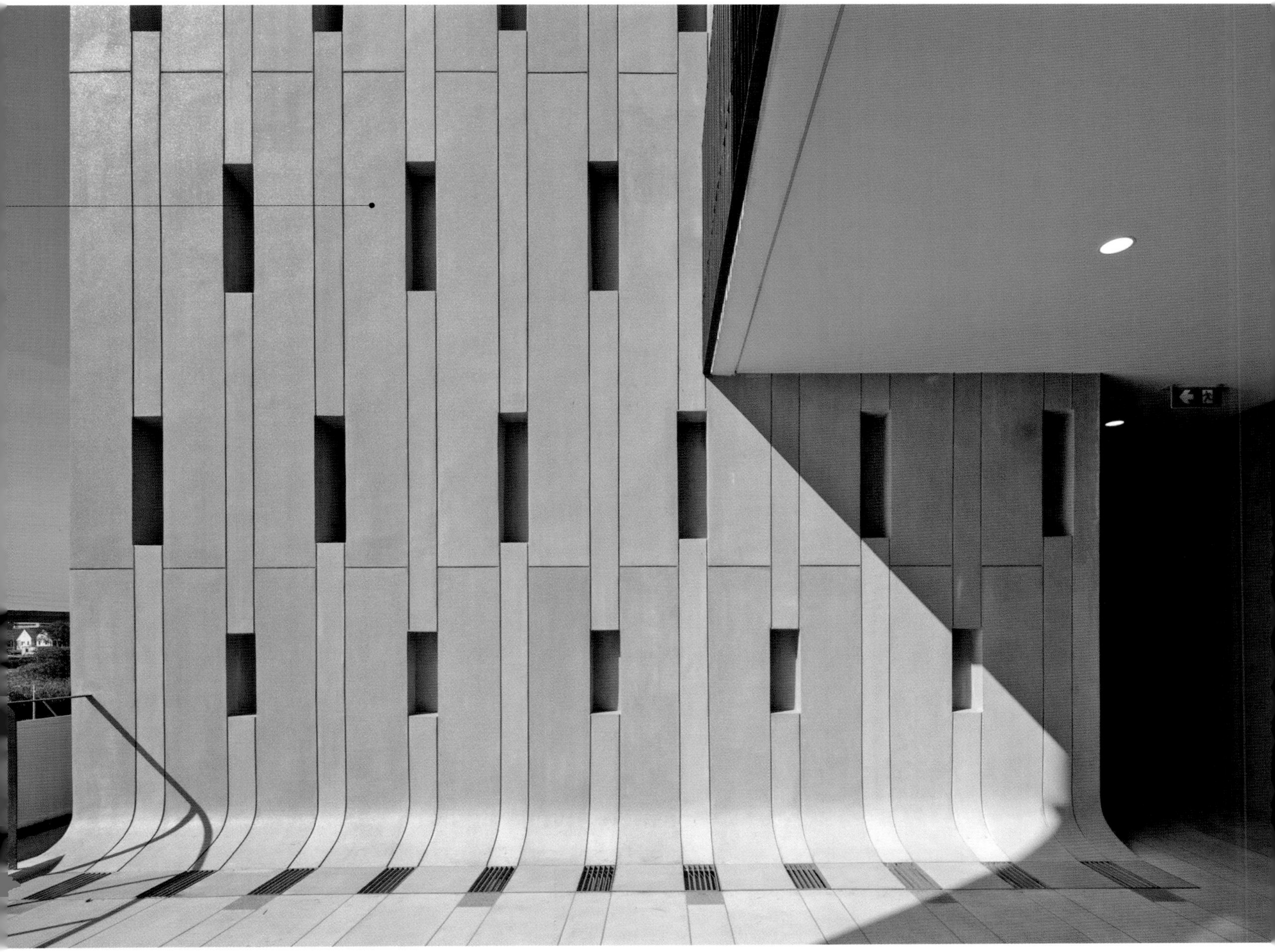

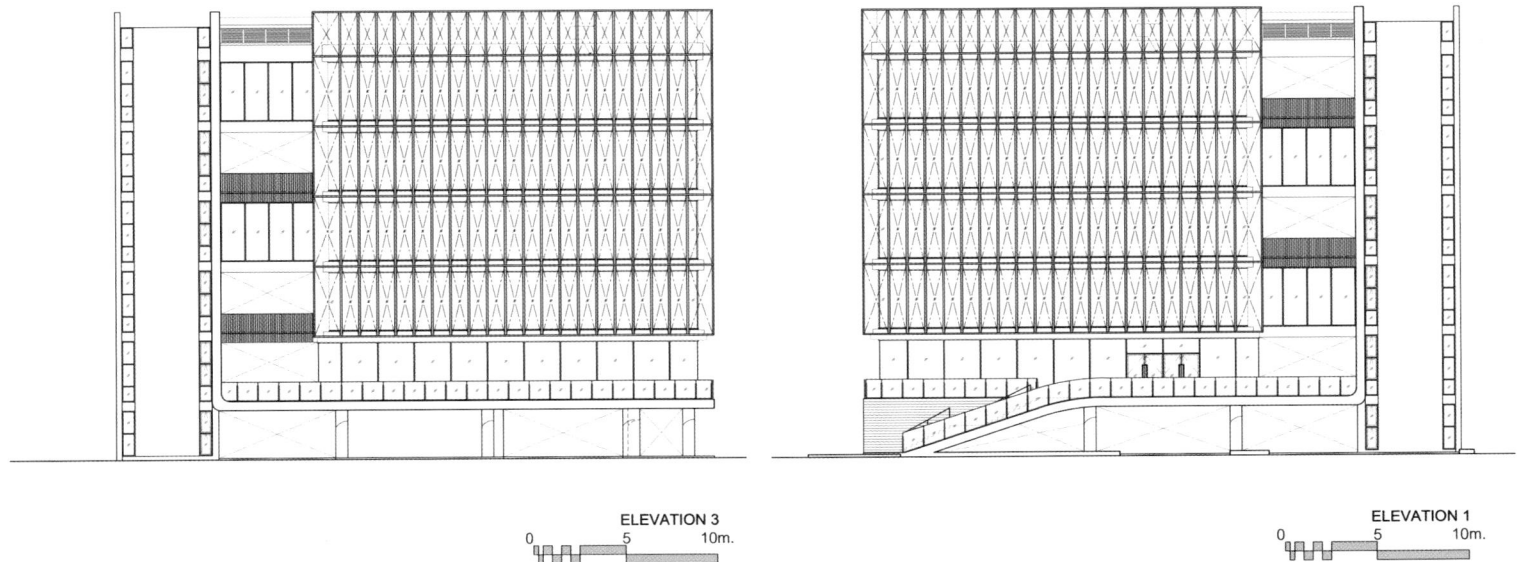

ELEVATION 3　0 5 10m.

ELEVATION 1　0 5 10m.

Material Type: **Steel & Metal**
(Vertical fins facade(wood-texture): Copper alloys; TECU® Classic)

Firstly, the prominent feature of the building is the large slab that transforms into various building components. Since the parking space is located on the ground floors, one corner of the slab touches the ground to create the stair that acts as main access from the ground level to the 2nd floor. The slab then transforms into the vertical wall that significantly reduces solar heat gain from the west during the afternoon. Secondly, vertical fins facade reduce glare from direct sunlight, eliminate the need for additional interior blinds. These wood-texture fins also fulfill the Feng shui requirement. Thirdly, the semi-outdoor atrium connected from the 2nd floor to the 6th floor facilitates passive ventilation and stack effect while provides a relaxing common area for users.

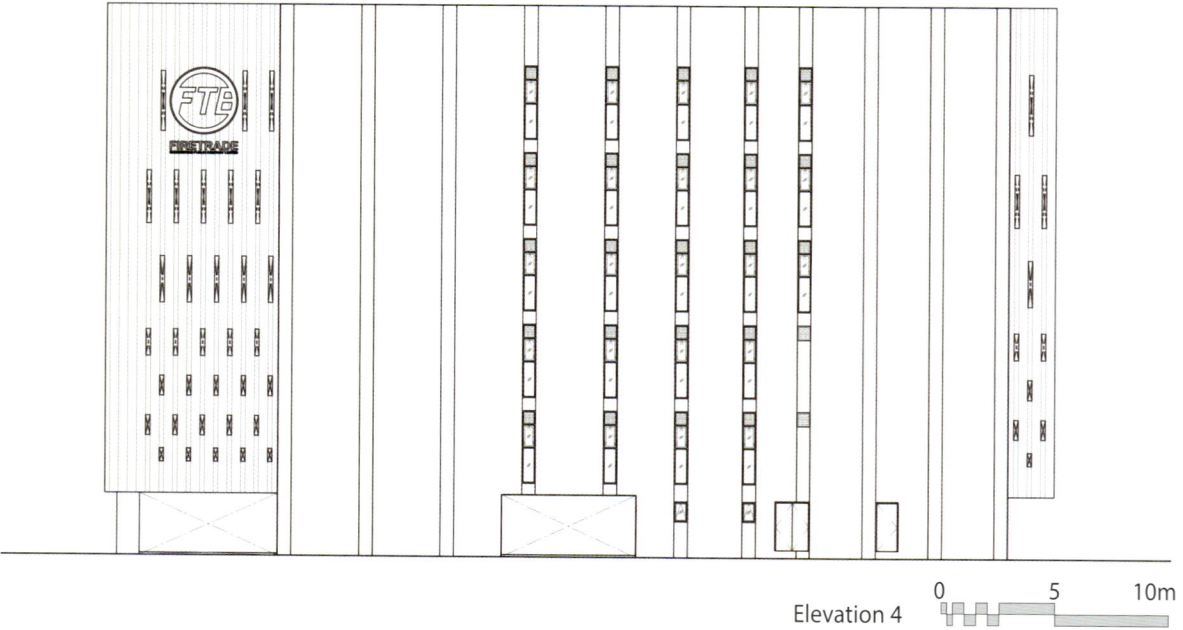

Elevation 4

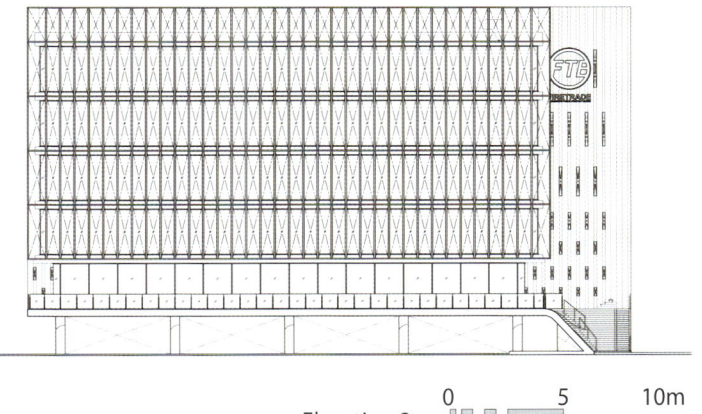
Elevation 2

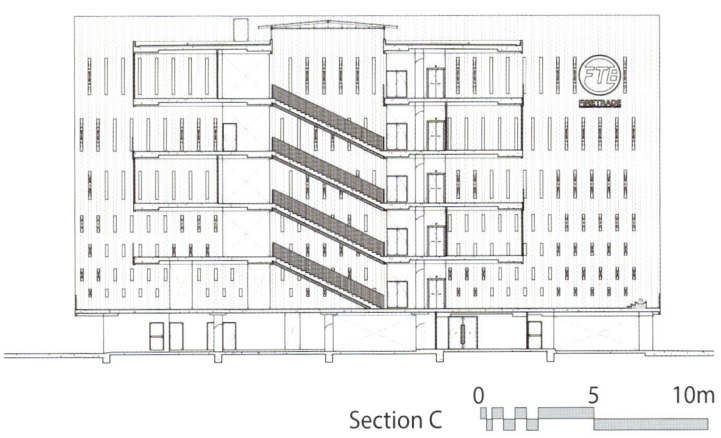
Section C

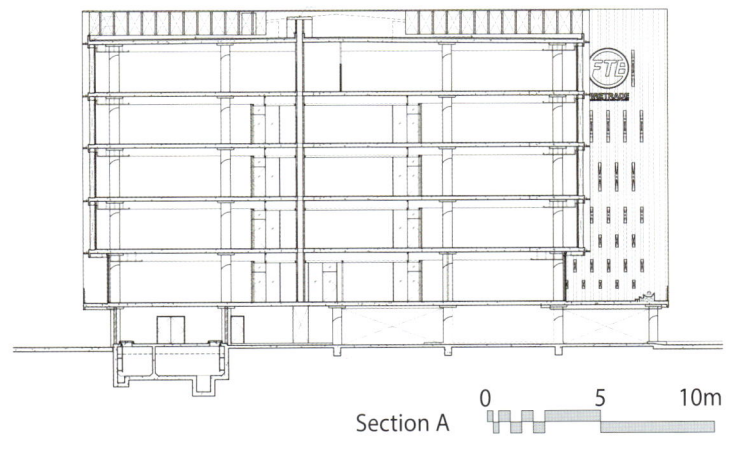
Section A

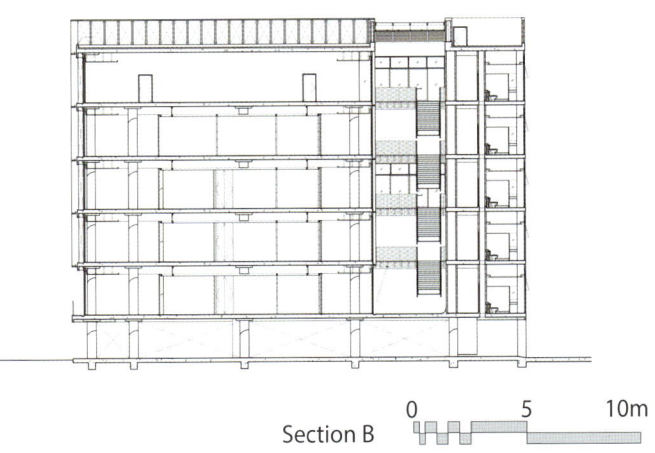
Section B

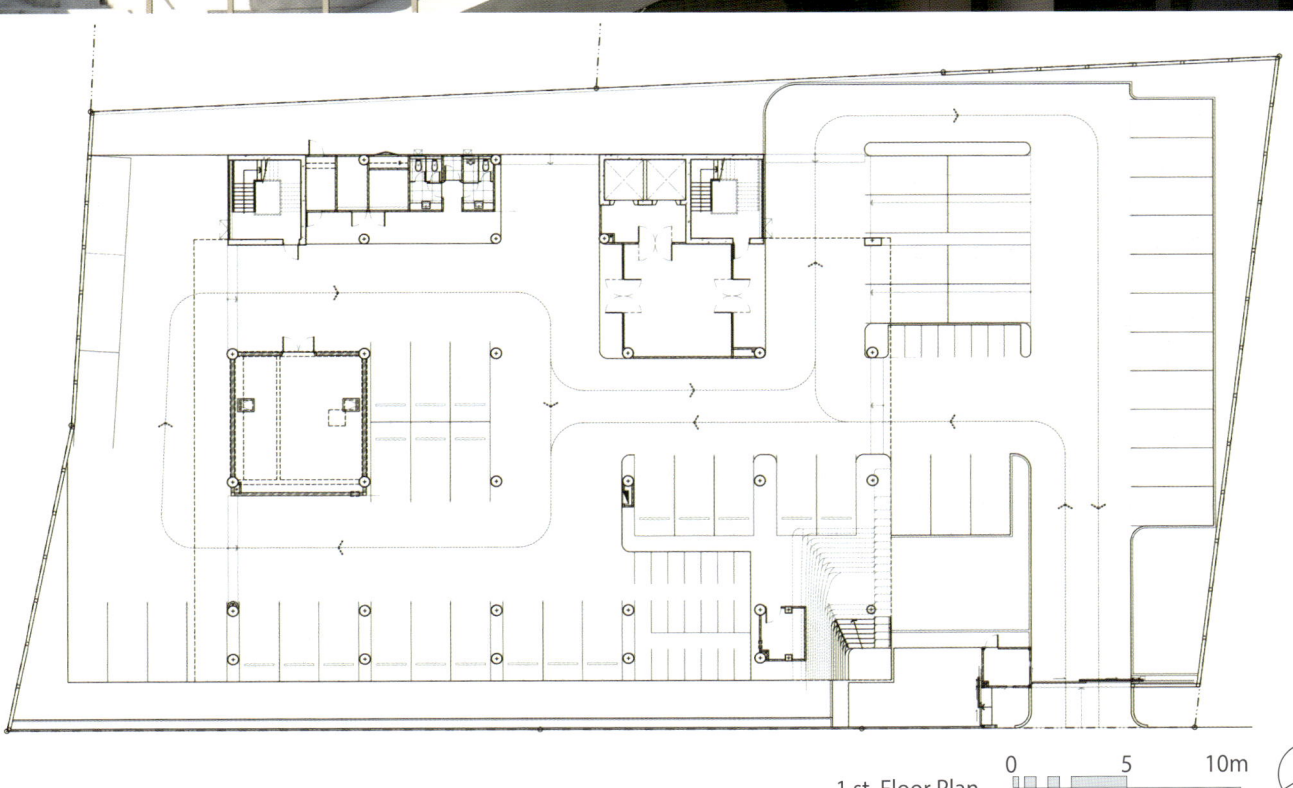

1 st. Floor Plan

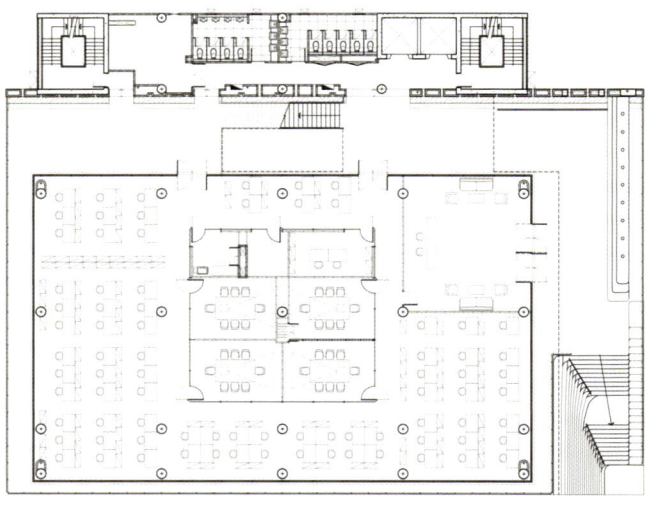
2nd. Floor Plan   0 5 10m
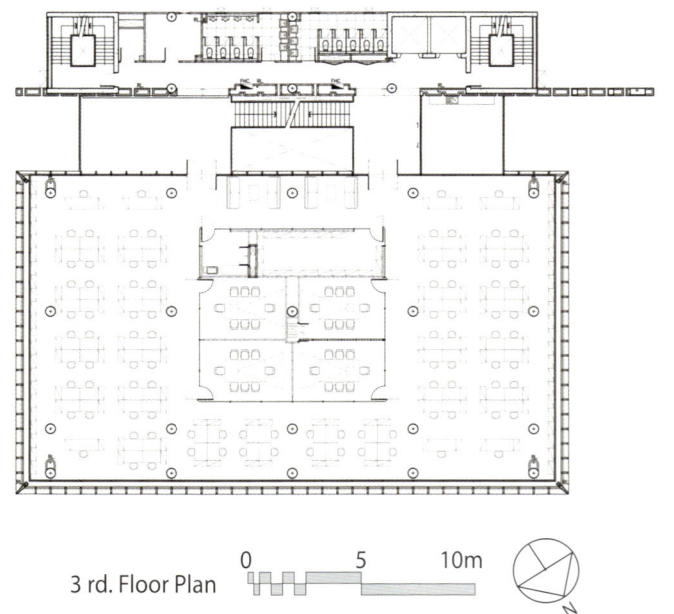
3rd. Floor Plan   0 5 10m

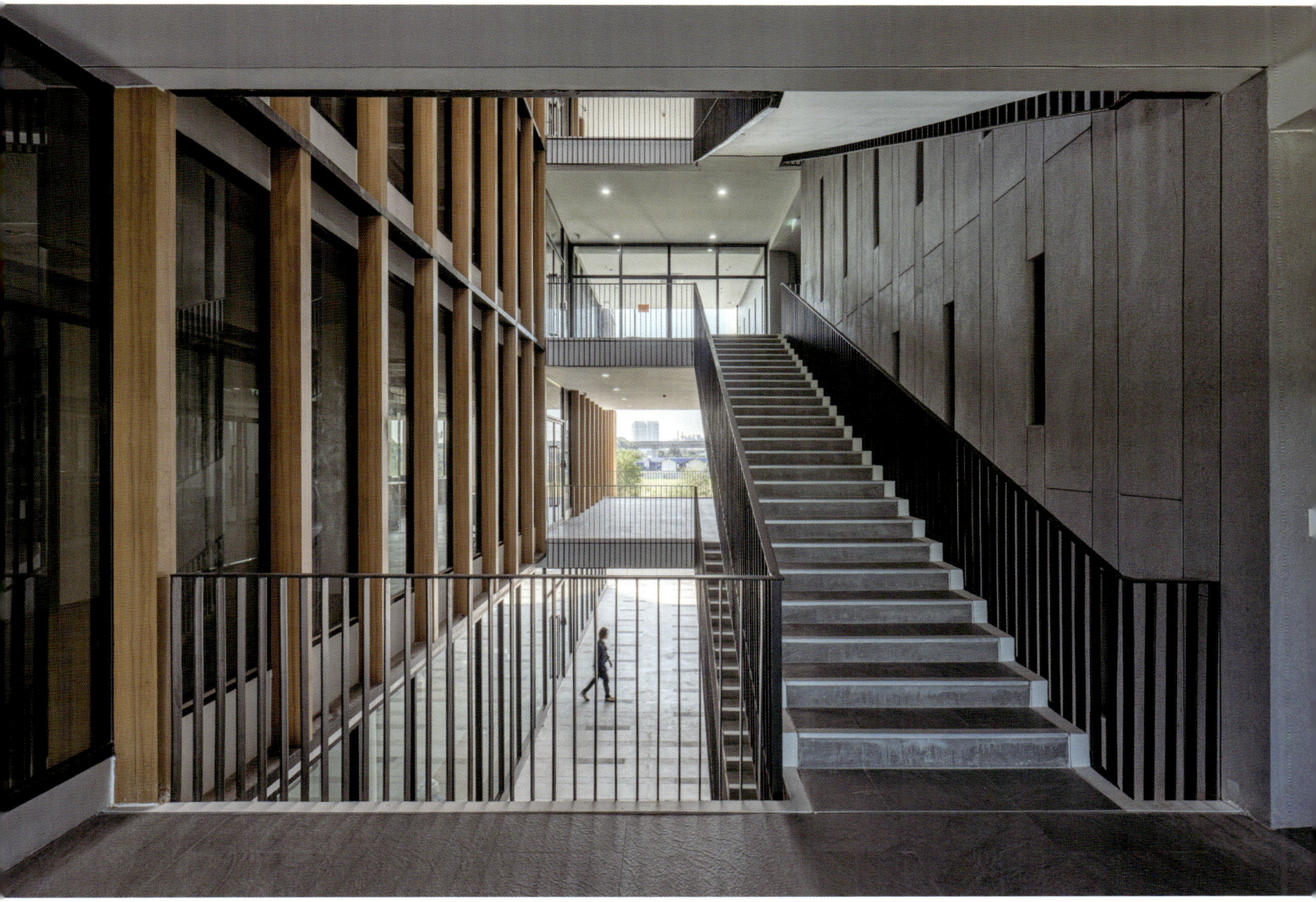

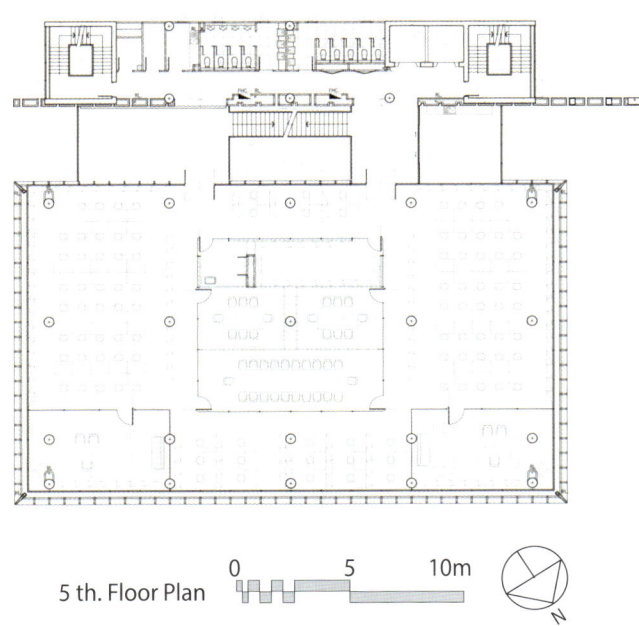

5 th. Floor Plan 0 5 10m

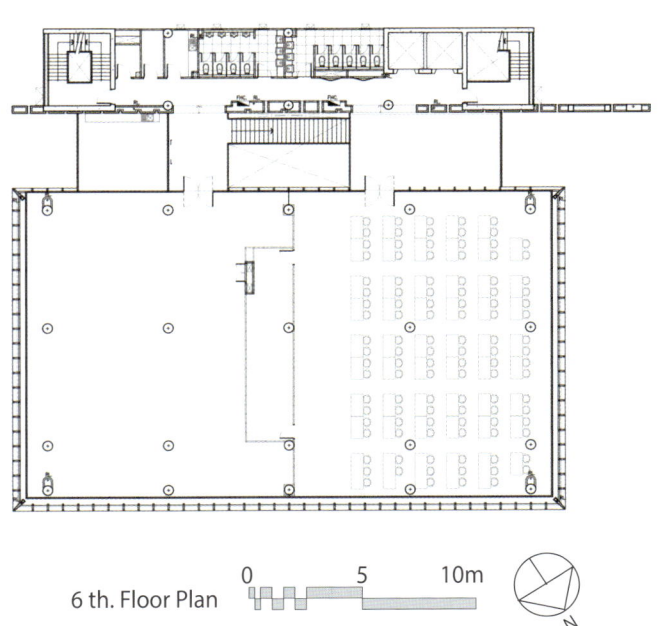

6 th. Floor Plan 0 5 10m

PG&E Embarcadero & Potrero Building

Architects: Stanley Saitowitz | Natoma Architects
Location: San Francisco, United States
Area: 15700 ft²
Photographs: Richard Barnes

Manufacturers: AutoDesk, Basalite, Winona Lighting, Rana, Structure Cast
Main Material: Concrete; Glassfibre Reinforced Concrete (Facade Panels - concrete skin | Rieder Group)

PG&E Substations at Embarcadero and Potrero. PG&E's earlier tradition of memorable urban substations which contribute to the fabric of the city is being revived. Two new buildings, one at Embarcadero and one at Potrero, have this goal. Both augment the existing substations they are adjacent to. They are set back 30' from the sidewalk to create public plazas. Their facades unfold onto the plazas and invite habitation. The new substations are compact compared to earlier buildings, with 30' height.

These 30' tall and 30' wide public surfaces are elaborated with a series of 4' wide concrete bands added to the facades. These bands are composed of five different profiles which combine to form 20 derivations which elaborate the vertical surfaces and continue onto the ground plane to become furniture. At the Embarcadero, in the midst of tall vertical structures, the bands are arrayed vertically; at the Potrero site, close to the bay and more linear forms, the bands are arrayed horizontally.

C Material Type: **Concrete**; Glassfibre Reinforced Concrete
(Facade Panels - concrete skin | Rieder Group)

Rieder presents concrete skin, facade panels made of fibreC glass fiber reinforced concrete.
It is an authentic material in line with the current trend towards natural, environmentally-friendly,
and sustainable materials that achieve an aesthetically appealing and modern effect.
The sturdy panels, only 13 mm thin, open up great freedom for the design of individual facades.
concrete skin pulls smoothly over buildings, corners, and edges like a skin and creates a unique material flow.

101

These bands continue onto the ground plane to become projecting surfaces on the plazas, extruding as seating and pedestals to encourage occupation. Interspersed among these linear habitable elements are trees for shade. Roof awnings mark the street edges.

EMBARCADERO ELEVATION

103

POTRERO ELEVATION

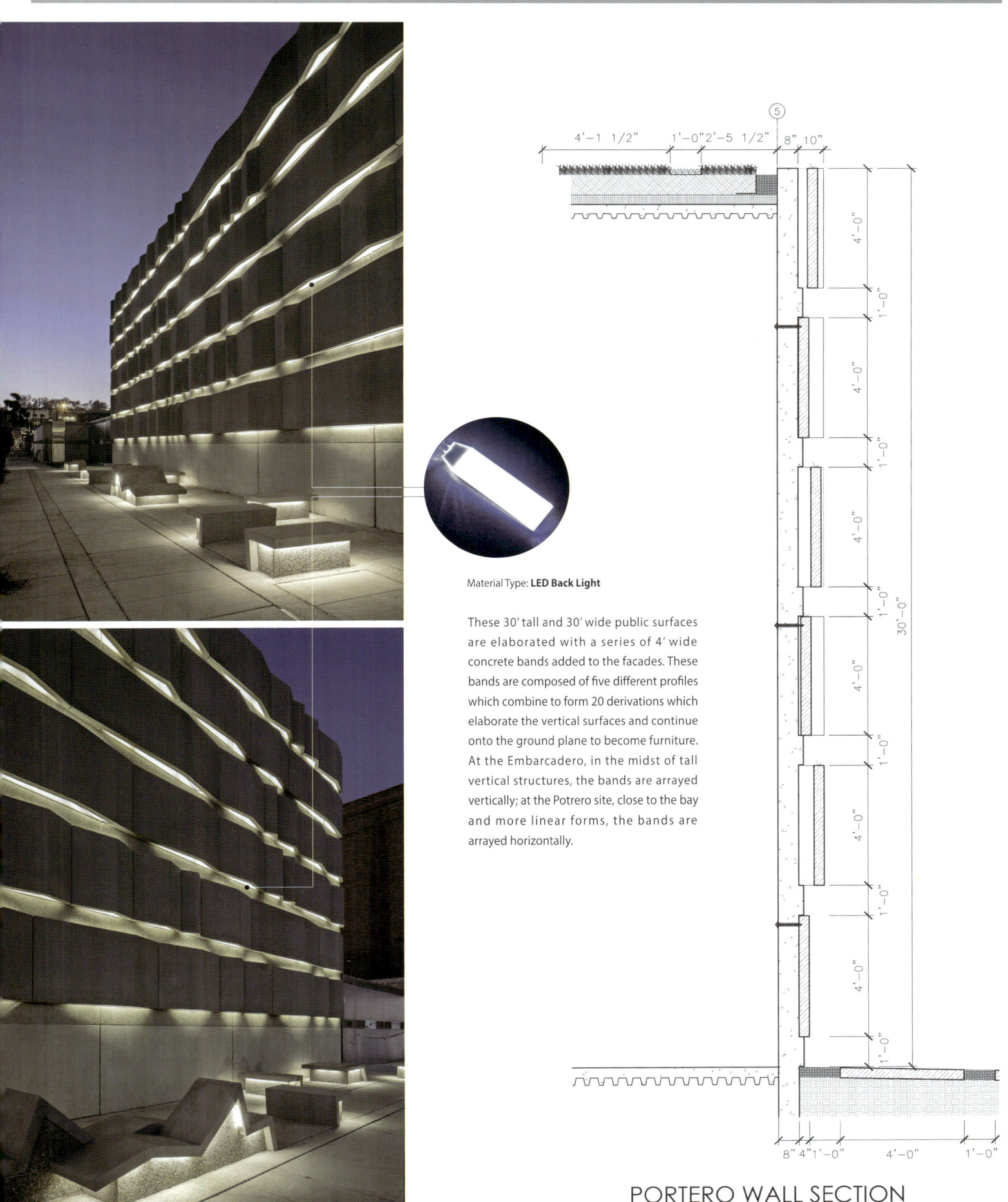

Material Type: **LED Back Light**

These 30' tall and 30' wide public surfaces are elaborated with a series of 4' wide concrete bands added to the facades. These bands are composed of five different profiles which combine to form 20 derivations which elaborate the vertical surfaces and continue onto the ground plane to become furniture. At the Embarcadero, in the midst of tall vertical structures, the bands are arrayed vertically; at the Potrero site, close to the bay and more linear forms, the bands are arrayed horizontally.

PORTERO WALL SECTION

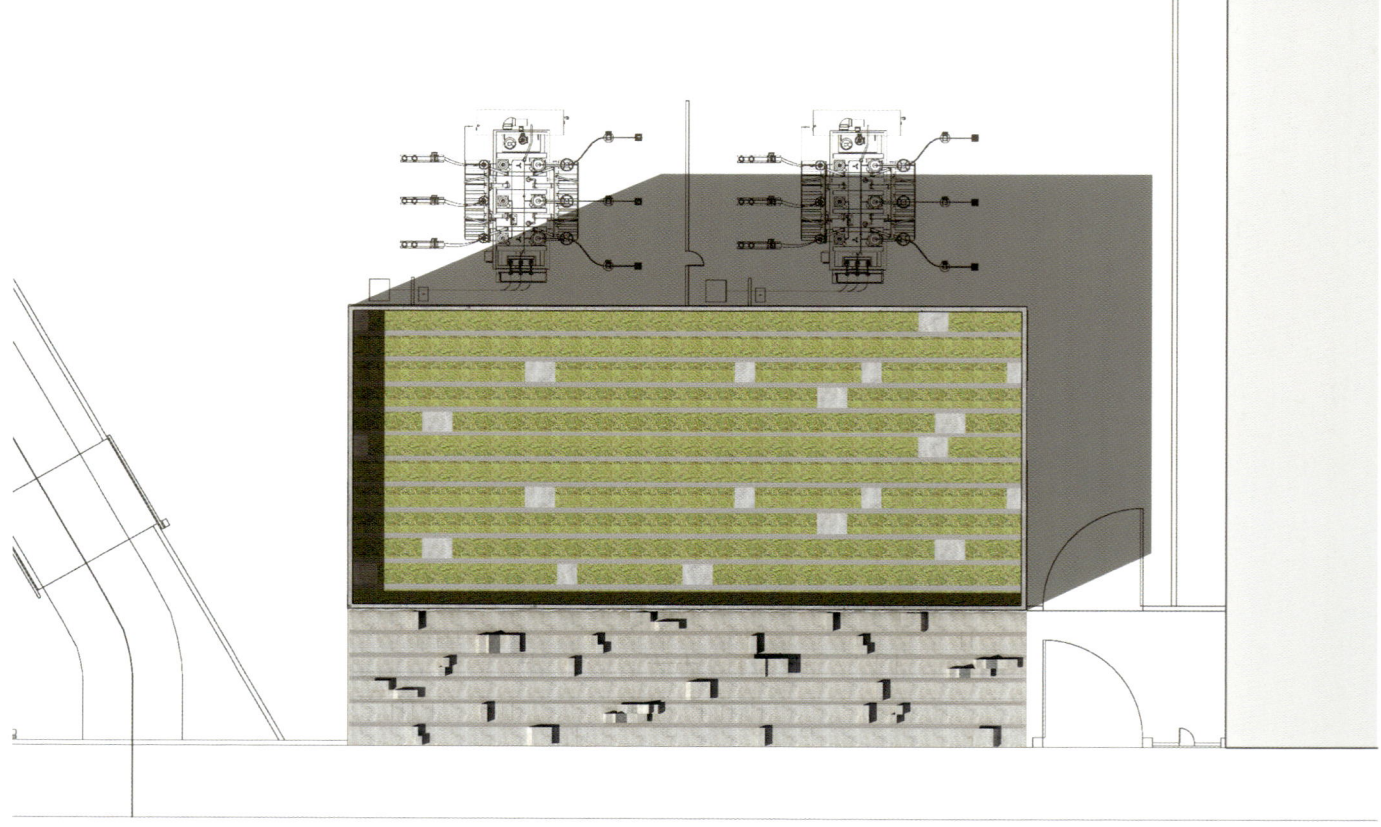

PORTERO SITE PLAN

PRE-CAST DERIVATIONS

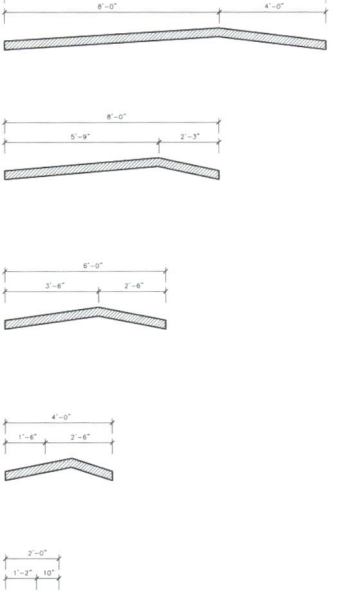

5 UNIQUE PIECES

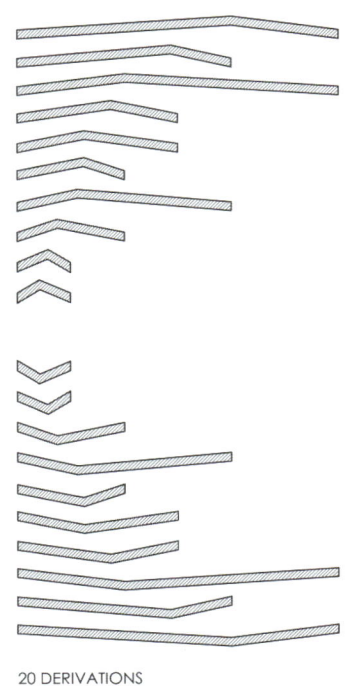

20 DERIVATIONS

EMBARCADERO ROOF PLAN

EMBARCADERO PLAN

EMBARCADERO ELEVATION

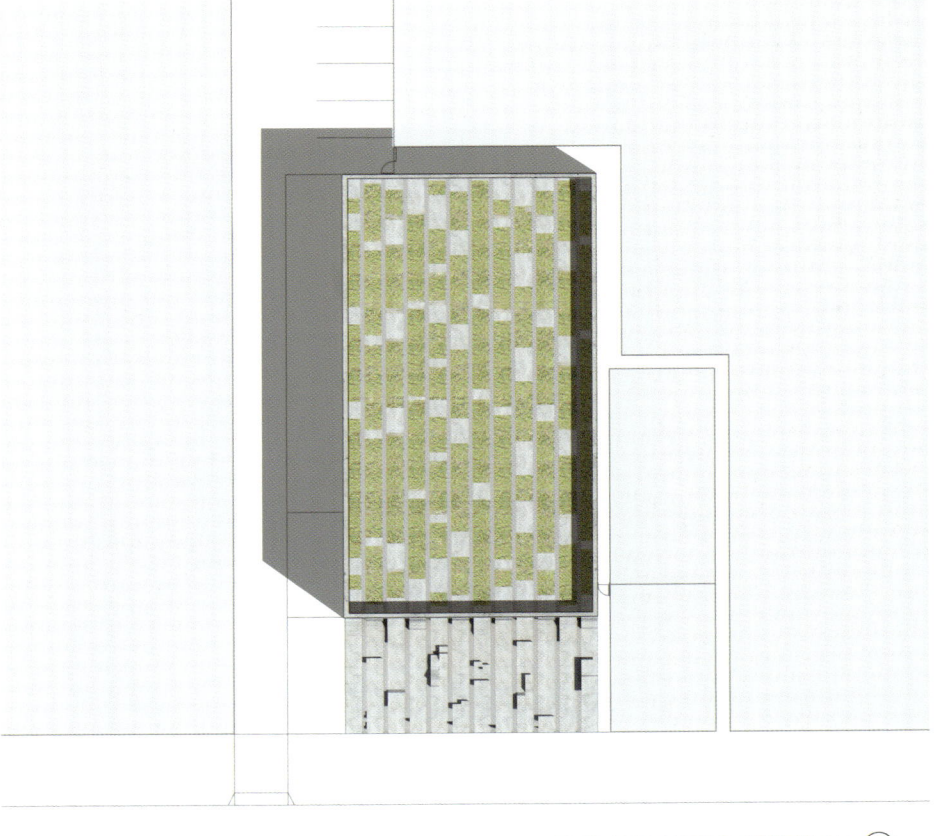

EMBARCADERO SITE PLAN

109

110

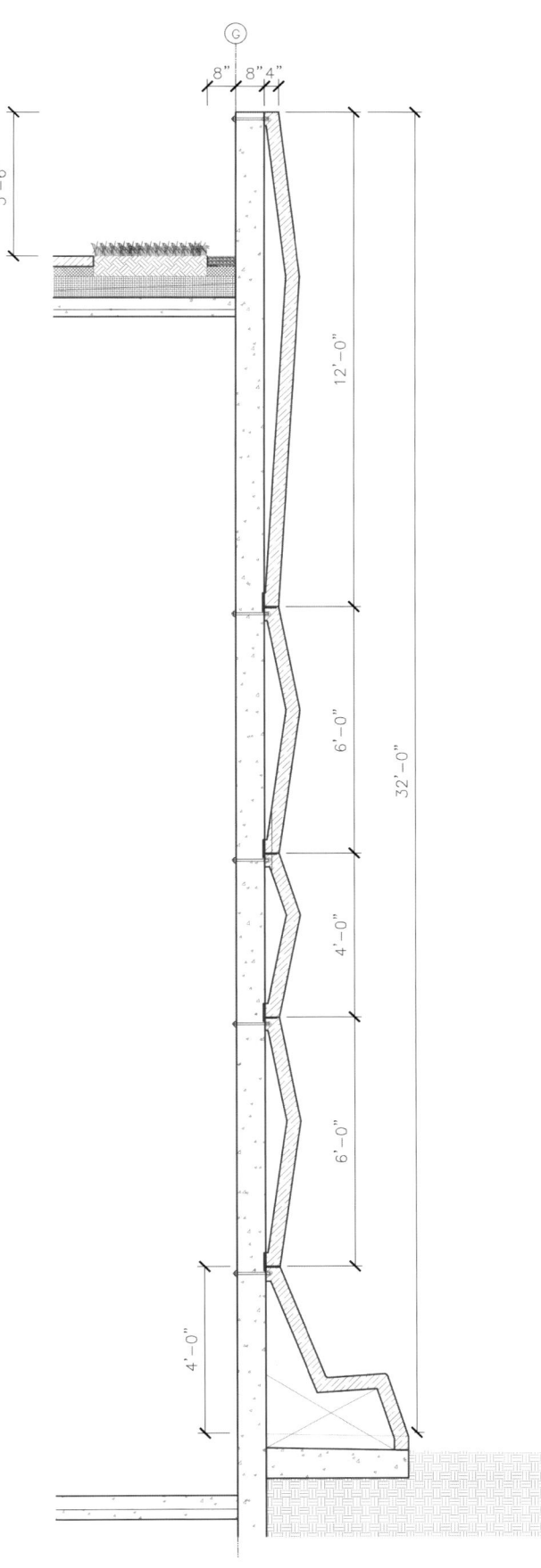

These bands continue onto the ground plane to become projecting surfaces on the plazas, extruding as seating and pedestals to encourage occupation. Interspersed among these linear habitable elements are trees for shade. Roof awnings mark the street edges. These plazas are free spaces for use by the public. Recessed between the rippling concrete bands, led lights illuminate the walls and ground planes. The color of this lighting can vary with the seasons and be used for events and celebrations. At the Embarcadero site, led lighting is added to the existing substation building to connect with the new plaza/façade. Creating these urban plaza/facades is a way for PG&E's substations which already provide vital services to the city, to also give back to the communities they are located in.

EMBARCADERO WALL SECTION

Greenhouse Orchid Punta del Este

Architects: Mateo Nunes Da Rosa
Location: Punta Del Este, Uruguay
Area: 20 m²
Photographs: Marcos Guiponi
Manufacturers: Plazit-Polygal, Elena Artagaveytia...
Main Material: Steel Chassis, compact polycarbonate sheets and Composite Aluminum

In times of pandemic, Ana, who is passionate about botany, specifically orchids, turns her hobby into an undertaking: Greenhouse Orchid, and entrusts us with a greenhouse that acts as an exhibition at the same time. The proposal is a transparent, transportable, modifiable prototype that generates the necessary climate for the survival of the orchids. Two are manufactured, one for exhibition and the other for flowering. Orchids require a microclimate within a very specific range. The environment must control temperature, lighting, humidity, ventilation, irrigation and nutrients, beyond the outside climate.

A double envelope is then formed that resolves it and formally exposes it. The translucent outer membrane protects from the wind and cold, directs the wind towards the air inlet, and depending on the orientation it allows you to see inside. The inner membrane stops direct sunlight, generates the required very bright environment and also allows you to see the environment from within. Between the membranes a laterally ventilated chamber is formed or through a retractable roof provided. The air intake is per floor and in case the passive systems are not enough, the same home automation system that controls the lighting activates air forcers.

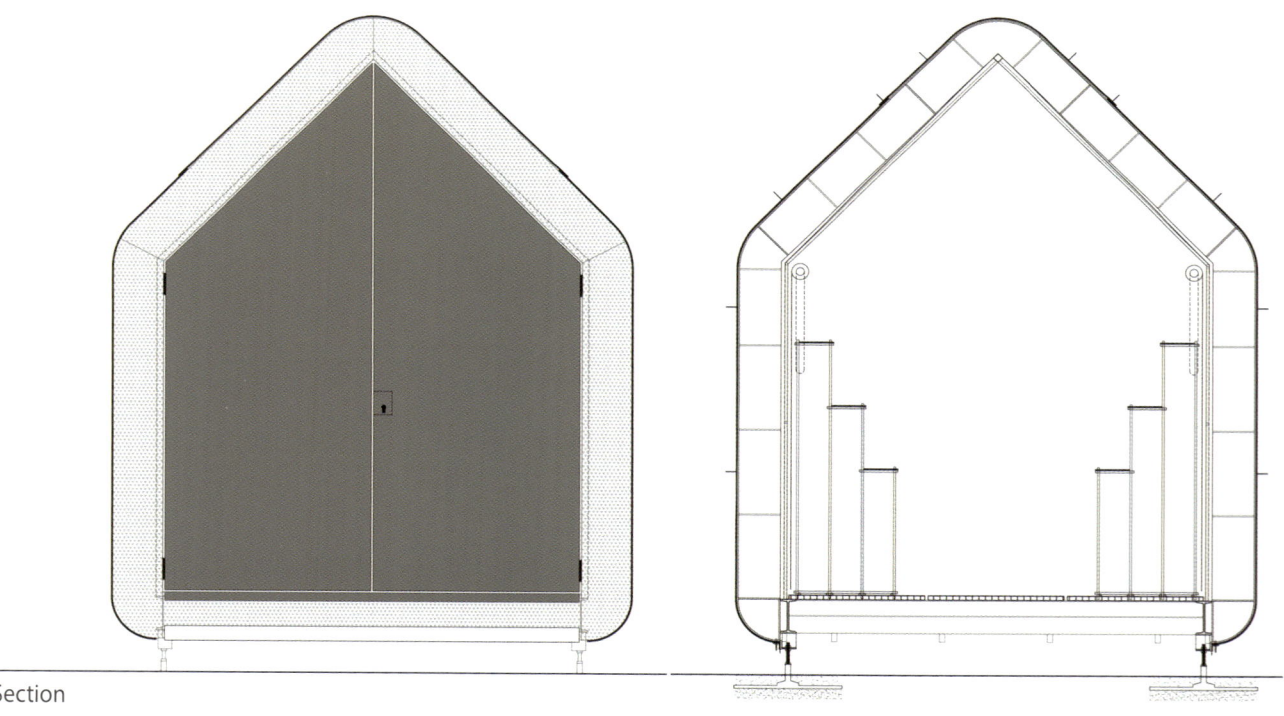

Section

Material Type: Steel & aluminum structures
(Stratlock range | Fastmount)

Fastmount presents its Stratlock range that allows the rapid installation of heavy panels in steel or aluminum structures.
Designed in collaboration with Poliform Contract Italia, the Stratlock range is the only product on the market that provides versatility for modular design with adjustable panel installation, thanks to its double locking function.
SL-M18 locks into the quick-fit female chassis (SL-FA or SL-FB) on the aluminum or steel frame. The panel is supported at the first click, with a tolerance of 4 mm in any direction to allow finding the correct position while the female supports the weight.
The second click locks the panel in position, secured with a pull-out load of 18kg per set of clips, making the system ideal for heavy decorative panels.

Material Type: **Other**
(compact polycarbonate sheets)

Two large doors are installed on the side that invite you to walk through it, generating a second natural ventilation tunnel. The dimensions of each greenhouse are given by the total volume of transport, its weight and the optimization of the materials with the highest incidence: Steel Chassis, compact polycarbonate sheets and Composite Aluminum.

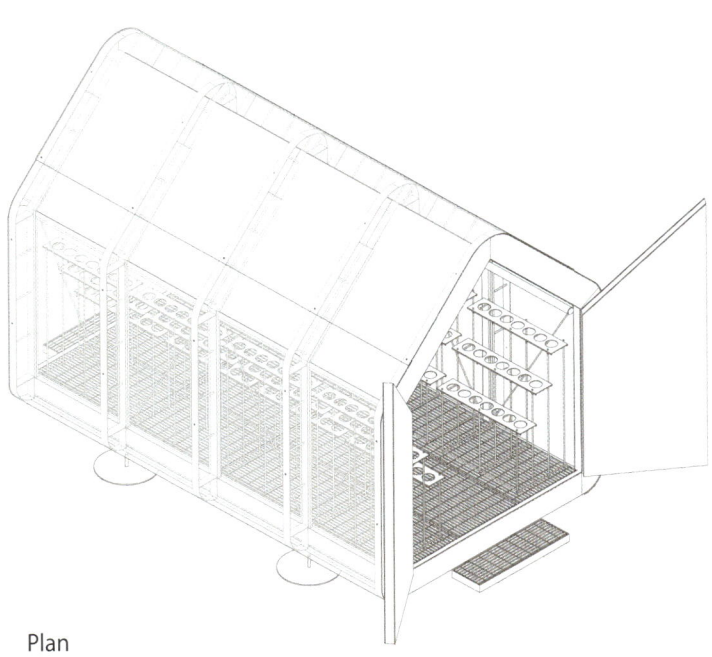

Plan

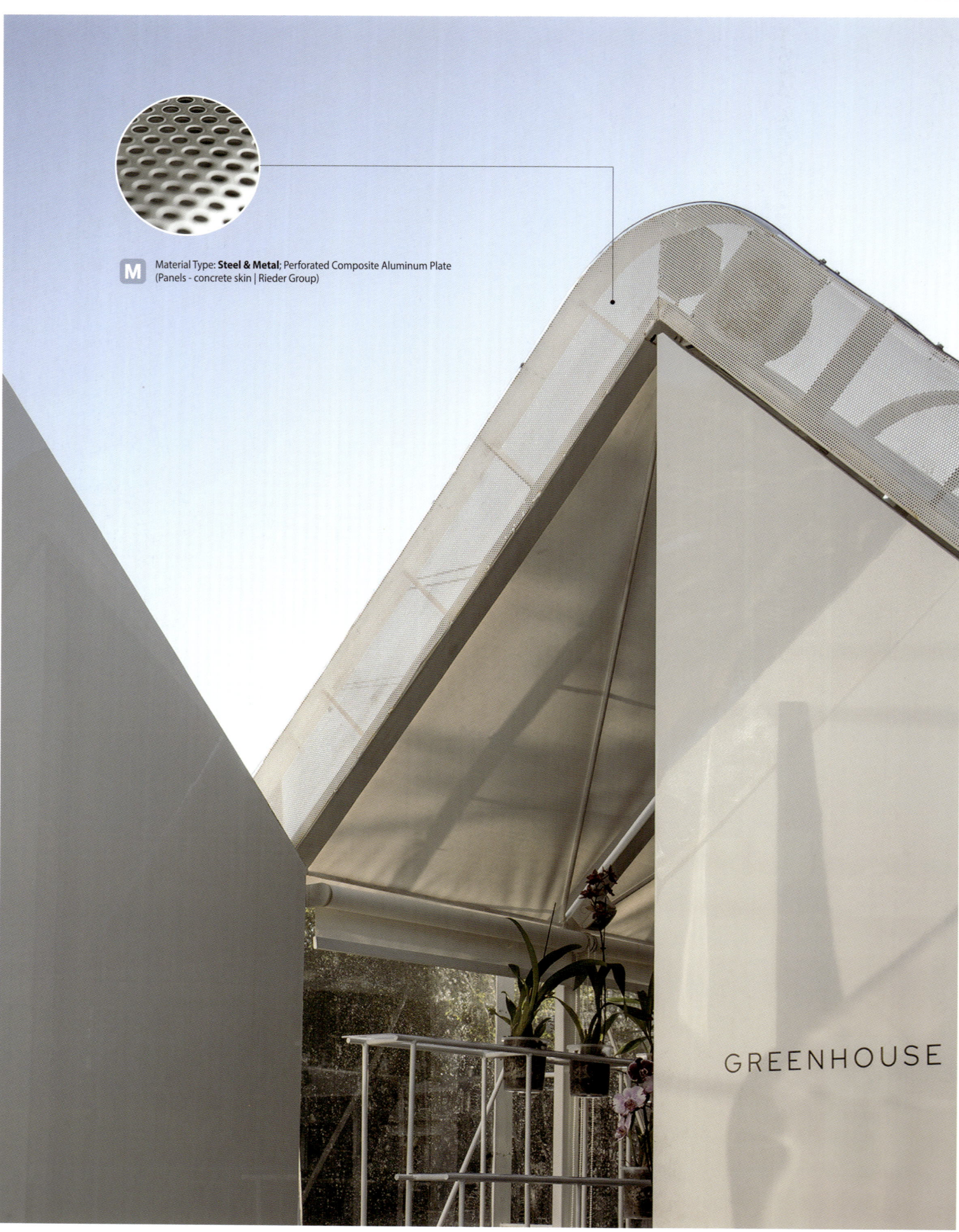

M Material Type: **Steel & Metal**; Perforated Composite Aluminum Plate
(Panels - concrete skin | Rieder Group)

GREENHOUSE

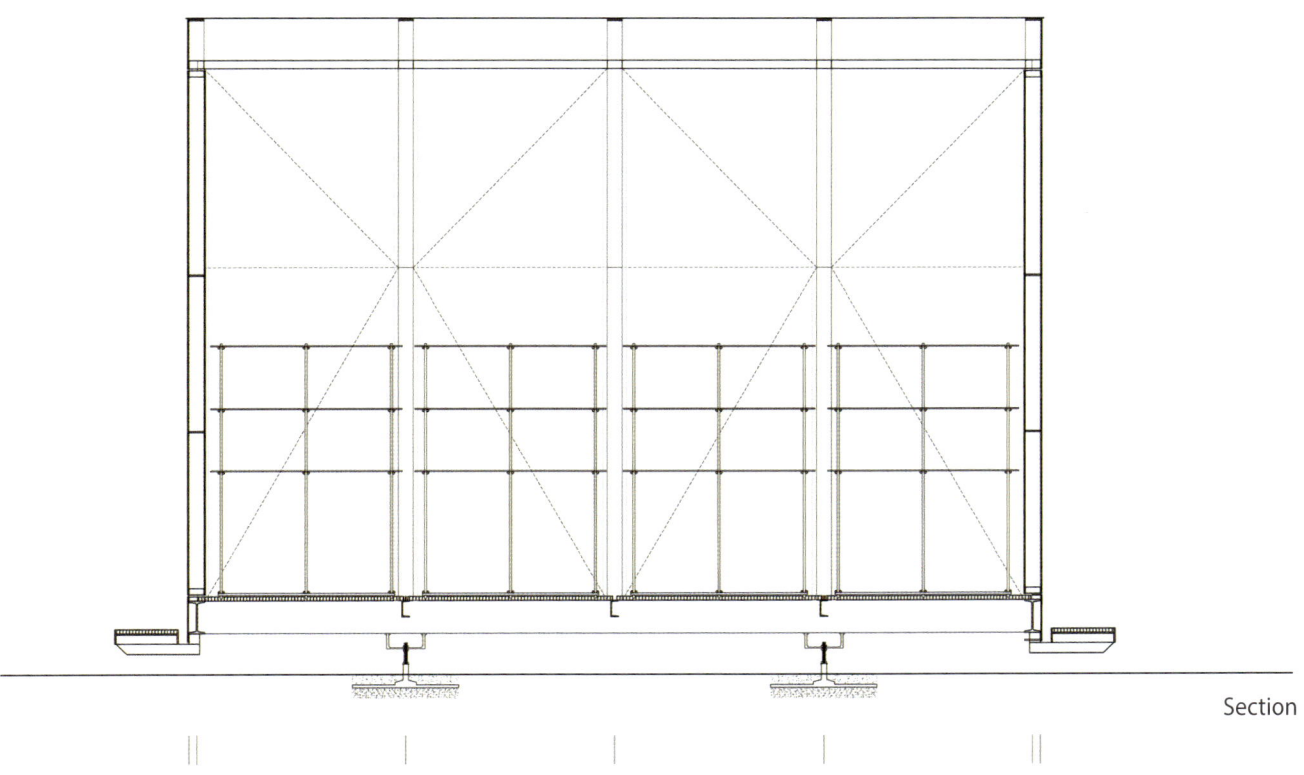

Section

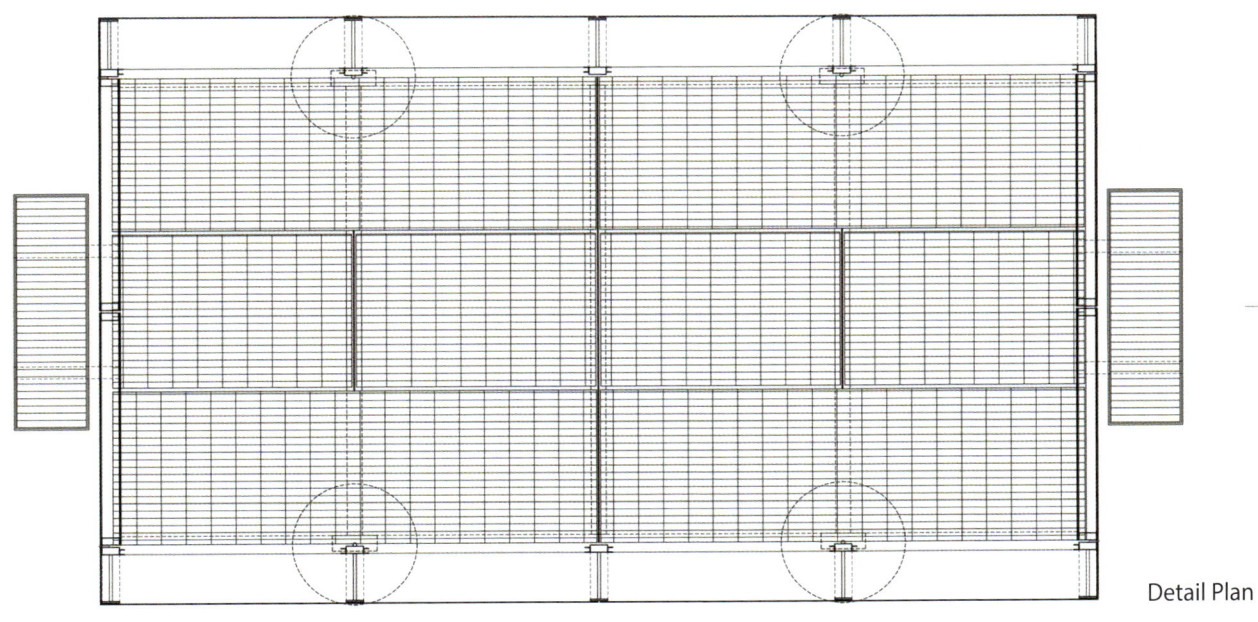

Detail Plan

123

Rural Hotel in an Olive Grove

Architects: GANA Arquitectura
Location: Villanueva Del Rosario, Spain
Area: 1006 m²
Photographs: Francisco Torreblanca Herrero

Manufacturers: Plazit-Polygal, Elena Artagaveytia...
Main Material: Concrete, Roofing Tile, Windows - Janisol Primo | Jansen

This project arises focused on the development of a hotel space around the original existing Cortijo (traditional cottage), ready to be part of the boom of Rural Tourism. Quite a challenge based on the rehabilitation of a building steeped in history and located in the heart of a plantation full of olive trees. When someone travels around Andalucia it is difficult not to pay attention to these green structures organized through streets where the trees are proud and convinced of their worth. That is why the hotel had to pay tribute to such a characteristic natural environment. The beauty behind the breathtaking environment of this plot is something that impresses from first sight.

In this sense, the key to the project was to achieve an interior-exterior connection that is as permeable as possible, without thereby damaging the essence of this peculiar building in any way. In contrast to this house, the plot also had a storehouse linked to the existing agricultural exploitation that was integrated into the project through the mandatory Proyecto de Actuacion (Urban Planning Development). A complex administrative procedure that nevertheless allowed the simple adaptation of these two buildings, which after the renovation and extension designed answer efficiently the requirements of this new activity.

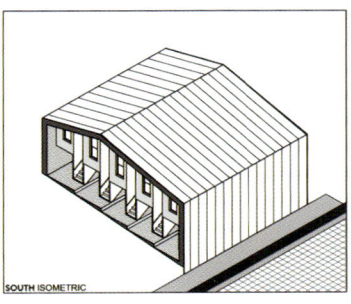 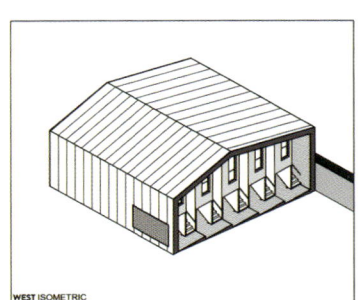 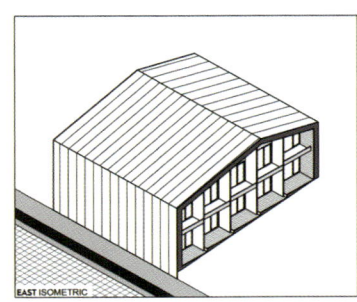 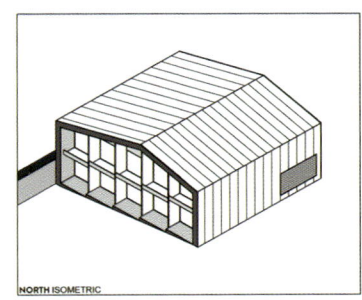

ADDITION ISOMETRIC

The result of the intervention is nothing but the perfect harmony between traditional and contemporary architecture, over the amazing influence of nature in its purest form. In this way, the common areas, together with half of the rooms, are located in the old farmhouse, giving these rooms and therefore the whole hotel its own indisputable identity, taking advantage of the heights and materials typical of that time.

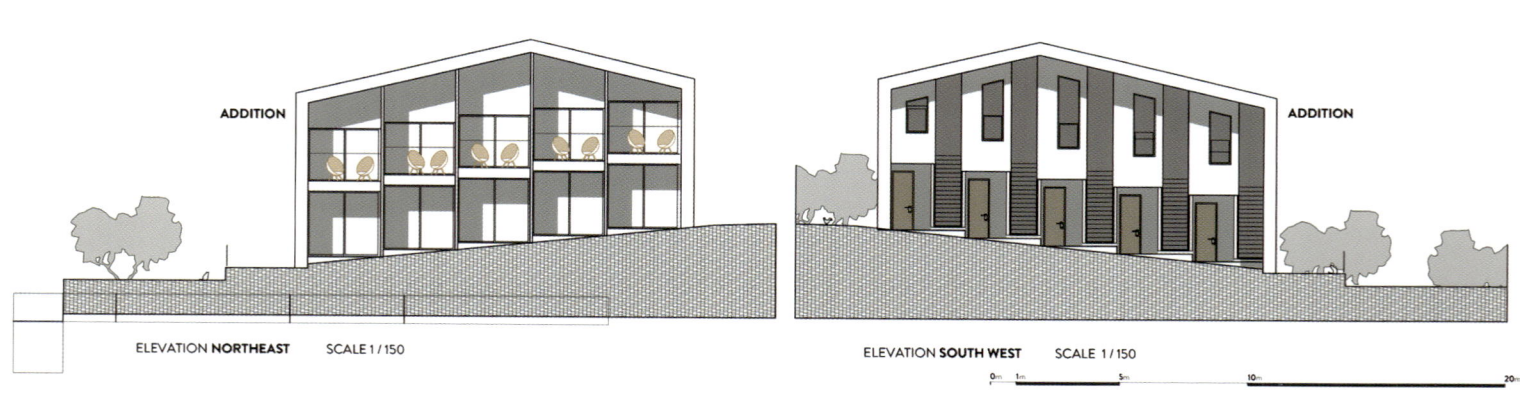

ELEVATION **NORTHEAST** SCALE 1 / 150

ELEVATION **SOUTH WEST** SCALE 1 / 150

SECTION **B-B'** SCALE 1 / 150

SECTION **C-C'** SCALE 1 / 150

0m 1m 5m 10m 20m

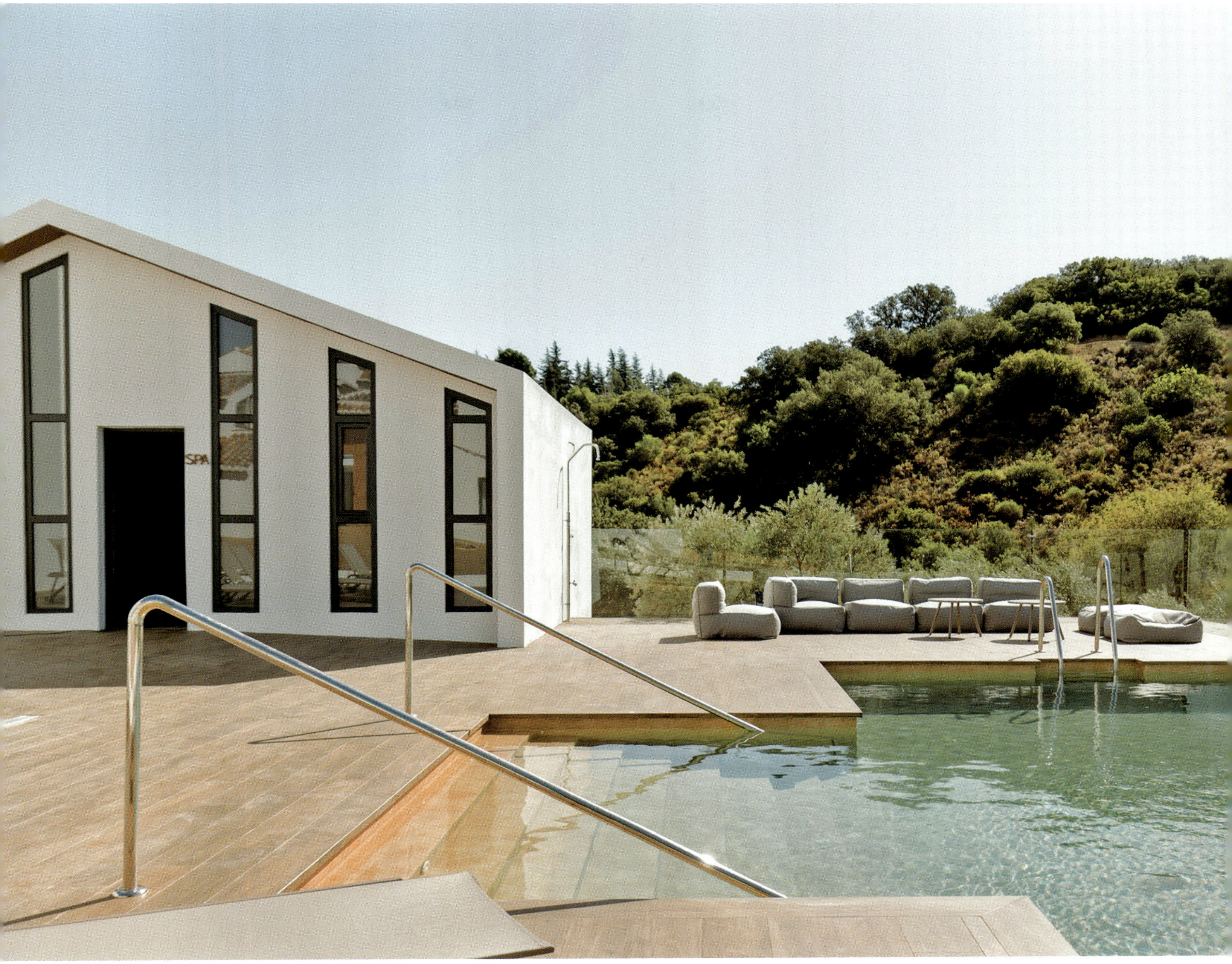

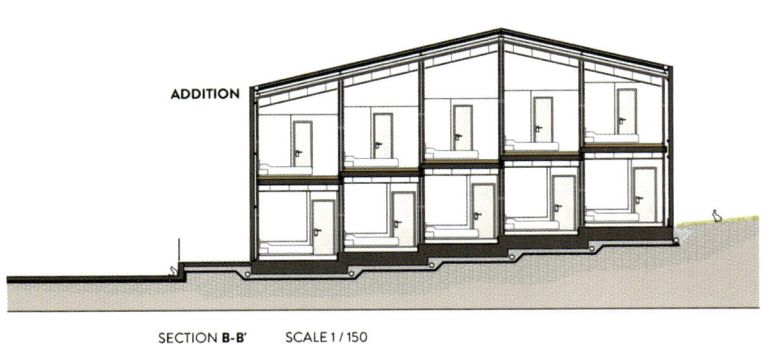

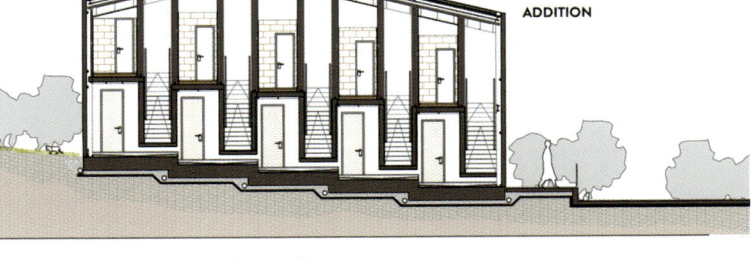

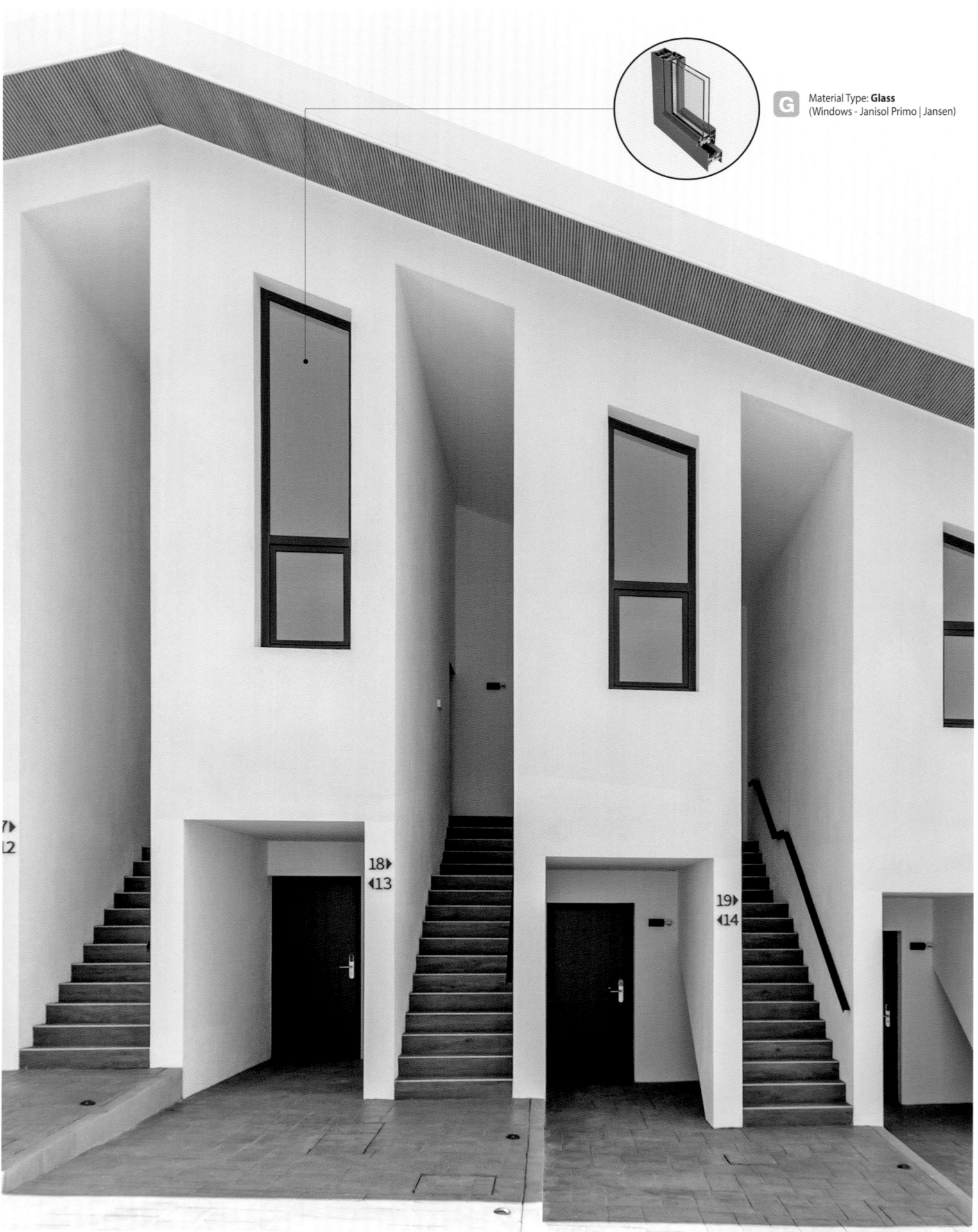

Material Type: **Glass**
(Windows - Janisol Primo | Jansen)

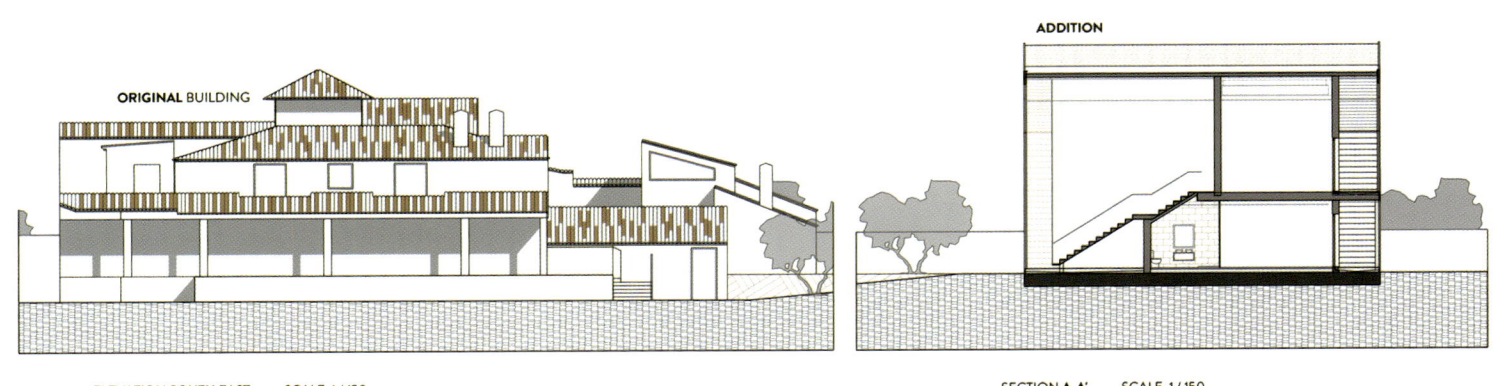

ELEVATION **SOUTH EAST**   SCALE 1/150

SECTION **A-A'**   SCALE 1/150

Material Type: **Concrete Roof Tiles**

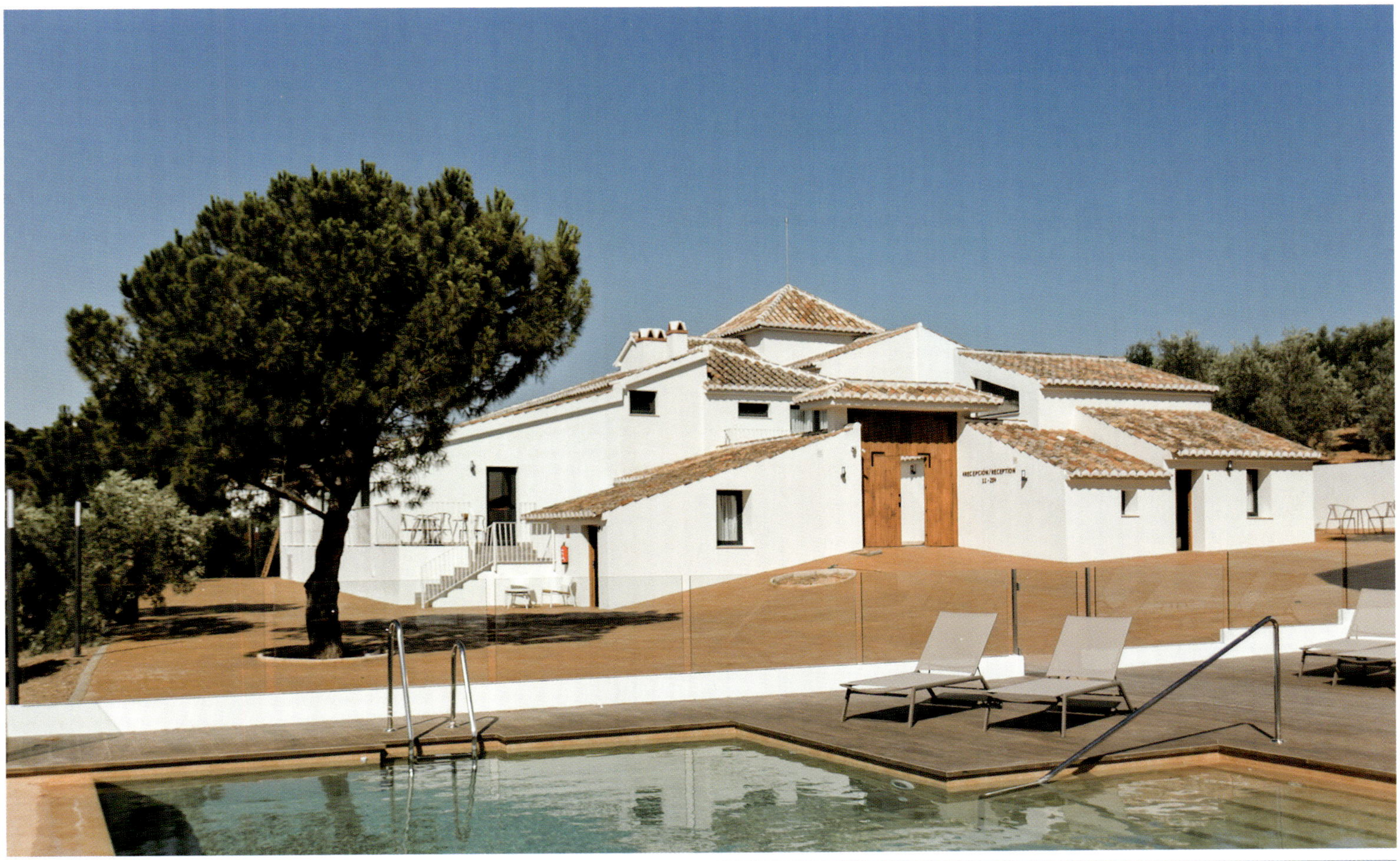

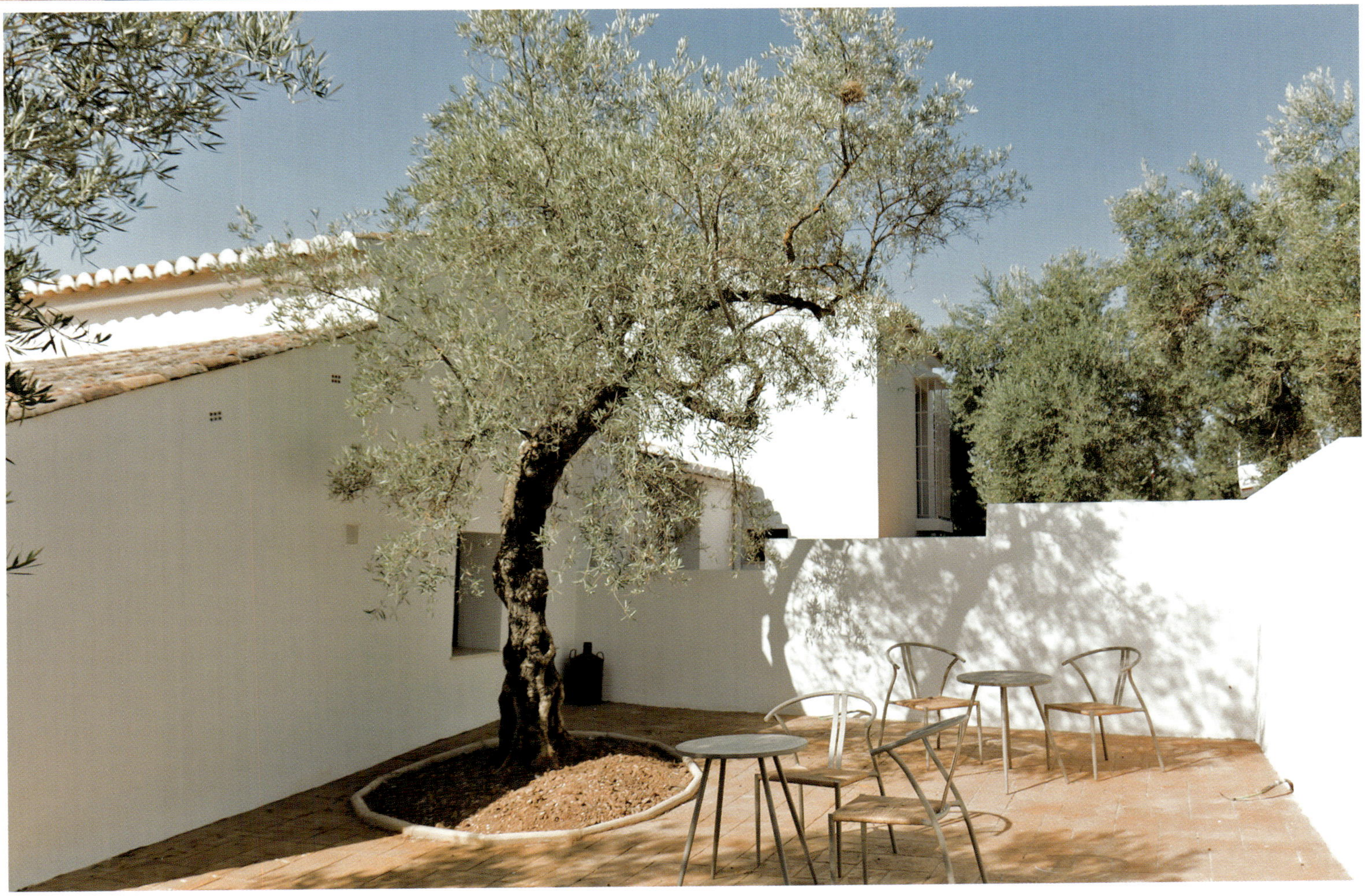

On the other hand, the remaining ten rooms are designed in a deliberately contemporary style, by the use of completely different materials and bigger slenderness. The outdoor access terrace, designed as a plaza, not only distributes the visitor through the different rooms but also allows the user to enjoy a pool area equipped with the best spa facilities, as well as recreational spaces where to disconnect from the current daily routine. This is how all the rooms converge towards the interstices generated in between the buildings so that the rooms can be opened exclusively towards the immediate natural environment. An unbeatable backdrop that makes this hotel a unique building ready to become a benchmark in the sector throughout the province.

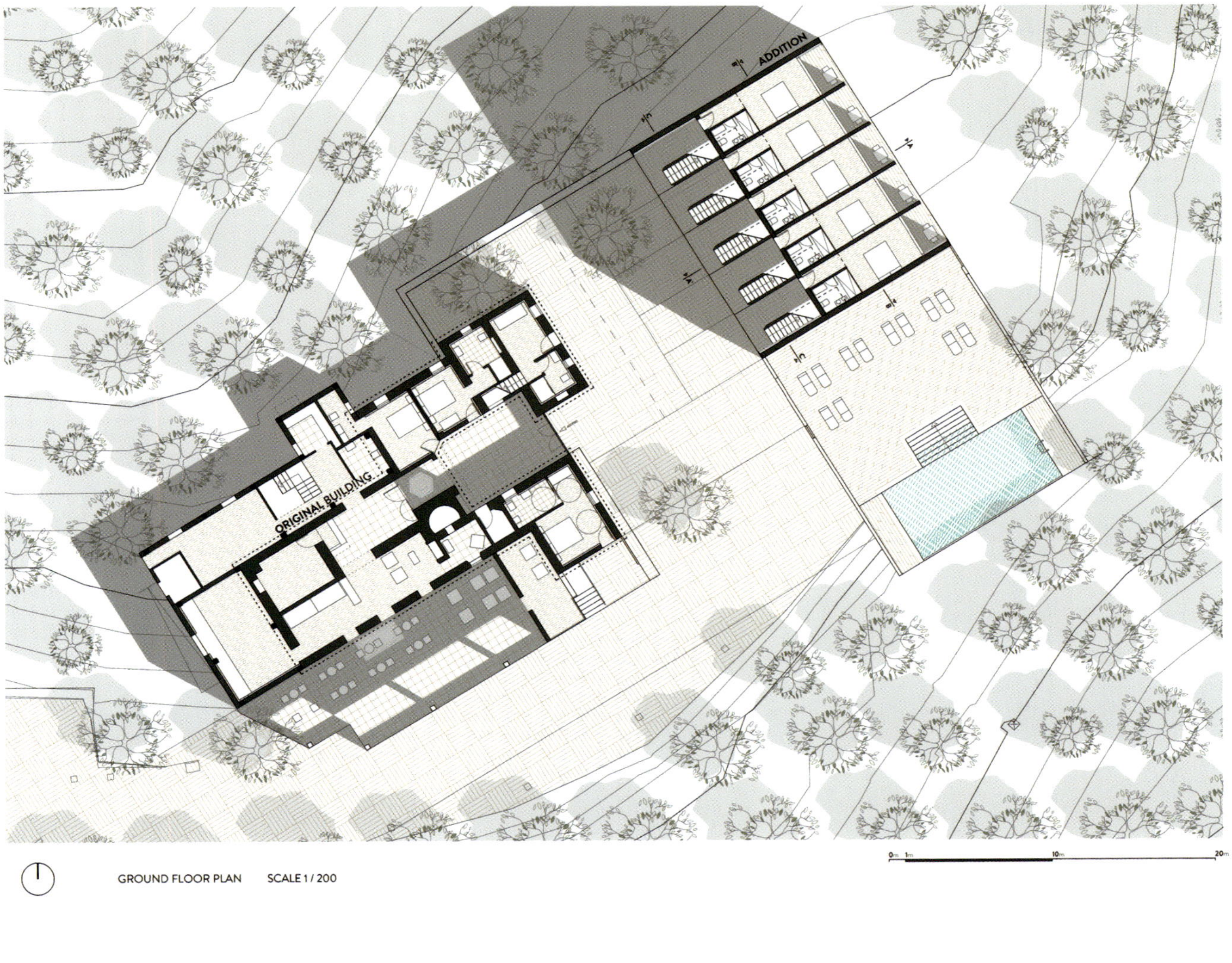

① GROUND FLOOR PLAN   SCALE 1 / 200

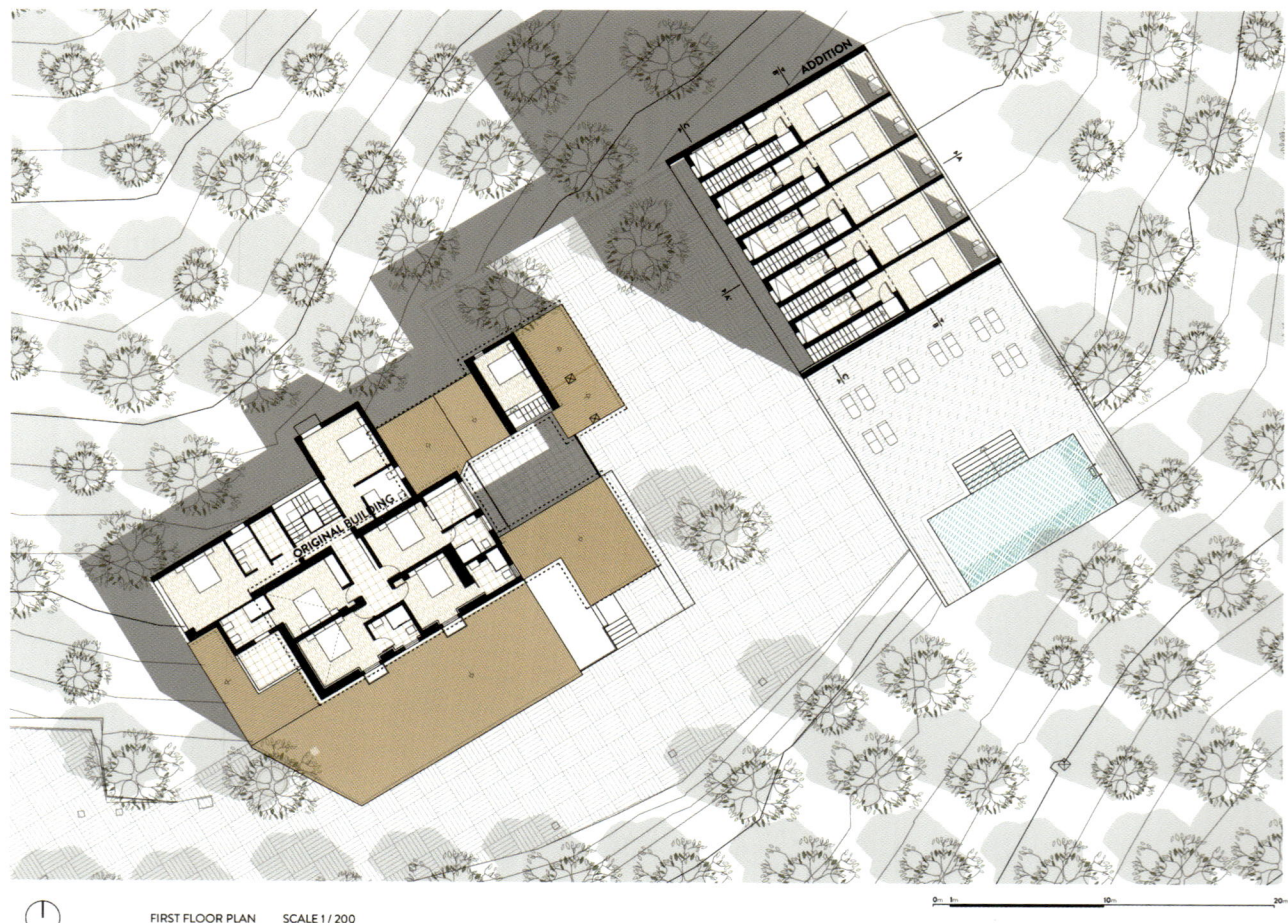

FIRST FLOOR PLAN   SCALE 1 / 200

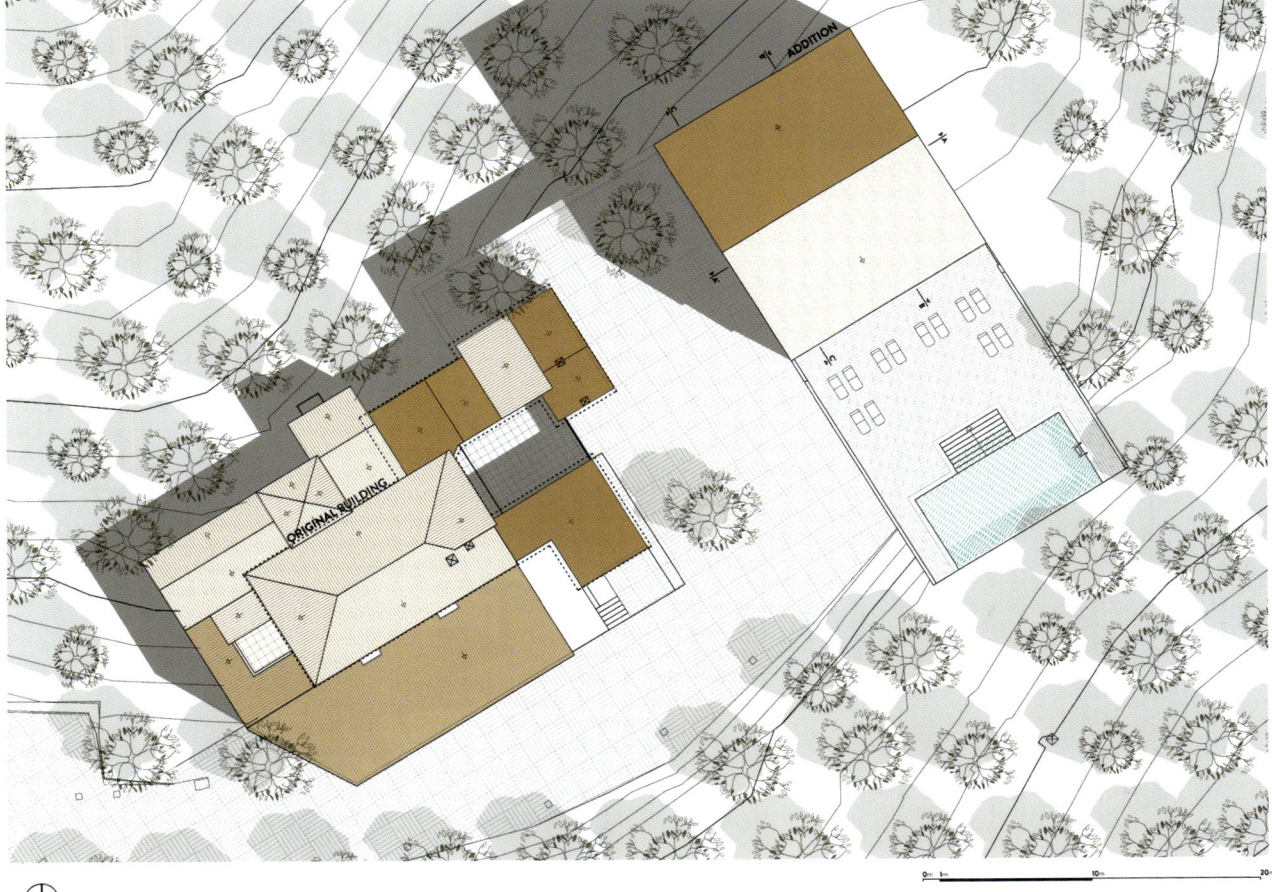

① ROOFTOP FLOOR PLAN   SCALE 1/200

Agency Giboire Morbihan Offices

Architects: a/LTA
Location: Vannes, France
Area: 270 m²
Photographs: Stephane Chalmeau

Main Material: Glass panels + Stainless steel rails

This project involves the heavy restructuring of a corner building located at the junction of rue Thiers and rue Carnot, in the heart of the historic center of the city of Vannes. The existing building, R+2 plus attic, was a bank that we transformed into offices for Groupe Giboire. The aim is to set up a new commercial and real estate development agency. The location of the project is remarkable, close to the port, it is backed by the city walls dating from the 4th century. The original structure of the building has been preserved, except for the attic which has been given a new style.

The first floor "catches up" with an alignment on the street, large openings are created to allow the reading of the ramparts located in the court. It is clad with a stainless steel cladding that is also found in the attic. The original configuration in angle (providing a real response to the articulation of the two streets), and the complexity of the structure of the existing facade, the bias was to choose the sobriety and finesse of the materiality. For this reason, the mineral facade was highlighted by a second curved glass skin located at the two intermediate levels.

Material Type: **Glass** (12mm curved glass panel)

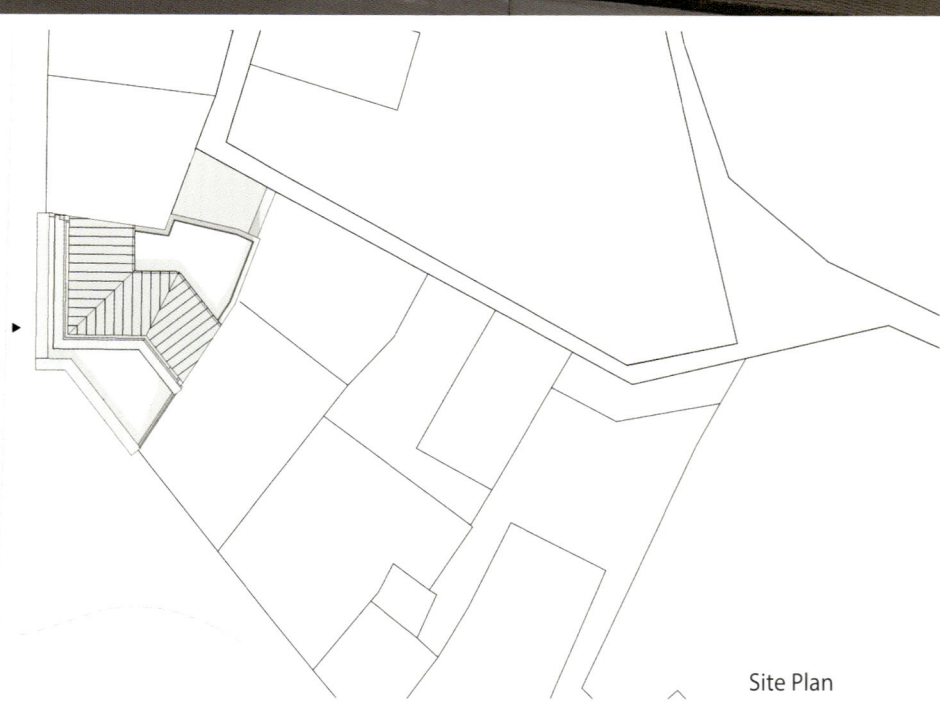

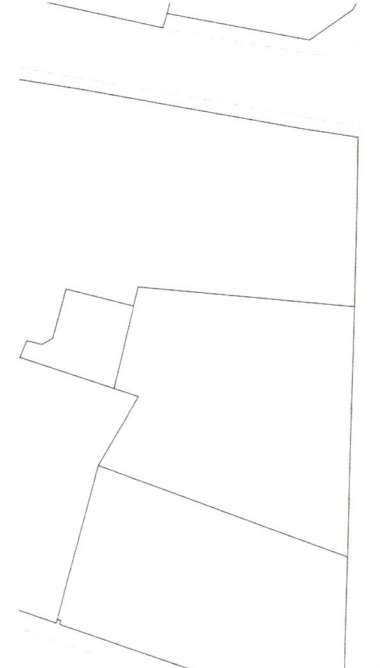

Site Plan

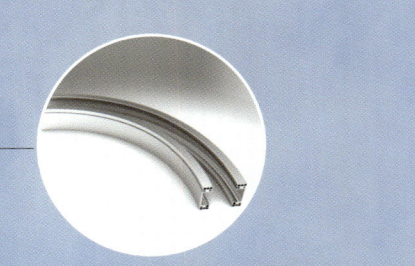

M Material Type: **Steel & Metal**
(stainless steel rails + anodized aluminum frame)

The combination of glass and stainless steel envelops the building and delicately slides it into the historic heart of the city. The redesigned facade plays with the reflections of the city as well as those of the sky, which vary according to the time of day.mThe glass panels, which make up this pleated structure, are supported by stainless steel rails. An anodized aluminum frame allows the natural ventilation of this double skin. The upper part of this frame, which constitutes the acroterion, is made up of removable panels which make its maintenance possible. The interior design was guided by the idea of transparency and luminosity. The attic of the building hosts a large meeting room that gives access to a terrace with a view of the city and the port.

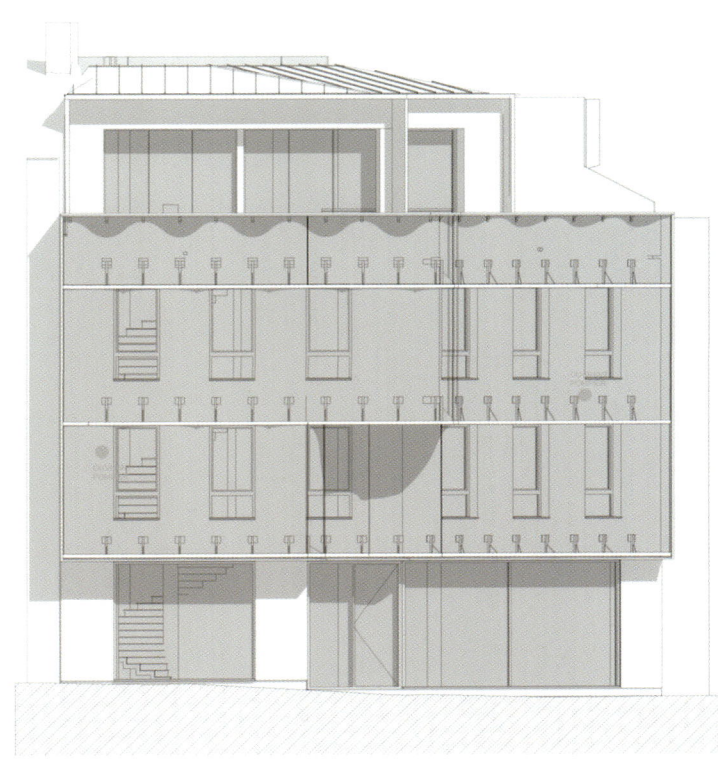

Elevation-A

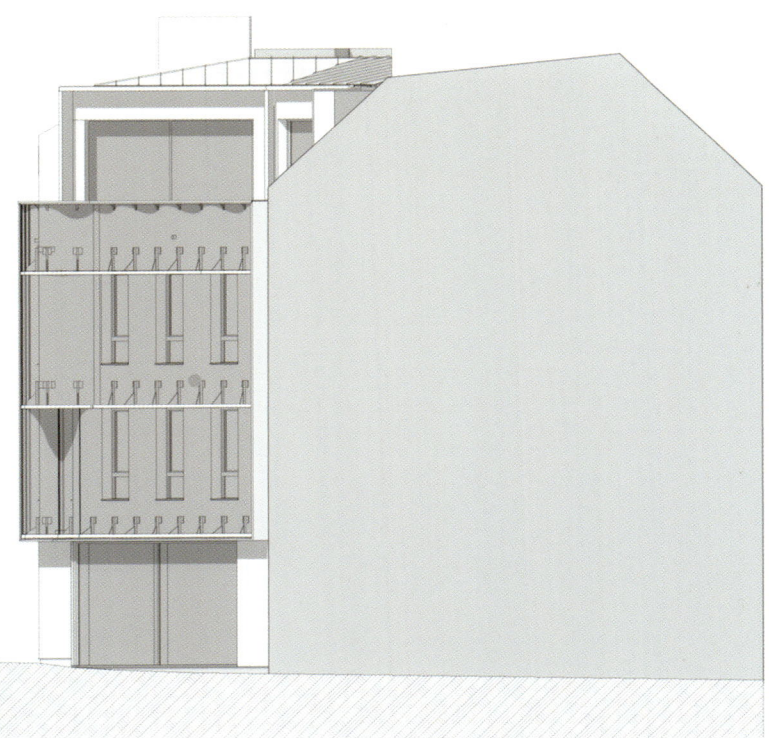

Elevation-B

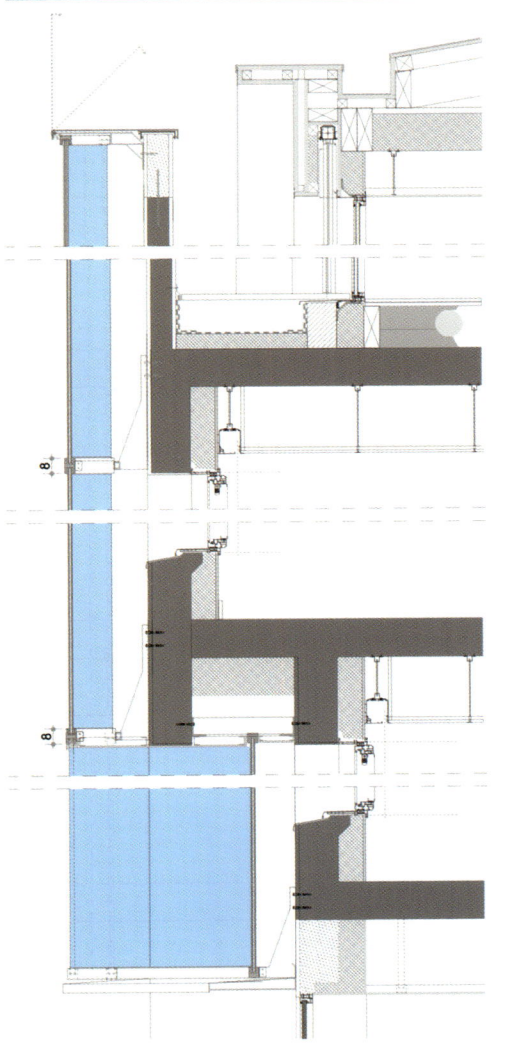
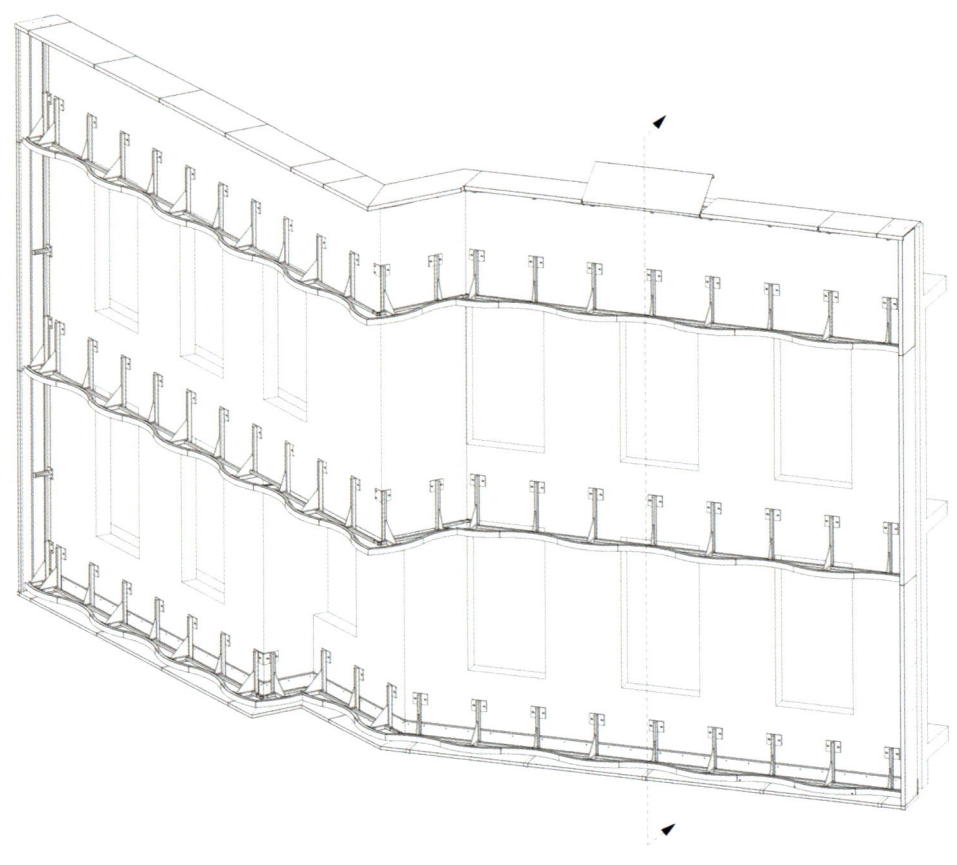

Detail Plan

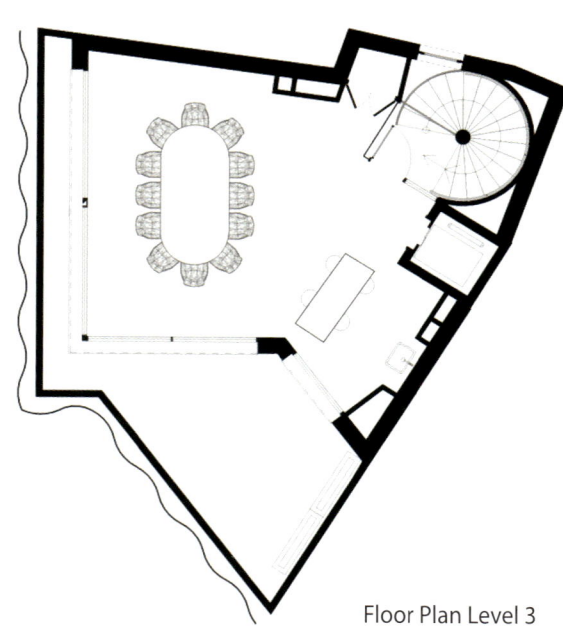

Floor Plan Level 3

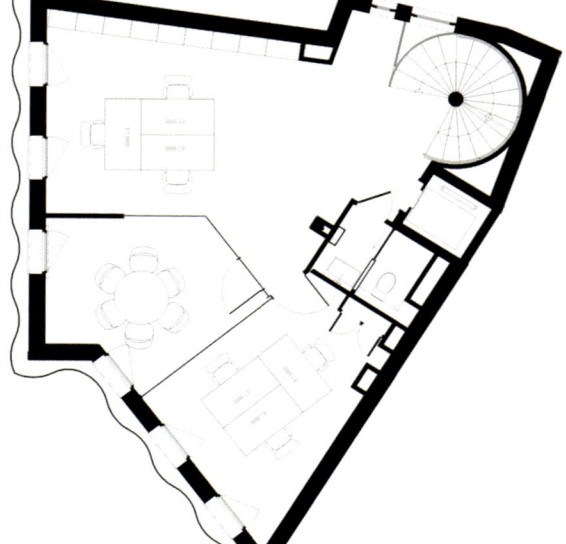

Floor Plan Level 2

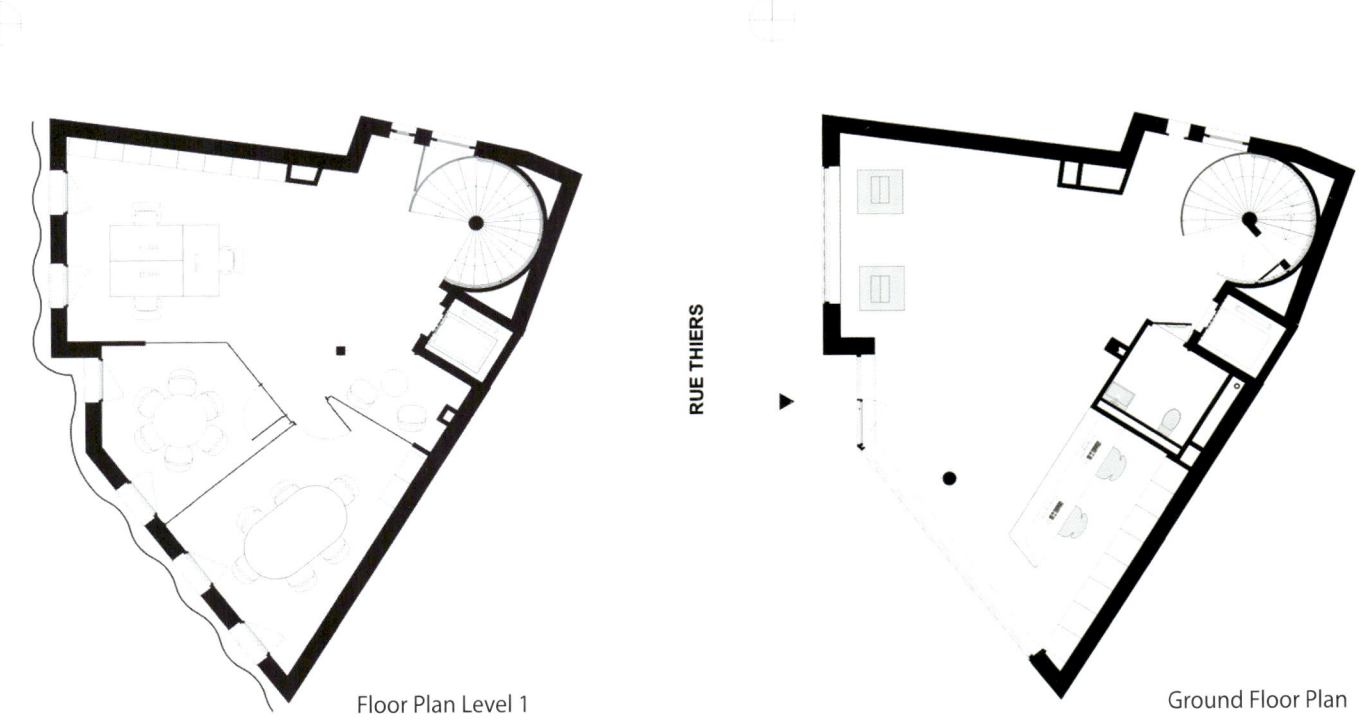

Floor Plan Level 1

Ground Floor Plan

RUE THIERS

The Centr'Al Building

Architects: B-architecten
Location: Forest, Belgium
Area: 1800 m²
Photographs: Lucid, Maxime Delvaux

Manufacturers: Geberit, Kingspan Insulated Panels, Rodeca, Aco, Led Linear, RENSON, CDM, DOX, Dauvister, Diamond, Labelfacade.......
Main Material: Concrete Hardening (Hard-Cem | Kryton), Translucent Glass facades.

Two buildings with amenities for the neighbourhood play a central part in the development of the Albertpool in Vorst (Brussels). These two buildings are the gateway to this district. The new Albertpool is an excellent opportunity to put this vibrant part of the city on the map in exemplary fashion. Big public areas, such as the auditorium and the sports halls are located on the buildings' corners. The transparent and translucent façades on the intersection liven up the buildings' functions.

The functional units are stacked as compactly as possible. Beneath the covered outdoor spaces is the entrance to the buildings. The local restaurant on the ground floor has a view of the new town square and is connected with the foyer inside. This foyer spreads over the different floors of the building and as such functions as a perfect meeting place for the users of the complex. Even the roof terrace can be used for different activities, such as sports.

Material Type: **Concrete Hardening** (Hard-Cem | Kryton)

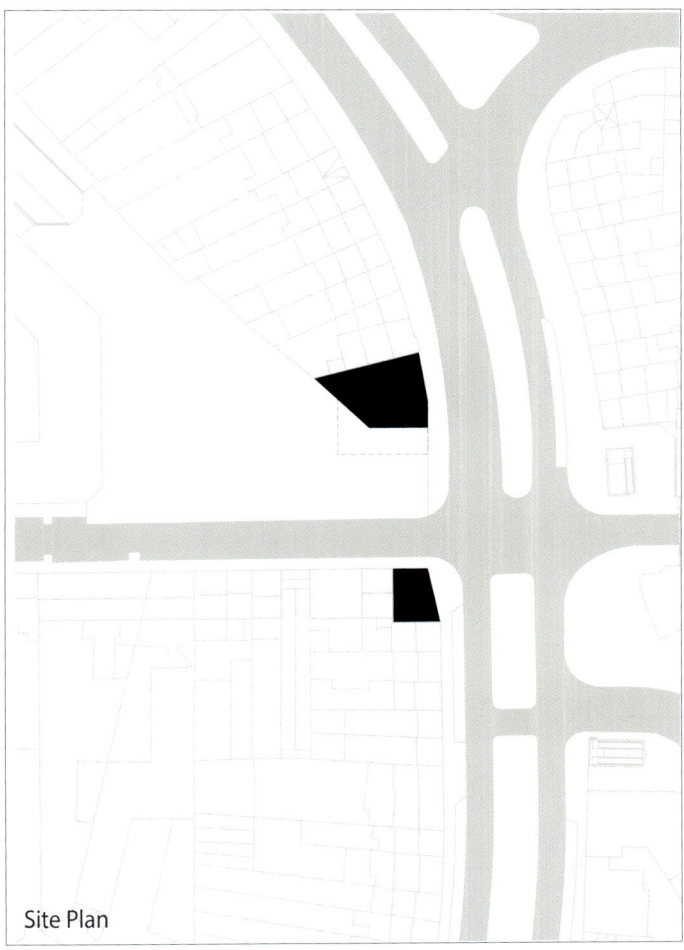

Site Plan

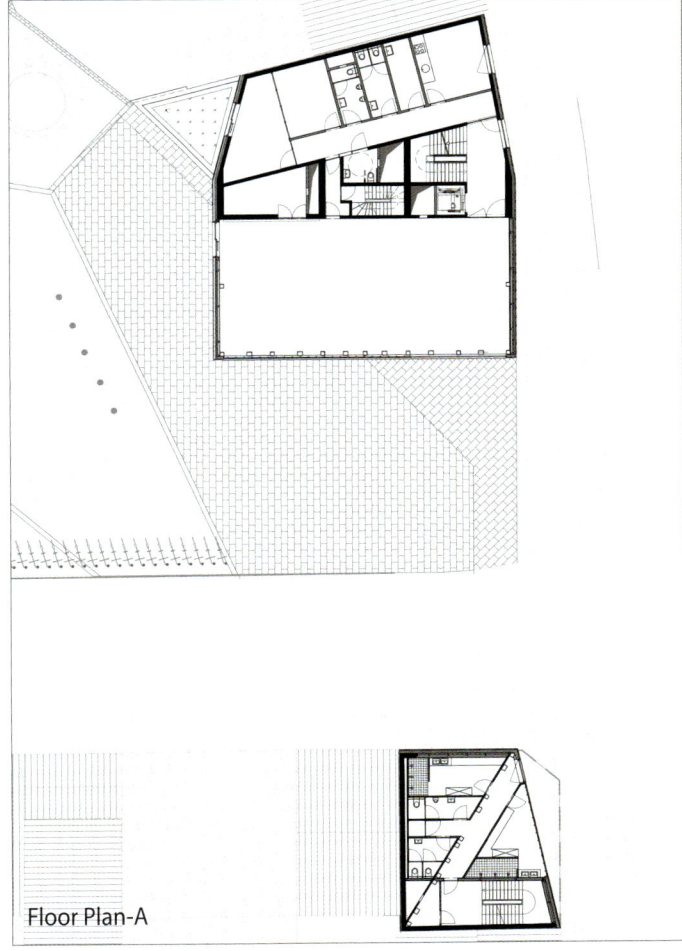

Floor Plan-A

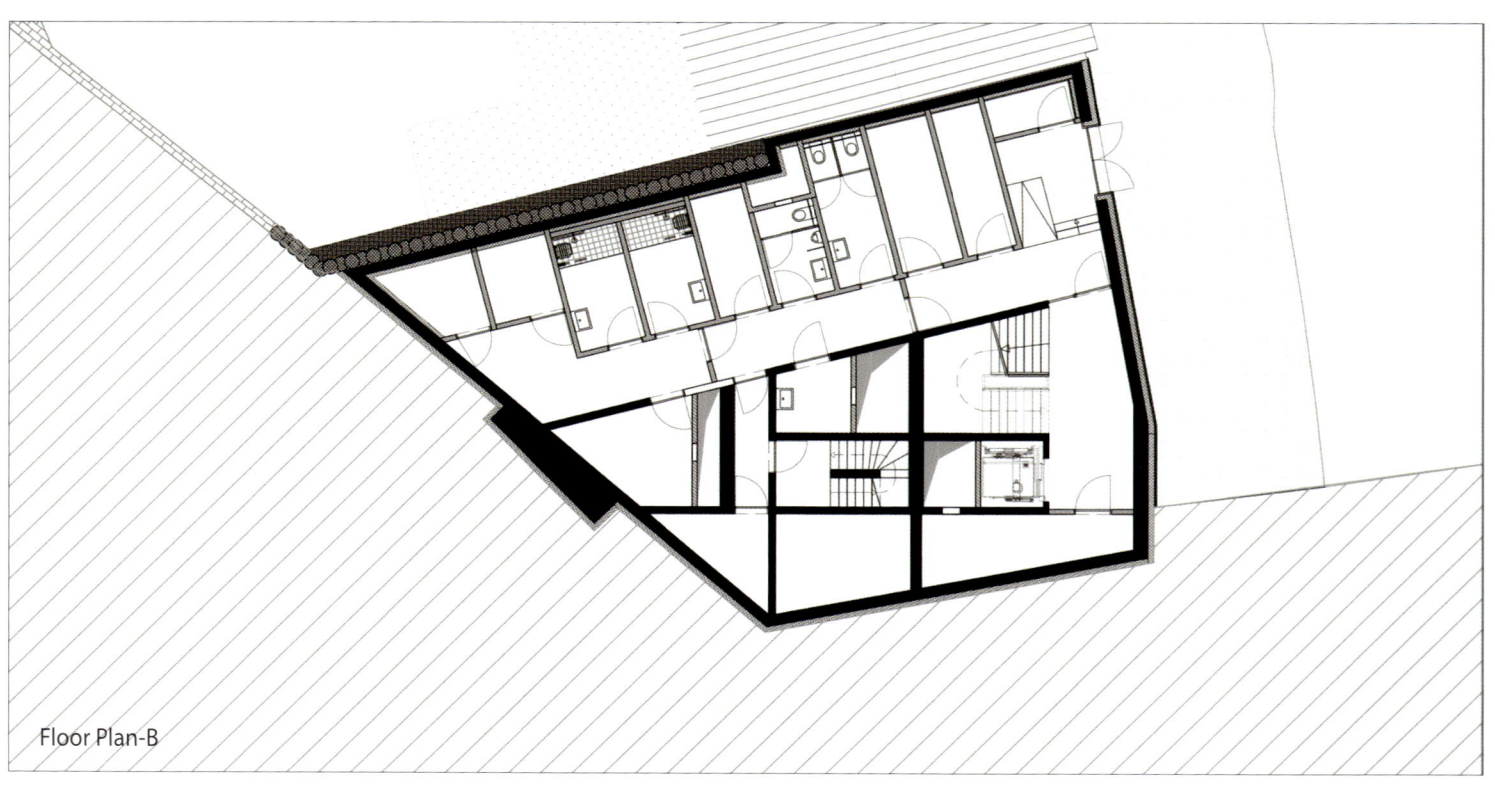

Floor Plan-B

Material Type: Glass
(Glass Curtain Wall)

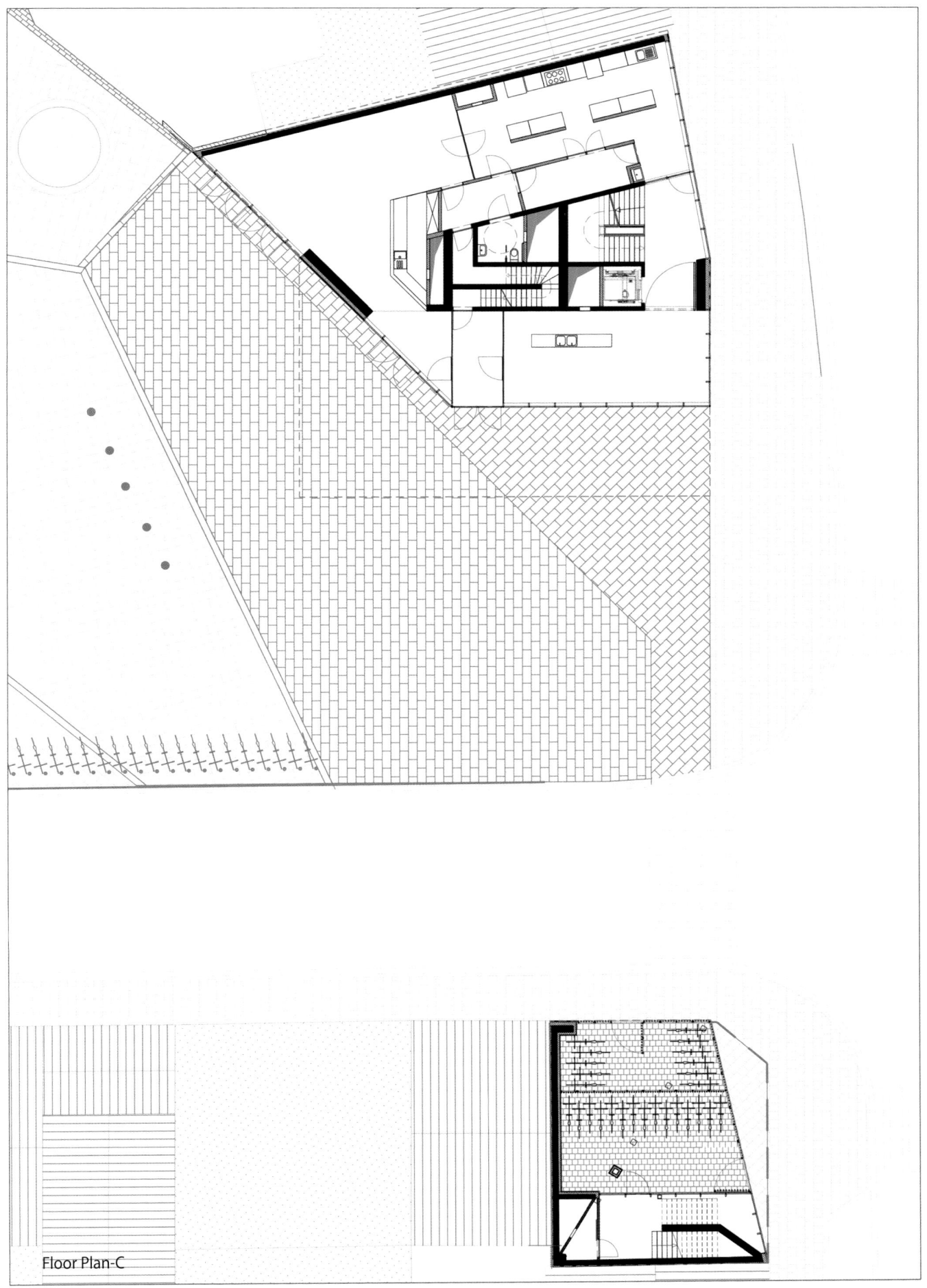

Floor Plan-C

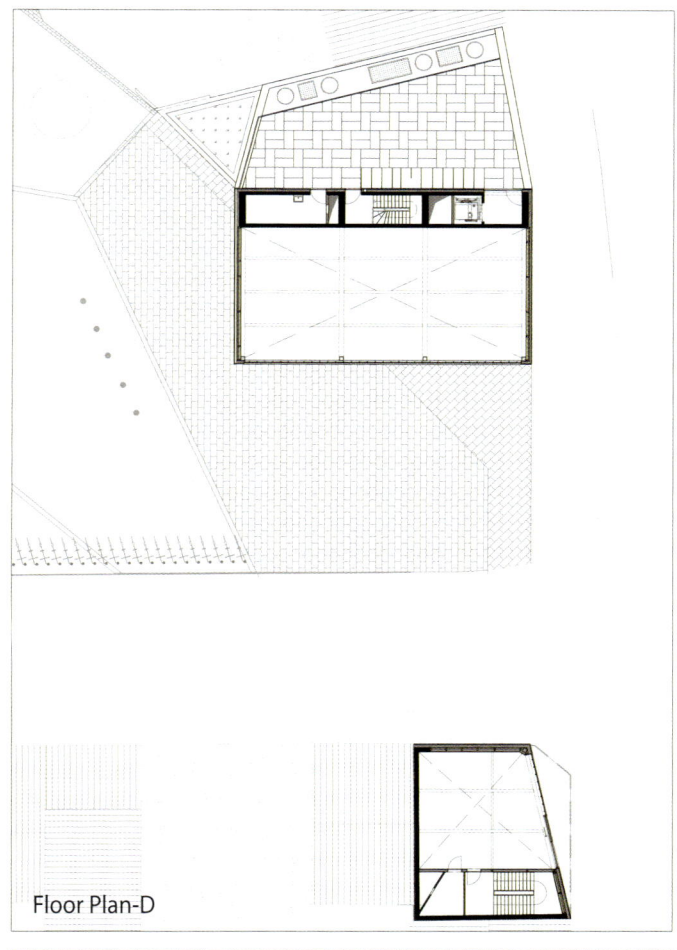

Floor Plan-D

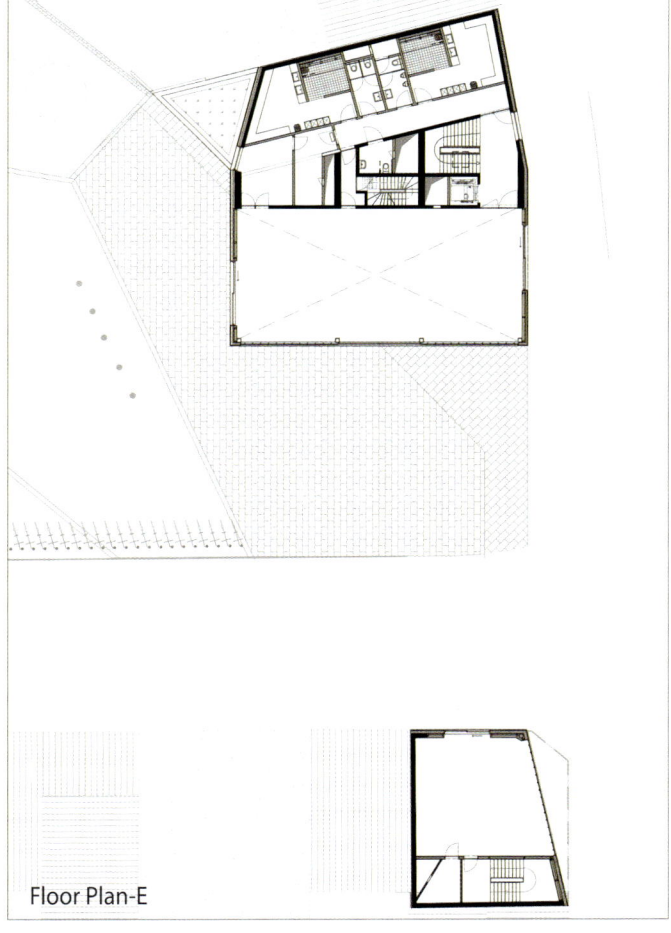

Floor Plan-E

The Framed House

Architects: Crest Architecture
Location: Forest, Belgium
Area: 2300 ft²
Photographs: Shamanth Patil J

Manufacturers: AutoDesk, Chaos Group, Grohe, Jaquar, Saint-Gobain, Adobe, Trimble....
Main Material: Concrete finish and white plastered walls

Located within a gated community in North Bangalore, the square shaped plot of this residence abuts the road on southern and western sides. Based on the client's requirements, our approach was to design a modest house with specific emphasis on natural light and ventilation. The design process and choice of materials for construction was influenced by a cost effective budget. Our approach was to create a compact layout that accommodated the requirements of the client while also establishing a sense of spaciousness to create a balance of connectivity and privacy. The south –western façade was kept almost entirely plain with narrow slit windows to avoid the harsh sun.

These windows are accentuated with Sadahalli stone chajjas which frames compelling views of the surrounding landscape. The open floor plan which consists of a living, dining, kitchen, puja and three bedrooms are spread across two floors. The double height living and dining spaces are opened up to get a view of the landscaped garden and glazed skylights allow ample natural light to infiltrate the open scheme of the house. Exposed concrete finish and white plastered walls induce a subtle ambience in both the interior and exterior. The design language throughout the house is intended to be simple that reflects the lifestyle of the client.

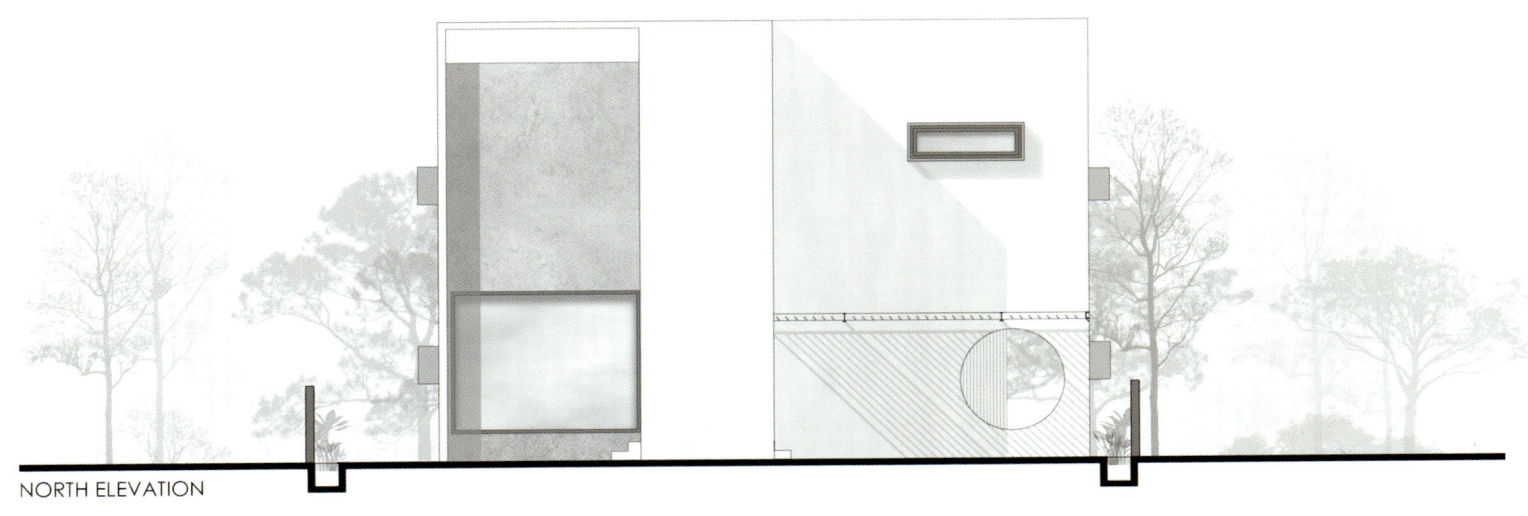

NORTH ELEVATION

155

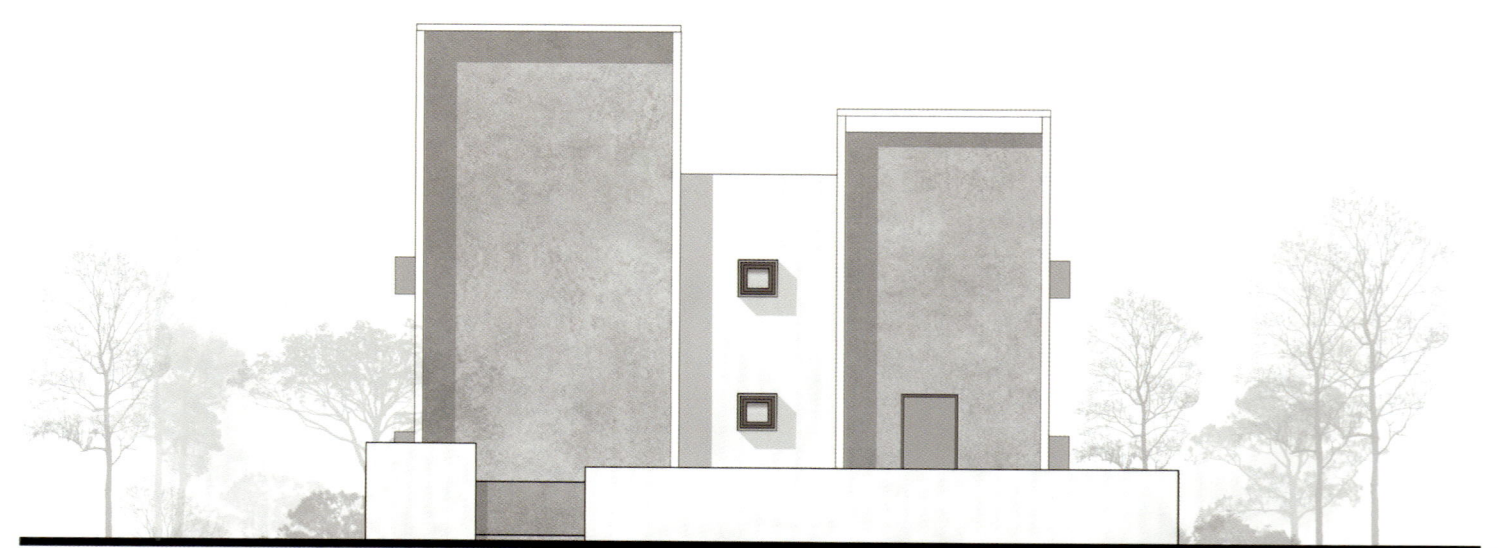

SOUTH ELEVATION

EAST ELEVATION

Material Type: **Concrete**
(Exposed concrete)

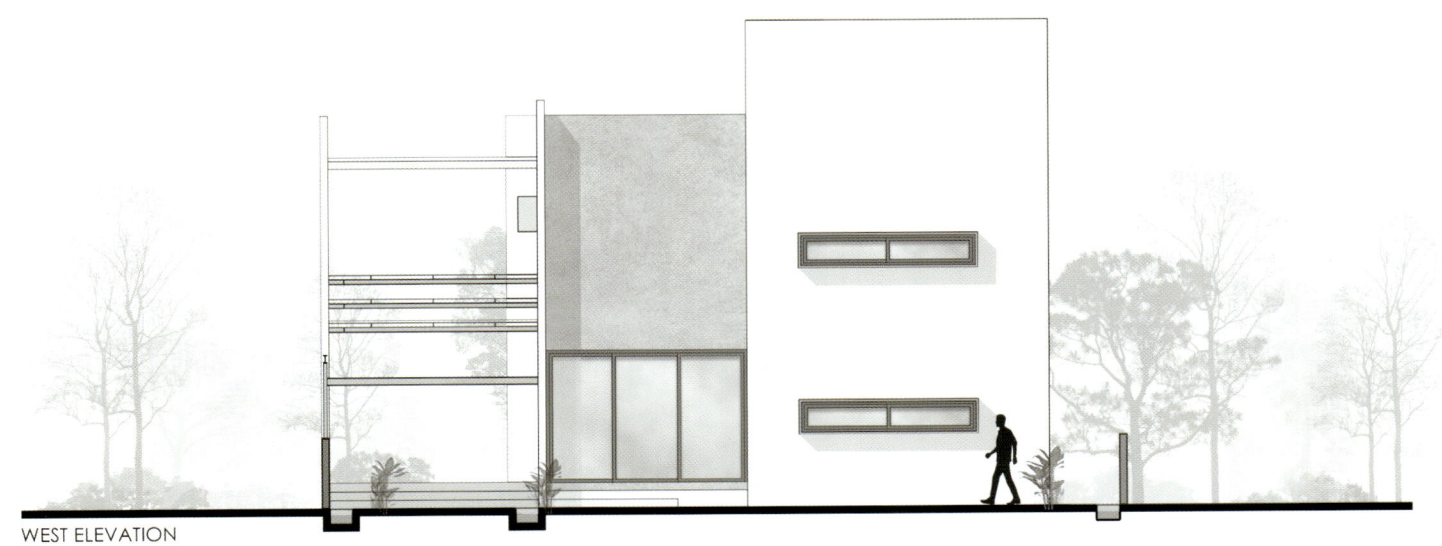

WEST ELEVATION

159

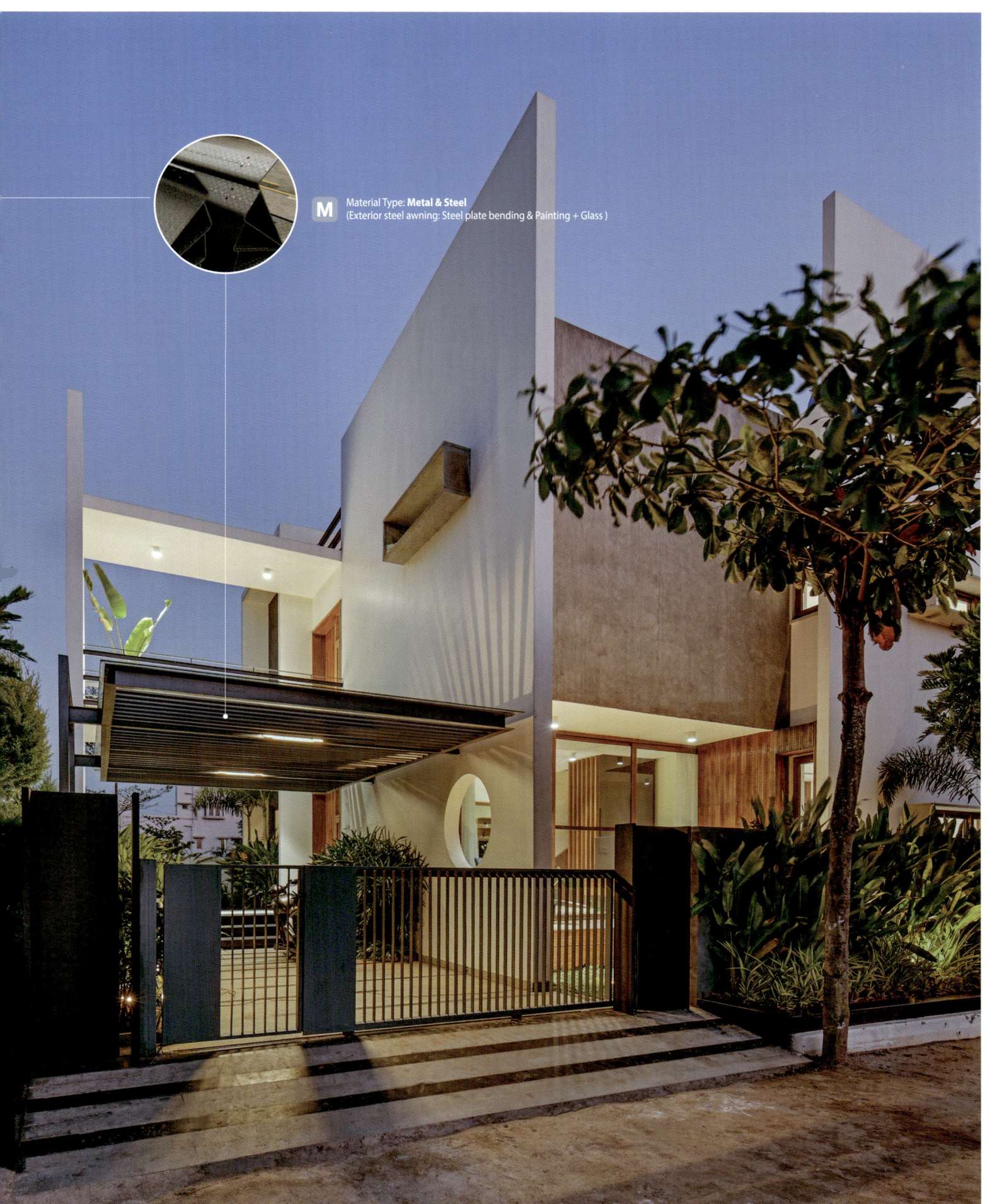

Material Type: **Metal & Steel**
(Exterior steel awning: Steel plate bending & Painting + Glass)

W Material Type: **Wood :** Thermowood from Scandinavian PEFC-certified pine from Finlan (Thermowood Battens | Lunawood) ① ② ③

The durable product from Lunawood used to create solar shades is LunaThermo-D; along with appearance, biological durability is a key property in the end-use applications of products in this treatment class. The tangential swelling and shrinkage due to moisture are low. Lunawood produces Thermowood from Scandinavian PEFC-certified pine from Finland. It can be used inside or outdoors, in any climate. Lunawood is an eco-friendly natural product that is easy to machine and install.

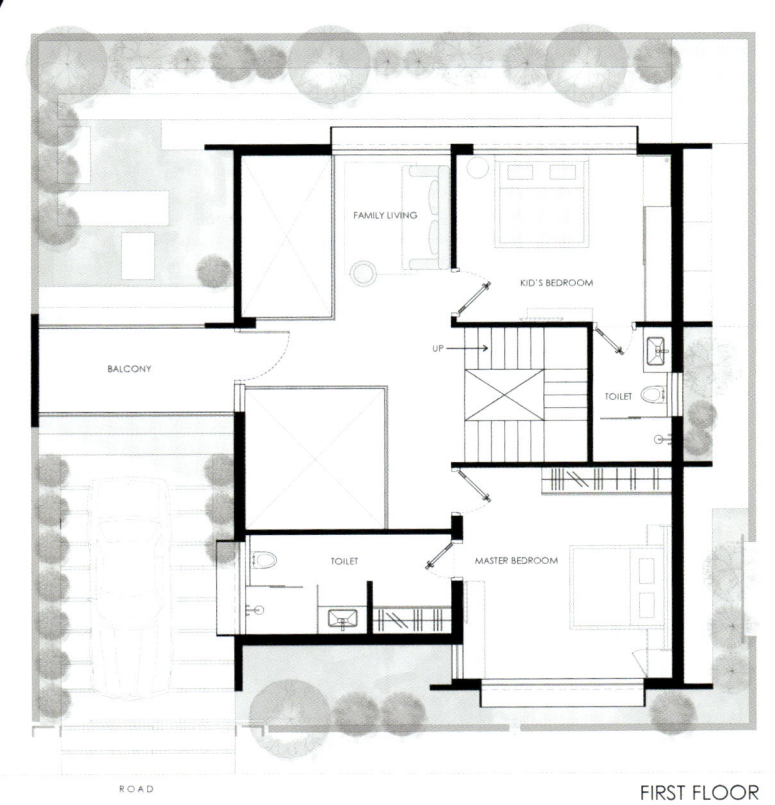

FIRST FLOOR

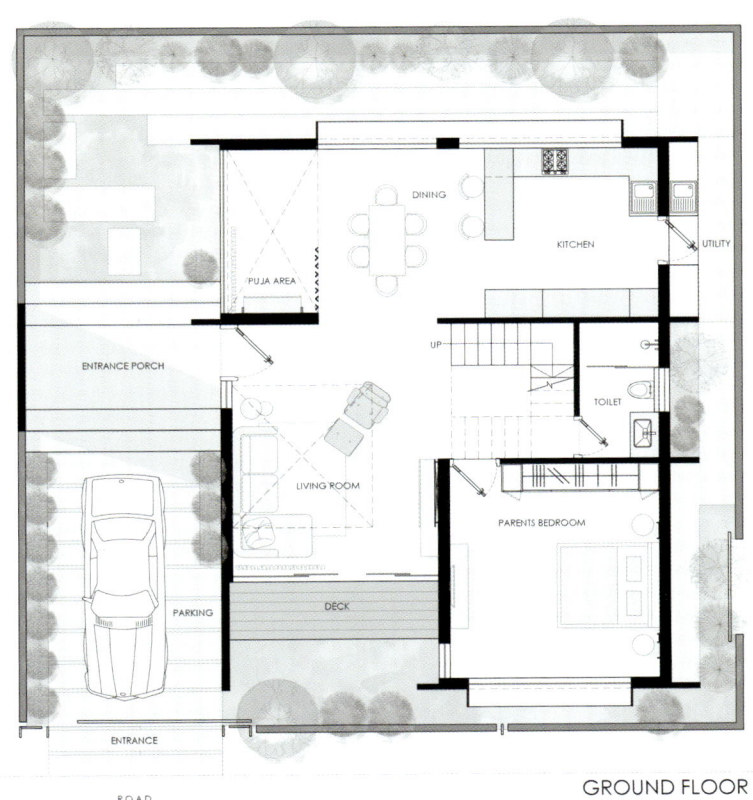

GROUND FLOOR

163

Travertine walls enclose Casa ZTG in Guadalajara

Architects: 1540 Arquitectura
Location: Guadalajara, Maxico
Area: 1,200-square-metre
Photographs: César Béjar

Main Material: Roman travertine marble(Brick + Concrete)

Mexican studio 1540 Arquitectura has created an inward-facing home for an older couple that features tall, marble-clad facades with limited openings. Casa ZTG is located in the metropolitan region of Guadalajara, within the western state of Jalisco. It was built in a private residential area with single-family dwellings. Local office 1540 Arquitectura aimed to create a residence well-suited for its occupants – an older couple. "The goal was to create a timeless and elegant atmosphere, both in form and in materials, that will reflect the personality and age of the owners," the firm told Dezeen.

For a flat site, the firm conceived a 1,200-square-metre building that is roughly rectangular in plan. Exterior walls are made of brick and are covered with Roman travertine marble. The front elevation consists of a tall, opaque wall that is lined with an L-shaped reflecting pool. In a lower corner of the facade, a rectangular opening provides access to the interior and offers a glimpse of a lush garden. Similar to the street-facing elevation, the home's side walls have a limited number of apertures. The architects decided to create an "introverted" home due to the lacklustre surroundings.

Corridors wrap an interior garden planted with trees. The home has three levels, one of which is below ground. In the front half of the main floor, lofty corridors wrap around the verdant garden. Arrayed along these hallways are a kitchen, a living room, a formal dining area, a small office and a master suite. The house is arranged over three levels. Retractable glass walls provide seamless access to the backyard, which features a terrace, swimming pool and hot tub. On the upper level, the team placed three bedrooms and a lounge. The basement holds a garage for the owners' car collection, service areas, and a sauna and steam room.

S Material Type: **Stone**
(Roman travertine marble)

Exterior walls are made of brick and are covered with Roman travertine marble.

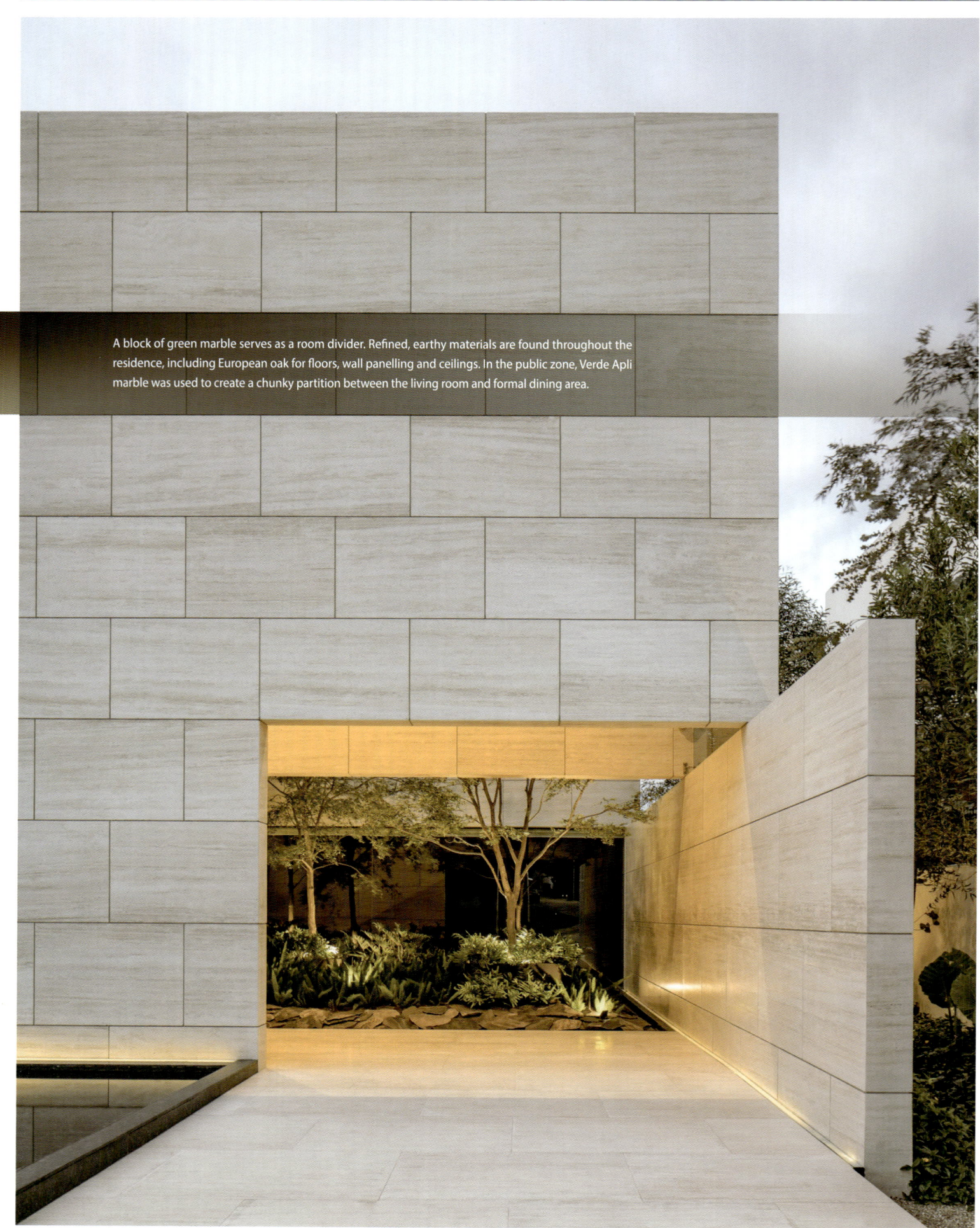

A block of green marble serves as a room divider. Refined, earthy materials are found throughout the residence, including European oak for floors, wall panelling and ceilings. In the public zone, Verde Apli marble was used to create a chunky partition between the living room and formal dining area.

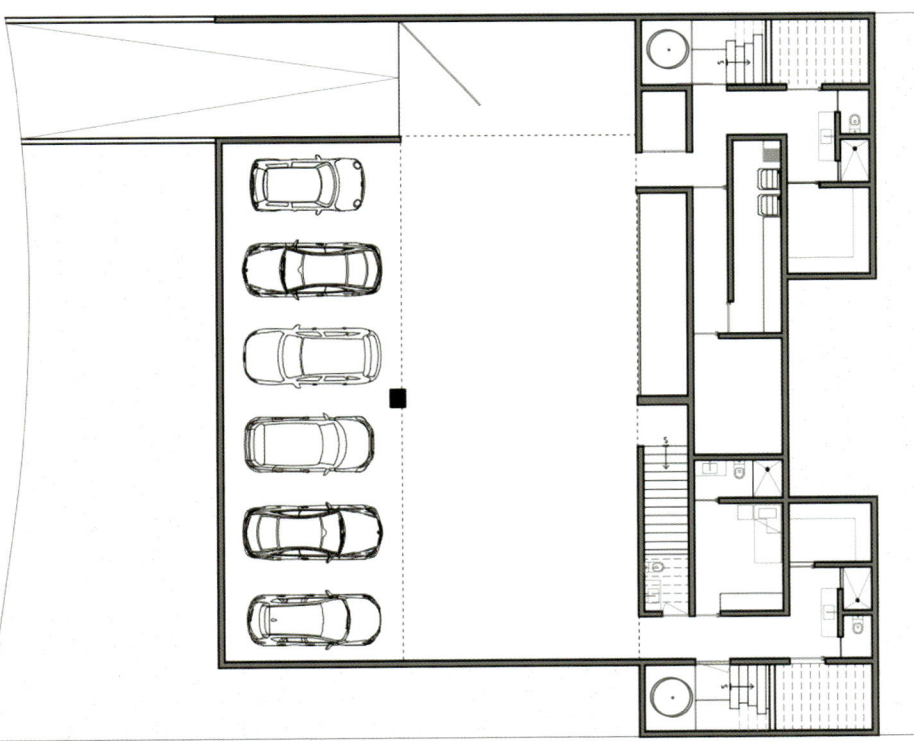

Grey marble forms a wall between the bedroom and the bathhroom. Marble was also used in the master suite. In this case, grey arabescato marble forms a wall behind the bed, which separates the sleeping area from the bathroom. In the dressing area, panels are wrapped with natural leather in a greenish hue. "In both the architecture and furniture, we decided to use a selection of materials, shapes and details that possess timeless sensorial, physical and aesthetic characteristics," the team said.

167

Travertine walls continue inside the house and Corridors wrap an interior garden planted with trees.

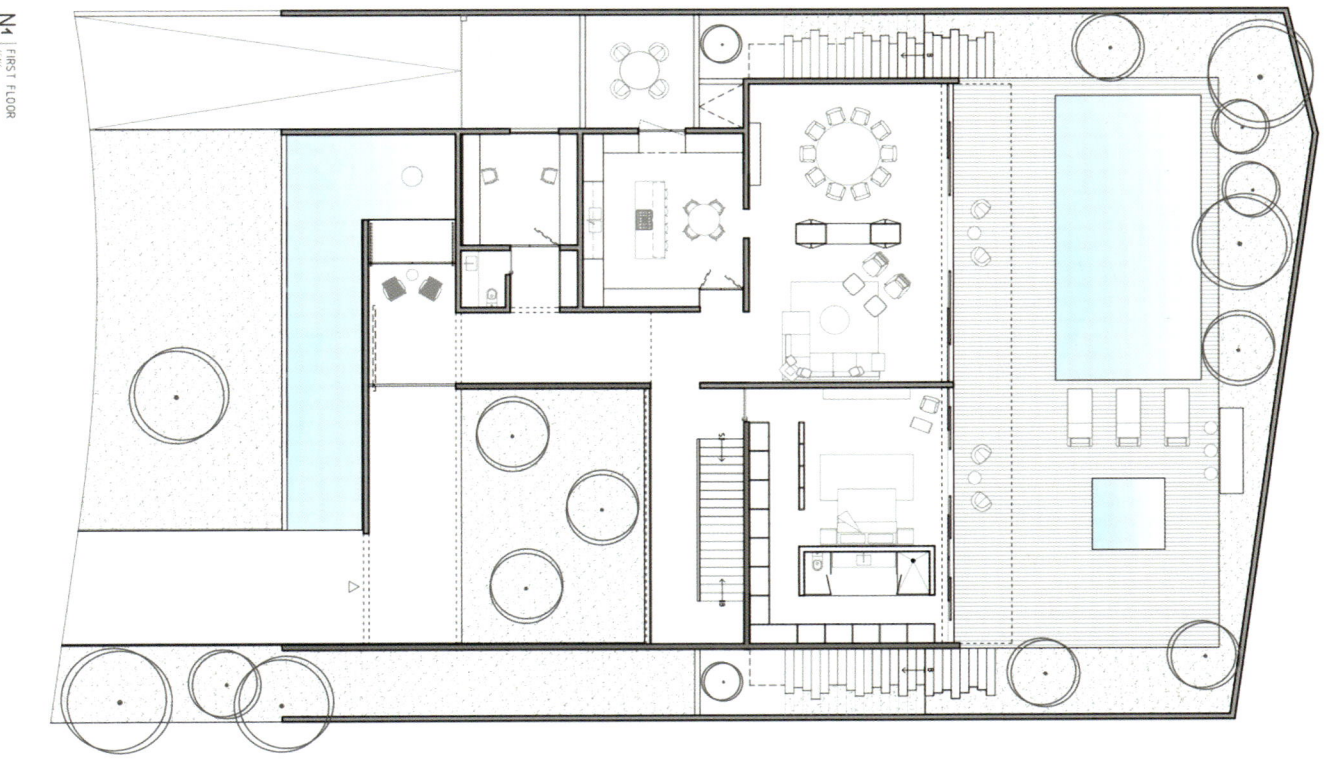

 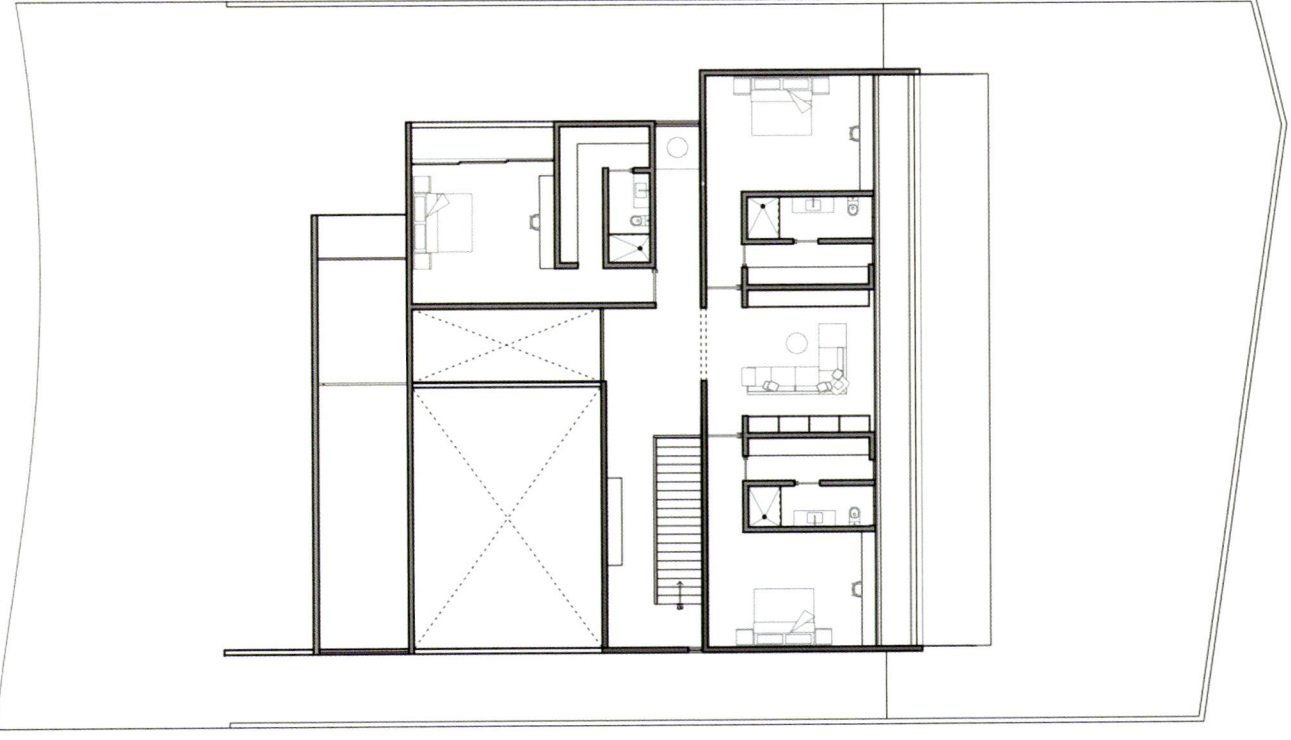

SECOND FLOOR

Renovation of No.1 Sinopec Gas Station

Architects: TJAD Original
Location: Shanghai, China
Area: 223 m²
Photographs: ZY Architectural Photography

Lead Architects: Ming Zhang, Zi Zhang
Project Architects: Xunan Wang, Chun Ding
Main Material: Concrete + Steel + Glass

No. 1 Sinopec Gas Station is located in the upper riverside along the Suzhou River, near its junction to Huangpu River. It was formerly China's first state-owned gas station built in 1948. The original building is used as supermarket and office with a steel structure that works as a canopy. The structure itself lacks publicity and transparency. Besides, there is a lack of proper diversion between motor vehicles that want to refuel and the public. Furthermore, the excellent river landscape cannot be absorbed by the supermarket and the gas station. Cultural resources are wasted. Therefore, the focus of our design is how to make a breakthrough of gas station type and form an infrastructure that is publicly transparent, suitable for circulation, complex in function, and fit for contemporary context. The design of the gas station starts from combing the riverside landscape, disassembling the mixed flow into two traffic lines: pedestrians bypassing the gas station from the north side near the river, and vehicles entering and exiting from the south side. Thus, the public side changes from the original south side to both the north and south sides, which fits public sphere of the city and riverside. It eliminates to the greatest extent the barrier the gas station causes to the riverside landscape.

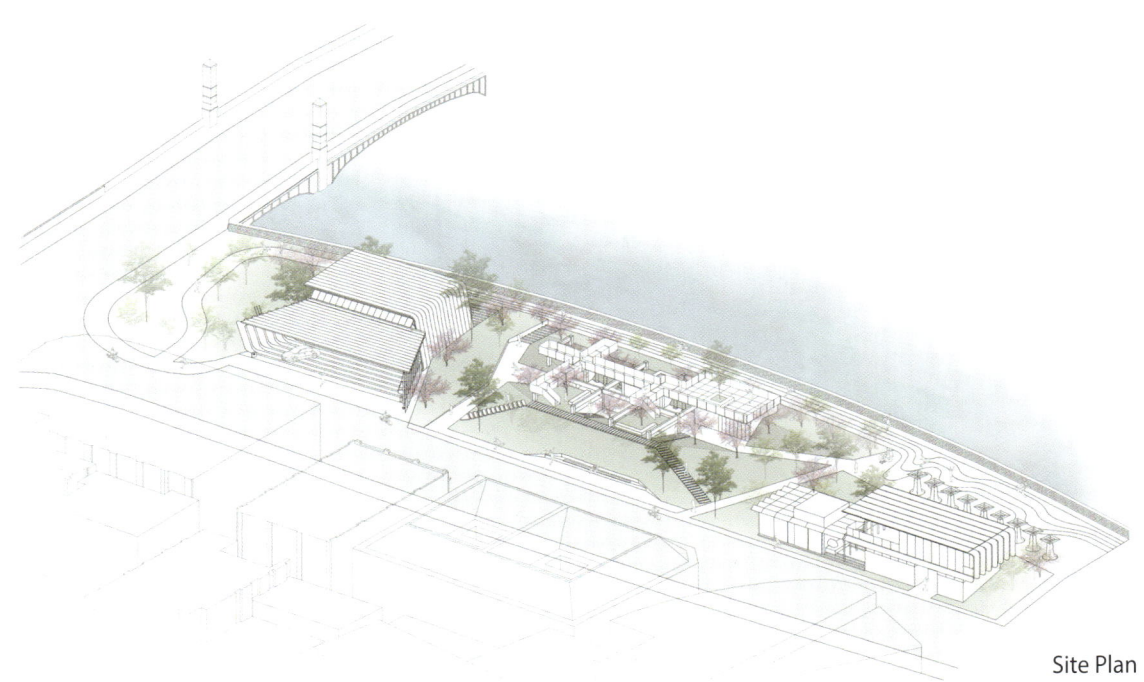

Site Plan

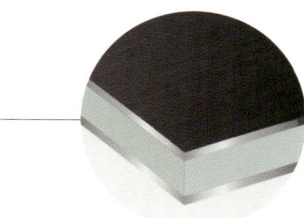

Material Type: **Metal & Steel**
(Roof a zinc panel)

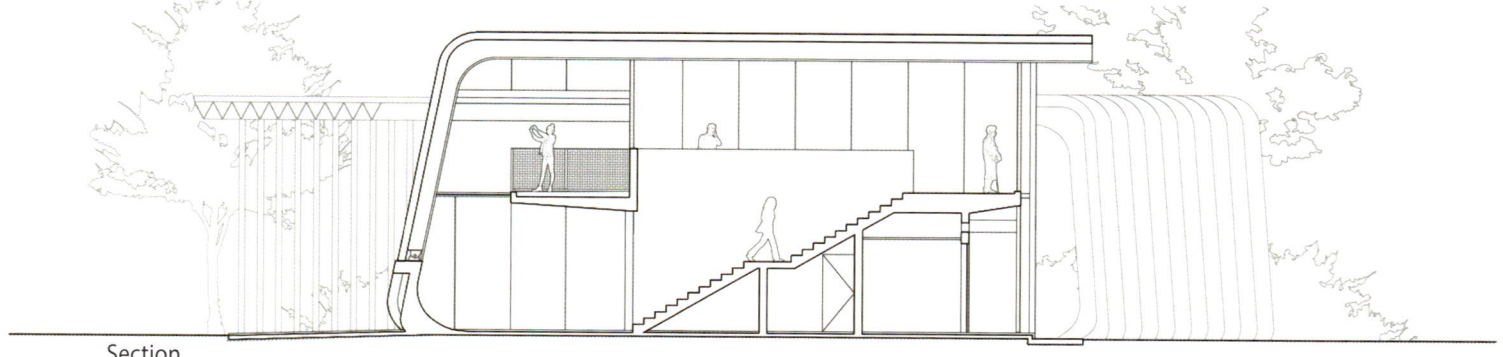

Section

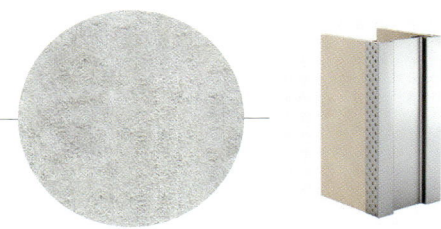

C Material Type: **Concrete**
(Wall Stop Ends - EzyCap | Studco)

Material is changed to fair-faced concrete, changing the panel into a combined structure of steel and concrete folded board.

The gas station is divided into two volumes, one virtual and one real. The virtual one is the refueling scaffold and the real one is the station building. Different from the previous heterogeneous blocks, we hope that the two volumes are more united, forming the shape of the building together. Due to the functional requirements of refueling, the scaffolding of the gas station has a certain span requirement. Considering that the folded plate form has the effect of increasing the span and has the characteristics of integration of structural rationality and visual characteristics, the building is conceived as a set of high and low folded panels that turns over from the ground. The high part of folded panels accommodates the functions of the two-story station building with a supermarket at the ground level and a café at the first level. The slightly lower part of the folded panels covers the refueling area.

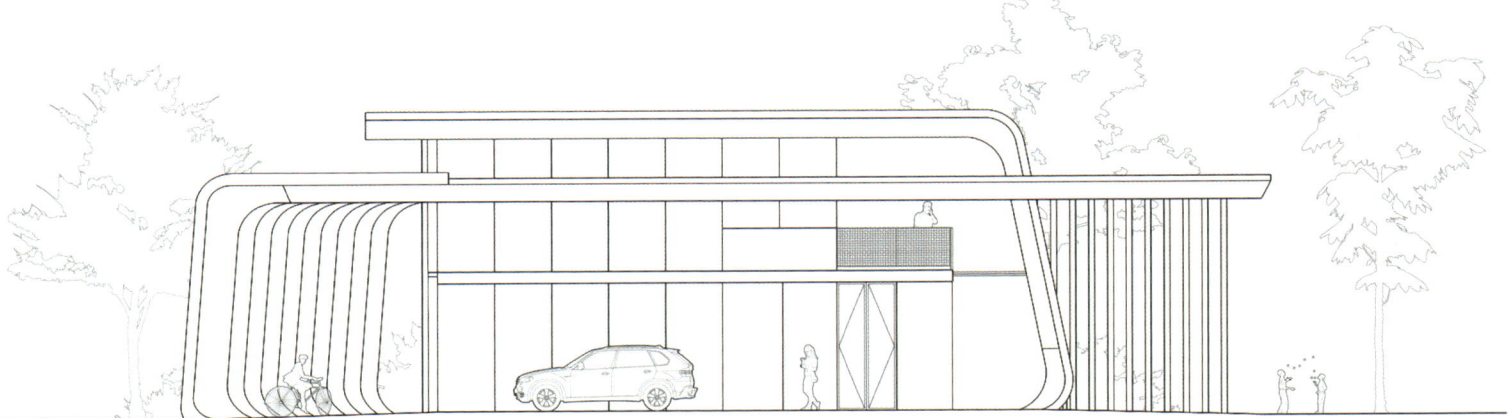

Elevation

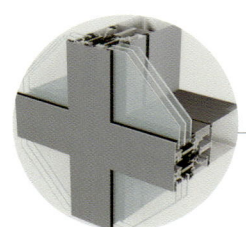

G Material Type: **Glass**
(3 layers Glass Curtain Wall)

1. 排水槽口　　9. 轨道射灯
2. 玻璃雨棚　　10. 加强筋
3. 加油口　　　11. 空调出风口
4. 加油管线　　12. 空调回风口
5. 栏杆　　　　14. 直立锁边屋面
6. 花架　　　　15. 射灯轨道
7. 设备管线　　16. 钢折板
8. 竖梃

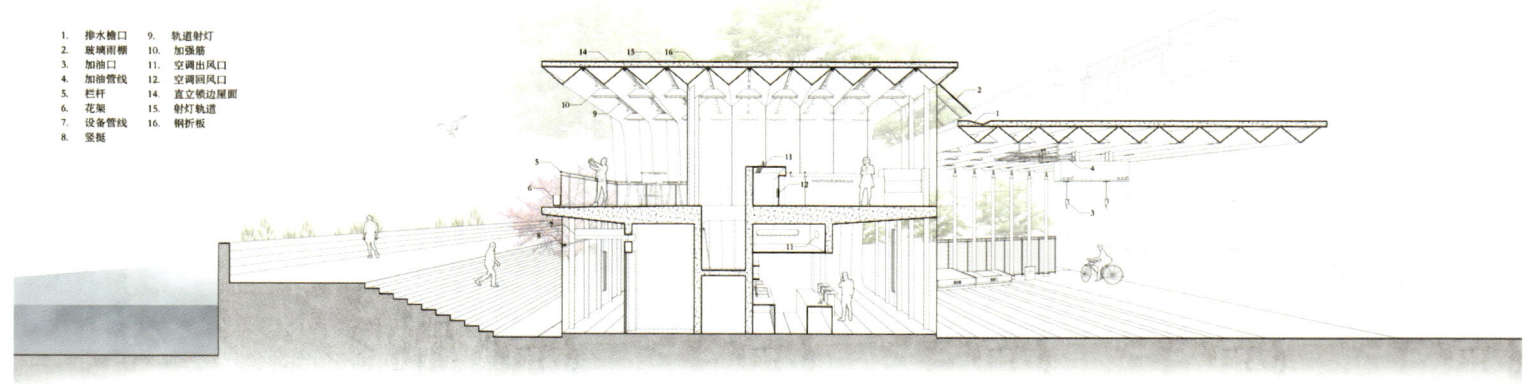

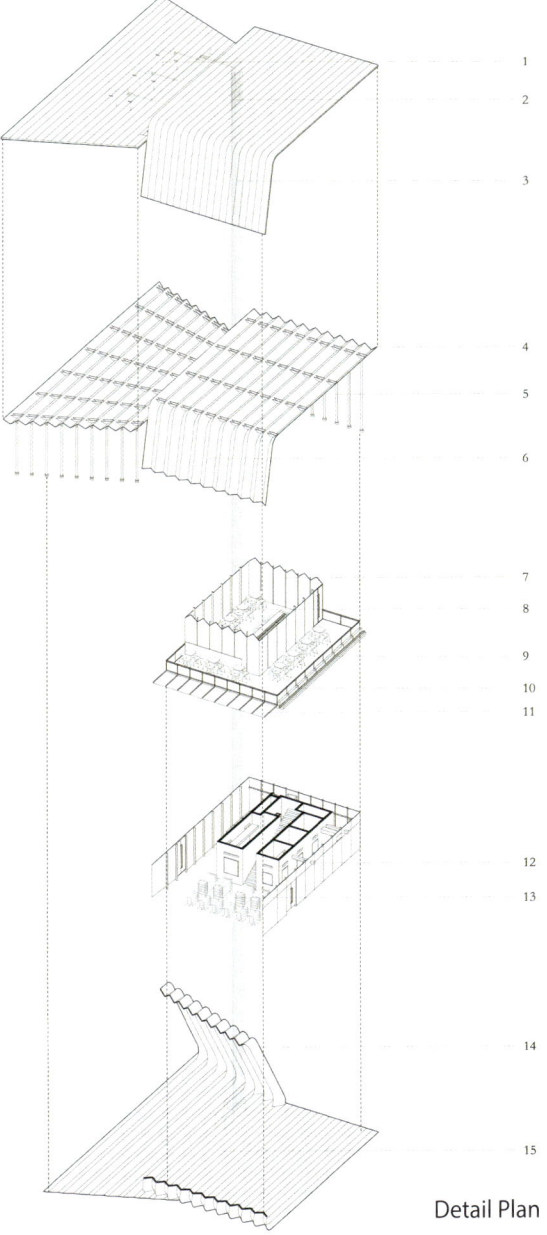

Detail Plan

The panels are inserted into the ground at one end and supported by a row of thin pillars at the other, which further highlights the morphological characteristics of the folded plates. The panels was conceived as steel structure in the early stage of the design. However, considering the risk of rust caused by the direct landing of the steel structure, the material is changed to fair-faced concrete, changing the panel into a combined structure of steel and concrete folded board. The volume of the two sets of folded boards changes at different heights. While ensuring the continuity of the form, the steel part is anchored on the concrete folded board by exposed hinges, which enhances the readability of the structure. The connection of the column and the folded panel also conforms to the way of the hinge point.

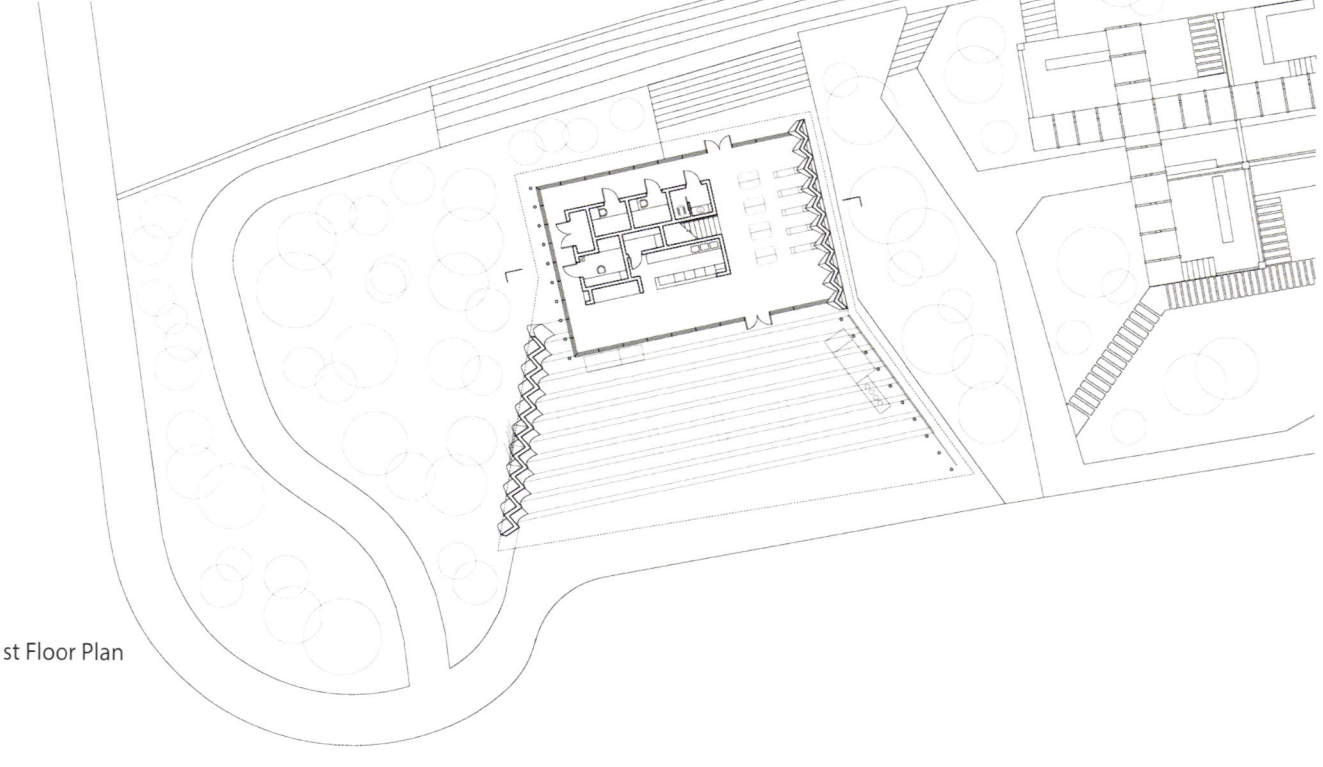

1st Floor Plan

On the one hand, the column is mainly subjected to axial force, which reduces the diameter of the column. On the other hand, the shape of the folded panel is more pure. The steel part shows a light and concise state, while the floor-to-ceiling concrete folded panel reveals an exquisite and powerful space experience. The gas station is named "Suhe Zhe". It means the integration of structure and building form, which makes the folded panels the most important feature of the building. Based on structural rationality, it combines a certain echo of the folded wall of the old building, the imagination of waves, and a kind of exquisiteness like a folding fan. The purity of the structure and the ambiguity of meaning endow the building with a contemporary era. In a place full of historical memory, this contemporaryity is undoubtedly the most reasonable annotation of "cultural gas station".

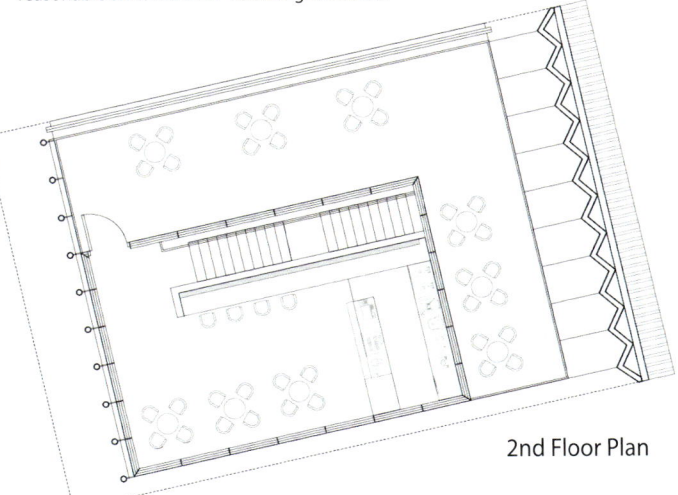

2nd Floor Plan

Kaolin Court Housing

Architects: Stolon Studio Ltd. + Baca Architects
Location: London, United Kingdom
Area: 1000 m²
Photographs: Robert Barker

Manufacturers: Johnson Tiles, Velfac, Wienerberger, Curtis Metal, GutterCentre, Krion, PWS, Sedum Supply, Velux
Main Material: Tiles (Johnson-tiles)

Kaolin Court, is small 'sociable' development of residential and live/work units, set around a shared courtyard garden. The client is a small private developer, specialising in unique projects. The brief was to redevelop the existing light industrial site with a private residential scheme that would appeal to families and professionals. The solution is a development of 4 houses, 2 live/work duplexes, and 3 apartments, set around a generous communal garden. The accommodation follows the boundary to maximise the available shared amenity space. The 2m slope was excavated to facilitate 2 storey houses without impacting adjoining owners, and to remove the contaminated land.

Whilst the solution may appear to be design driven, it was developed within the strict budget and programme set by the client. Therefore, design decisions were made with the view of adding value, not just cost, and Stolon worked to identify suppliers and sub-contractors that fit the budget. The sculptural forms were designed to maximise sunlight into the courtyard and minimise the effect on neighbours. The architectural angles are softened by the planting, reflecting pools, and the dappled light on the tiled facades. A 15° geometry was used throughout the development, in a plan to maintain outlook from neighbours and in section to maintain daylight/sunlight, as well as meet suit the minimum pitch for tiles and roof-lights.

Elevation-A

The front block is divided in two with a dramatic pedestrian route-way through to the new homes at the rear, which also allows the morning sun to stretch into the rear of the site.

Elavation-B

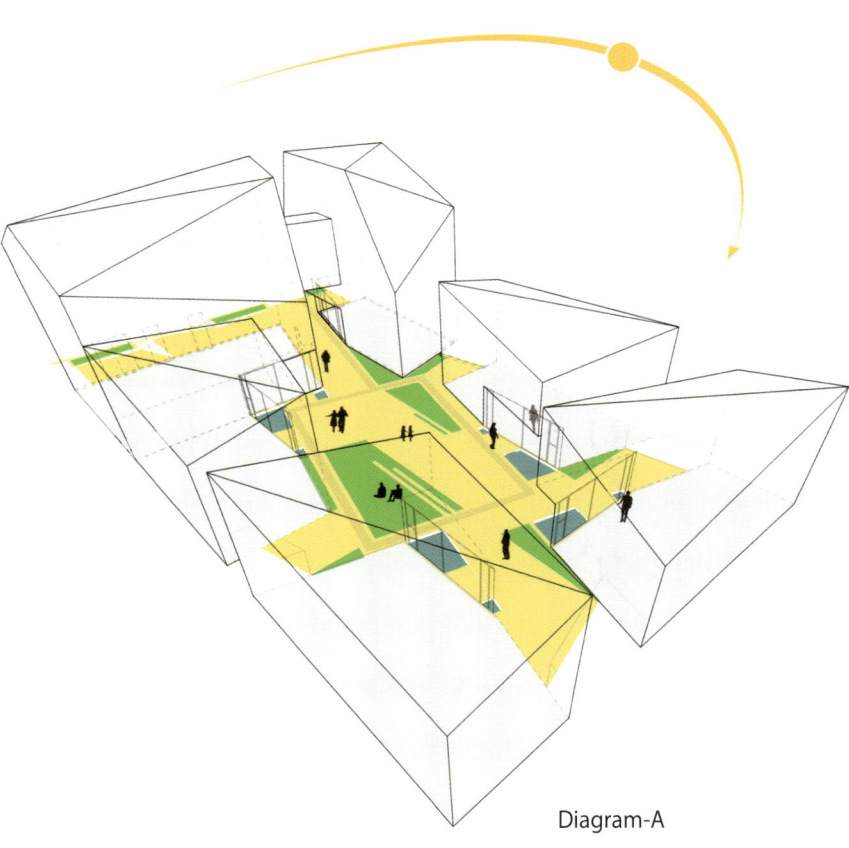

Diagram-A

Material Type: **Stone & Tiles**
(Johnson Tiles)

Johnson Tiles!

Proud to have set the environmental standard for ceramics in the white wares sector Johnson Tiles is known for high quality products and first class service. Our track record in committing to making positive change happen has helped us to uphold this reputation. We're constantly making and setting the industry standards for health and safety, quality assurance and environmental improvement ahead of time.

Landscape Plan

The shared garden is a place to relax, socialize, and play. Each house has its own courtyard which opens onto the central garden. The interchange between public and private space has been planned meticulously - from the most public street-front through to the most private spaces in the homes. The large picture windows onto the kitchen/dining space are separated from the shared courtyard by a reflecting pool. Houses are staggered in plan so that the more open dining areas are opposite the more private living areas. On the first and second floors, the building's form shelters the entrances and outdoor spaces, providing seclusion and protection, leading through to the delightful discovery of the garden beyond.

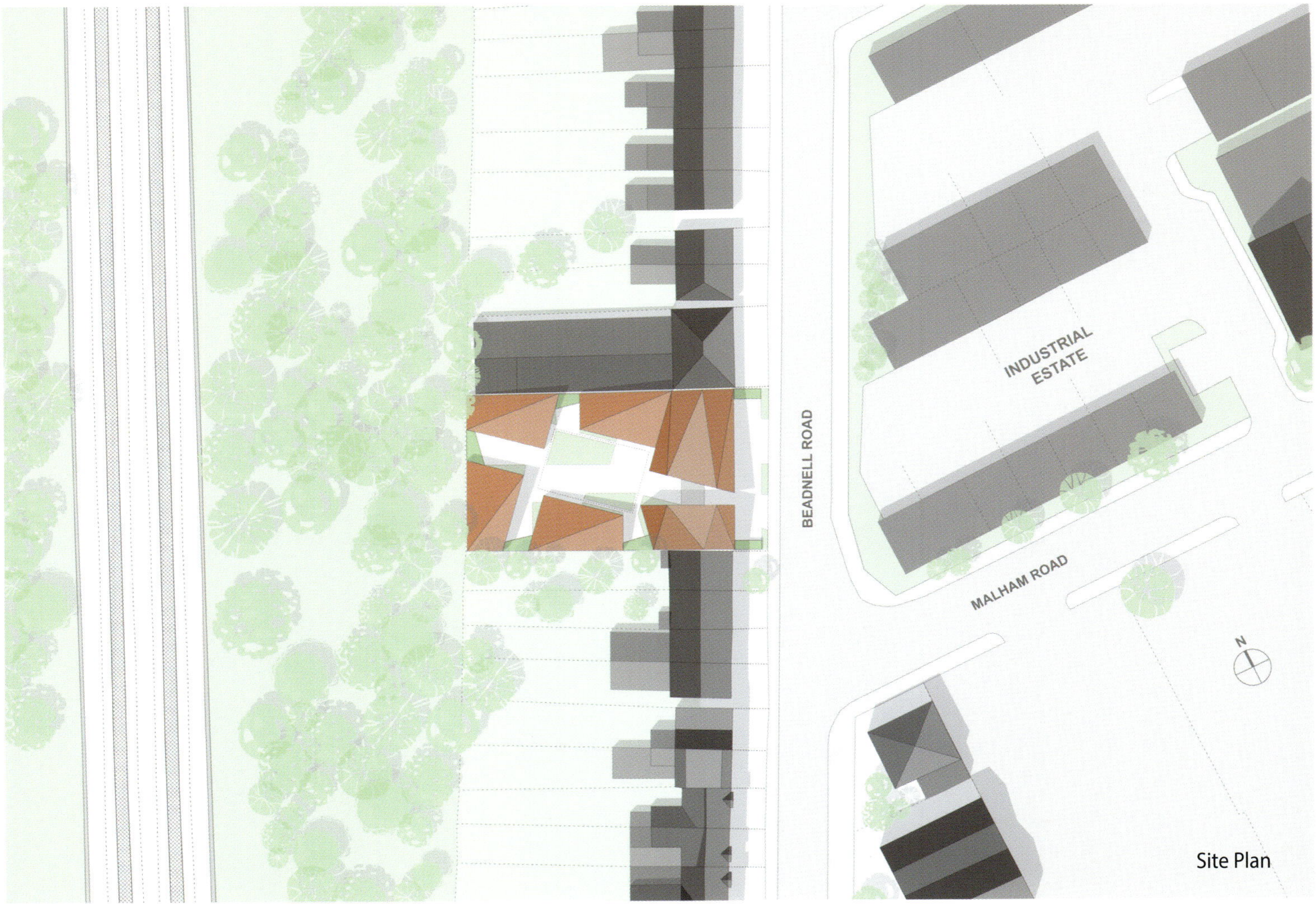

Site Plan

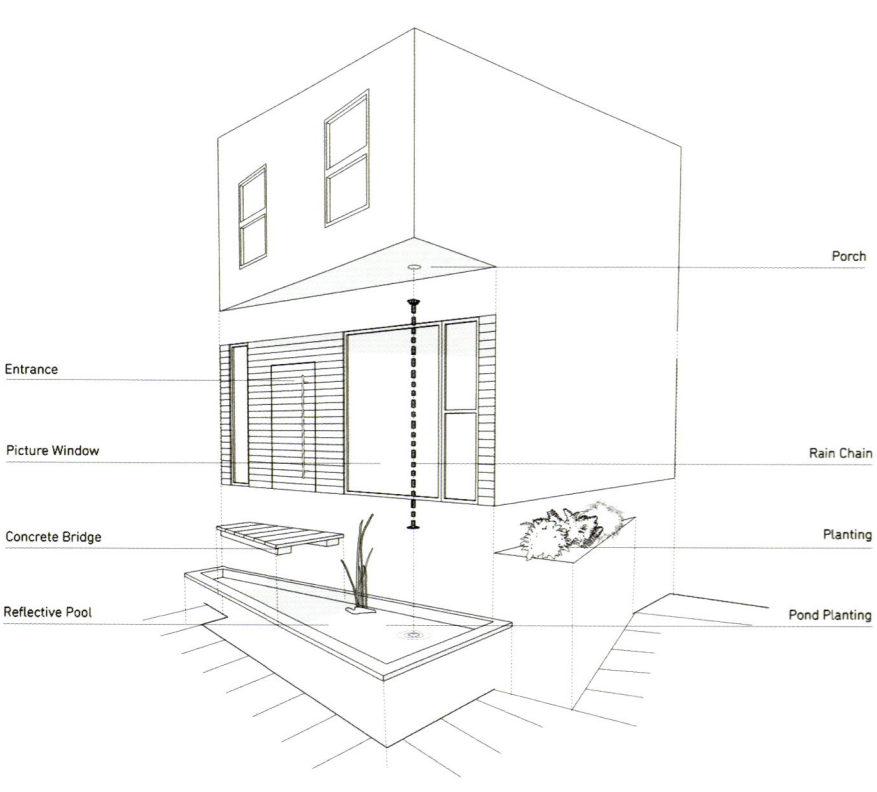

- Porch
- Entrance
- Picture Window
- Concrete Bridge
- Reflective Pool
- Rain Chain
- Planting
- Pond Planting

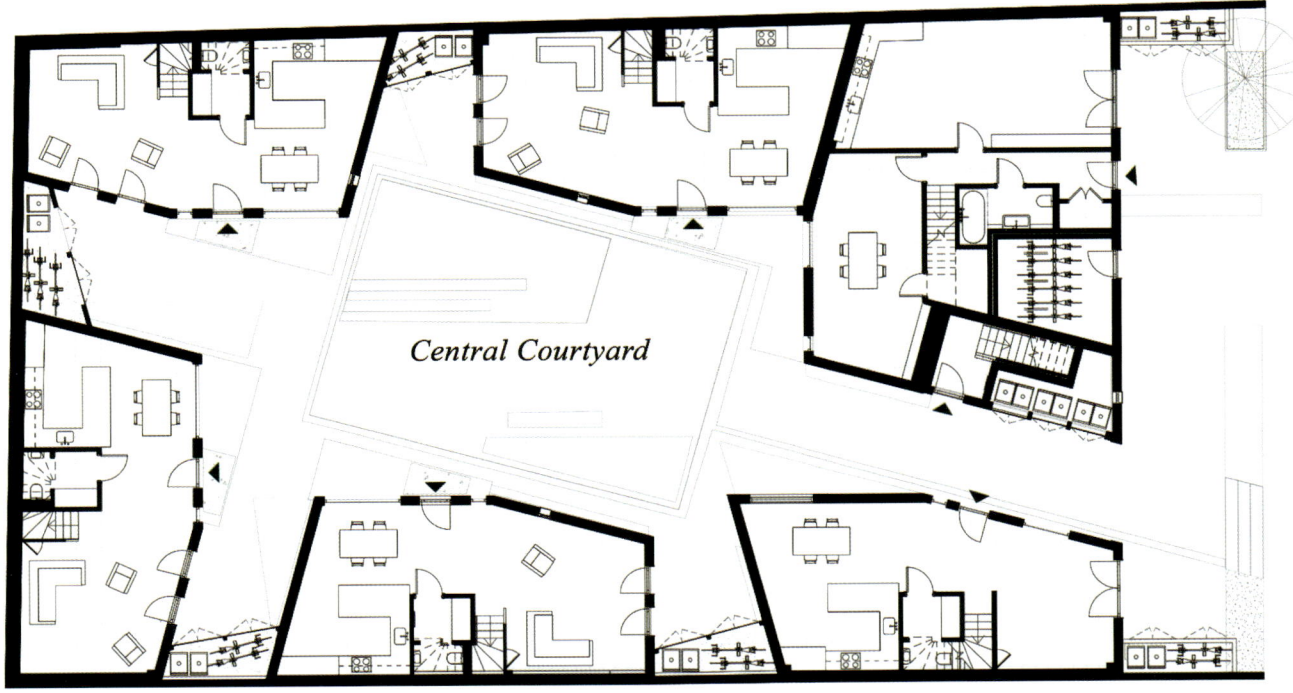

Ground Floor Plan

Together with the design, landscaping, 'neighbouryard', and reflection pools, the development has a clear drive to enhance the well-being and 'sociability' of the residents.

The Sao Felix da Marinha House

Architects: Raulino Silva
Location: Vila Nova de Gaia, Portugal
Area: 3200 ft²
Photographs: João Morgado

Main Material: Concrete + Glass + Wood

We offer you two villas built by the Raulino Silva Arquitecto studio in Portugal. Two elegant and essential houses, characterized by the white color of the typical Portuguese houses, which hide a balanced interior design made of glass, wood and light. Casa Aldoar is located in Oporto, Portugal, the lot in which it is located is long and narrow, but this has not prevented the designers from designing an elegant and minimal house including two floors and a basement. The main front, candidly white, like the whole building, overlooks a small pedestrian garden, while on the opposite side there is a large garden with swimming pool and outdoor living room.

The ground floor includes the whole living area. The kitchen is adjacent to the street, while the living room is directly connected to the rear swimming pool; to separate them is only a large sliding window. The solid pine stairs, in continuity with the floor finish, mark a focal point of the house and lead to the upper sleeping area. Here there are three suites and a large office with a large terrace. A single-family villa located in São Félix da Marinha in Portugal that rises among the pines and oaks in the well-known seaside resort at the southern end of the Porto district.

C Material Type: **Concrete**
(Concrete + painting)

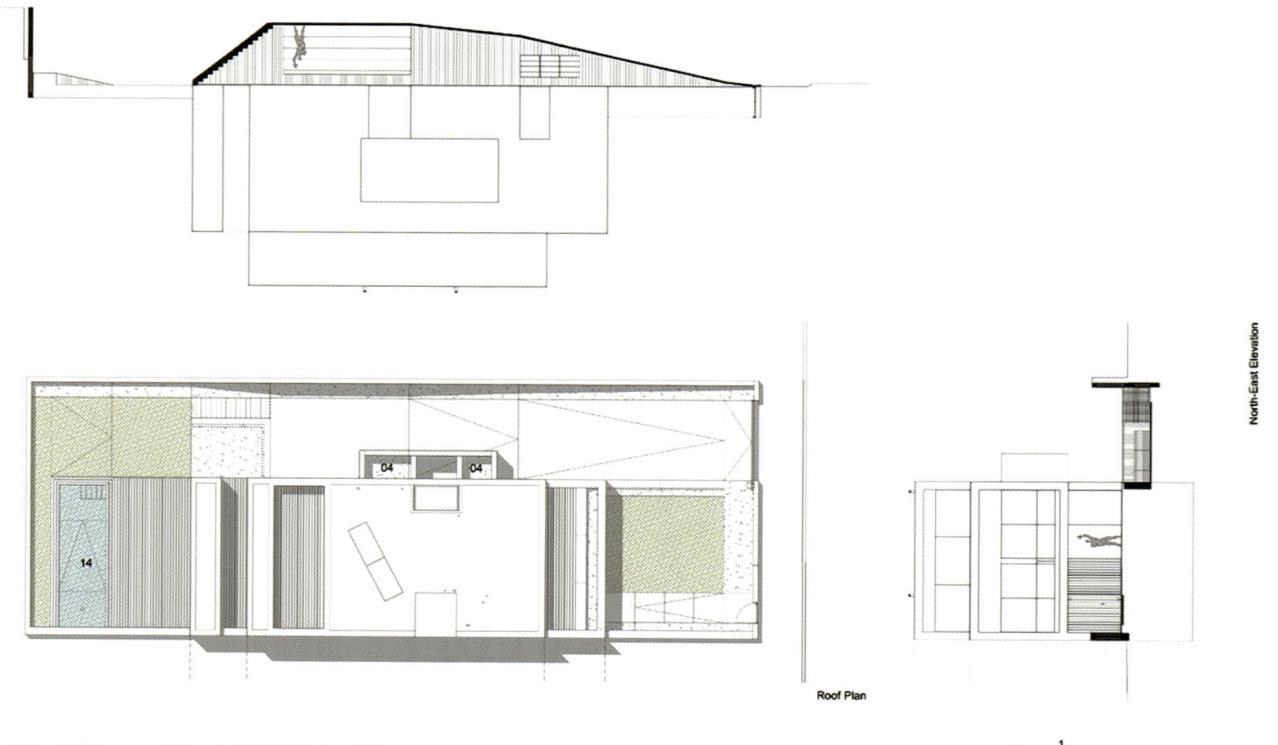

North-West Elevation

North-East Elevation

Roof Plan

01 Garage | 02 Storage room | 03 Laundry | 04 Patio | 05 Bedroom | 06 Bathroom | 07 Technical area | 08 Elevator | 09 Entry | 10 Kitchen | 11 Living room | 12 Closet | 13 Office | 14 Pool

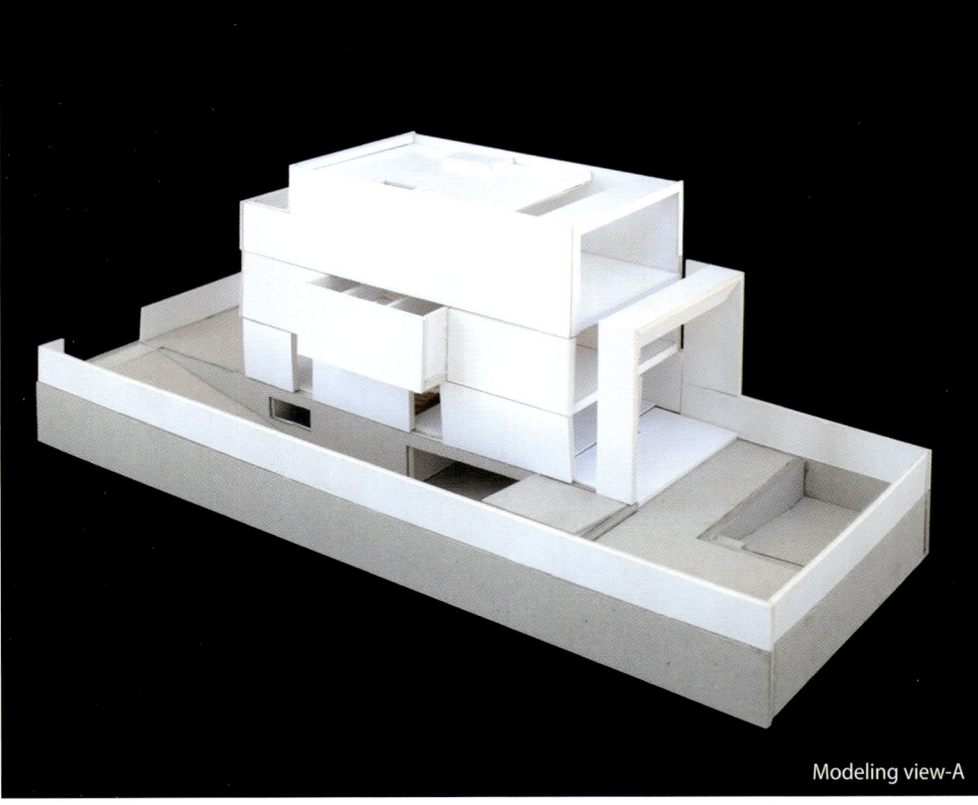

Modeling view-A

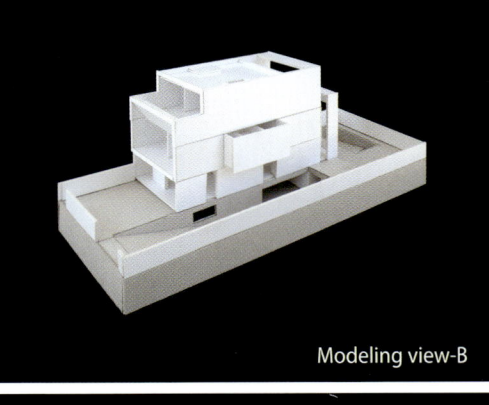

Modeling view-B

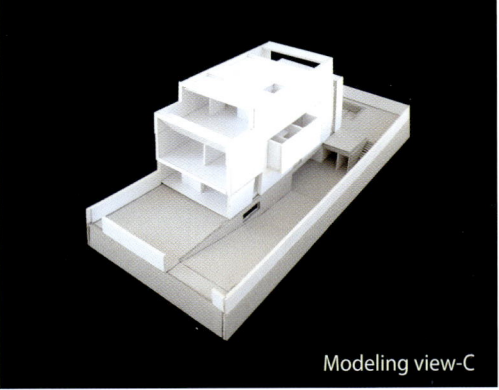

Modeling view-C

A private architecture that combines the tranquility of rural life with the pleasantness of the maritime environment. The villa is oriented to enjoy the well-being lavished by the natural light of the sun that invests the upper floor from which to admire the suggestive sea view. On the ground floor, the main entrance leads into the atrium which opens onto the courtyard to emphasize the contact between architecture and nature. From the lobby you can access all the internal spaces: on the east side, adjacent to the road, there are the garage and the laundry, while on the west side there is the kitchen.

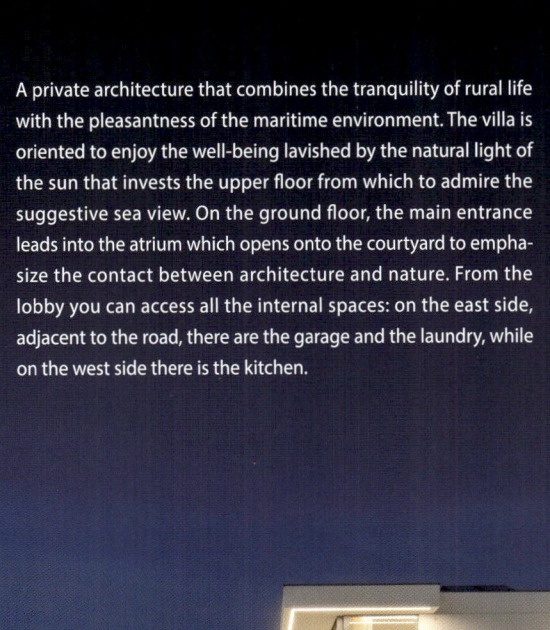

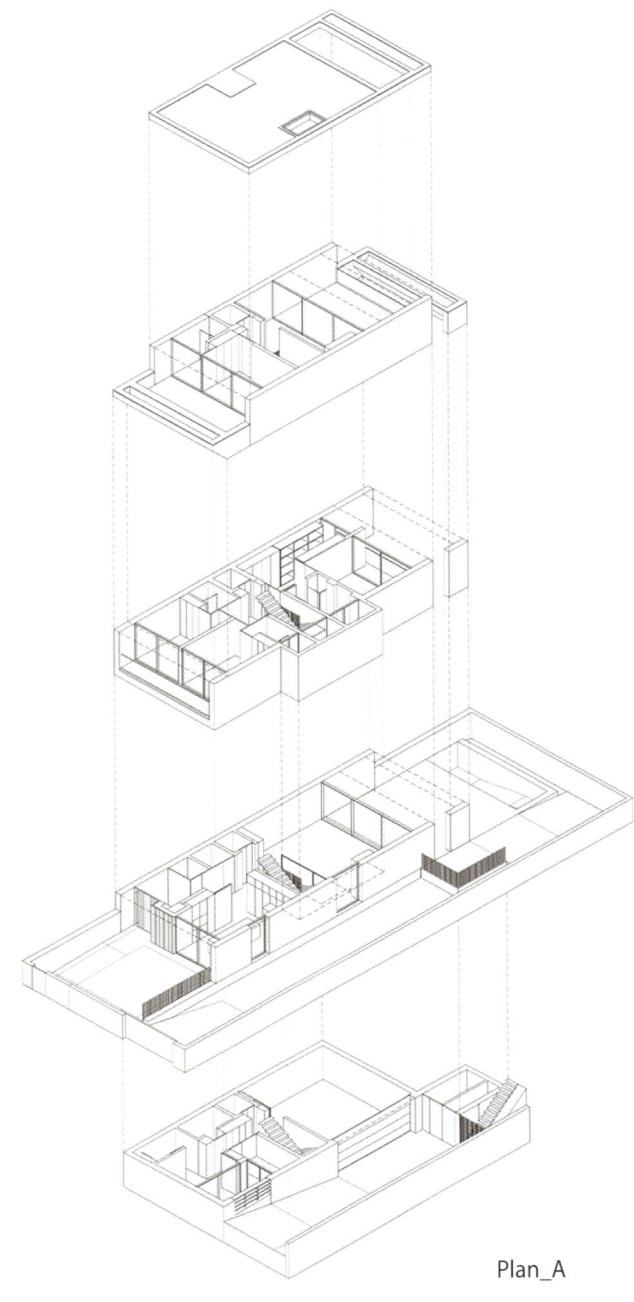

Plan_A

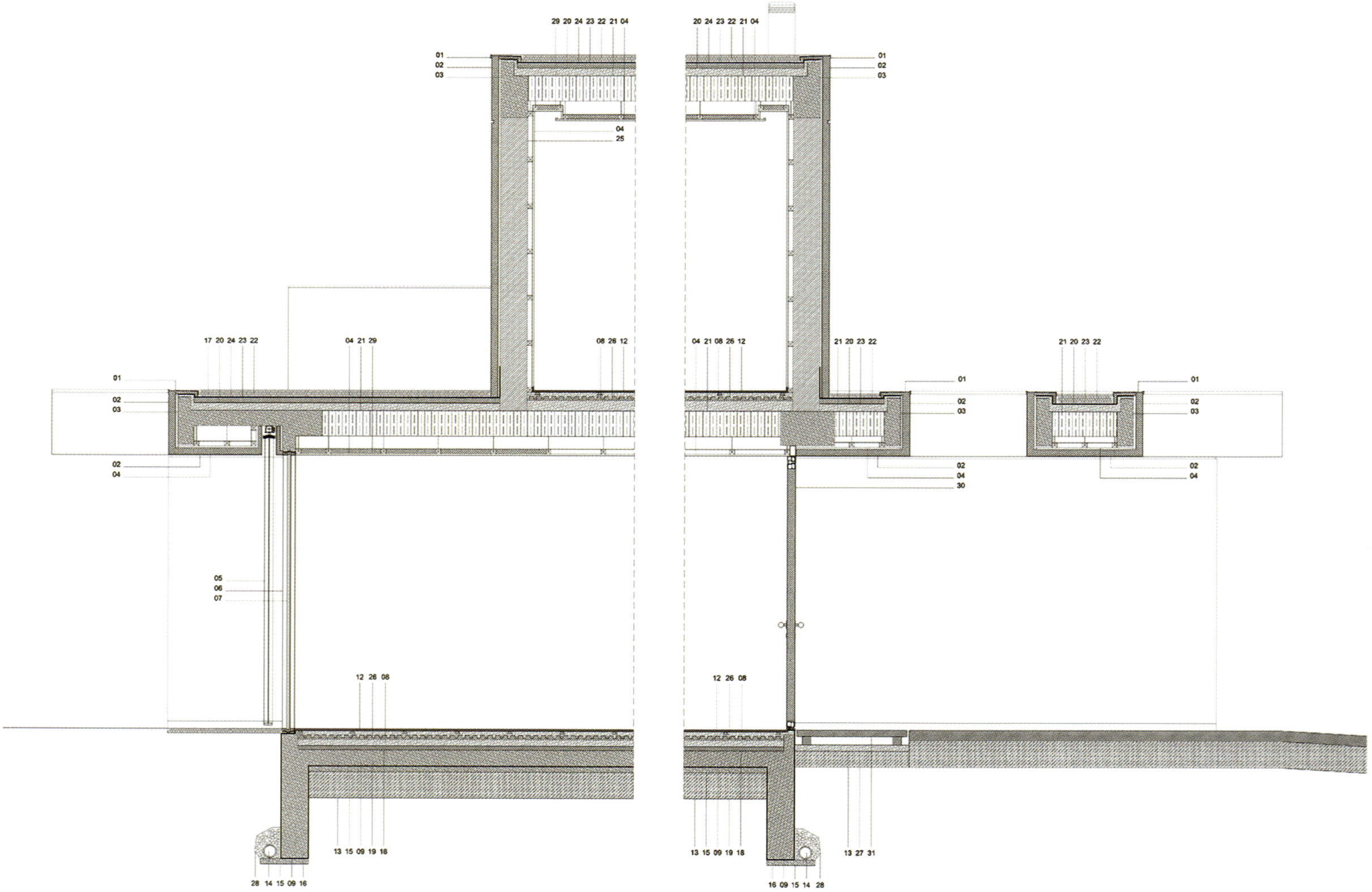

01 Zinc capping | 02 Etics system | 03 Reinforced concrete beam | 04 Plasterboard | 05 Adjustable shutters | 06 Aluminium window frame | 07 Clear double glazing | 08 Lightweight concrete layer | 09 Waterproofing membrane | 10 Stone flooring | 11 Glass railing
12 Linkfloor vinyl flooring | 13 Tout-venant | 14 Drain pipe | 15 Concrete blinding | 16 Concrete lintel | 17 Reinforced concrete slab | 18 Insulation | 19 Reinforced concrete slab with steel grid | 20 Shape layer (lightweight concrete) | 21 Lightened slab | 22 Pebbles
23 Waterproofing membrane + textile | 24 Thermal insulation (pir system) | 25 Thermal block | 26 Water underfloor heating system | 27 Draining concrete | 28 Gravels/chippings wrapped in geotextile felt | 29 Stone wool | 30 Aluminium entrance door | 31 Black concrete flooring

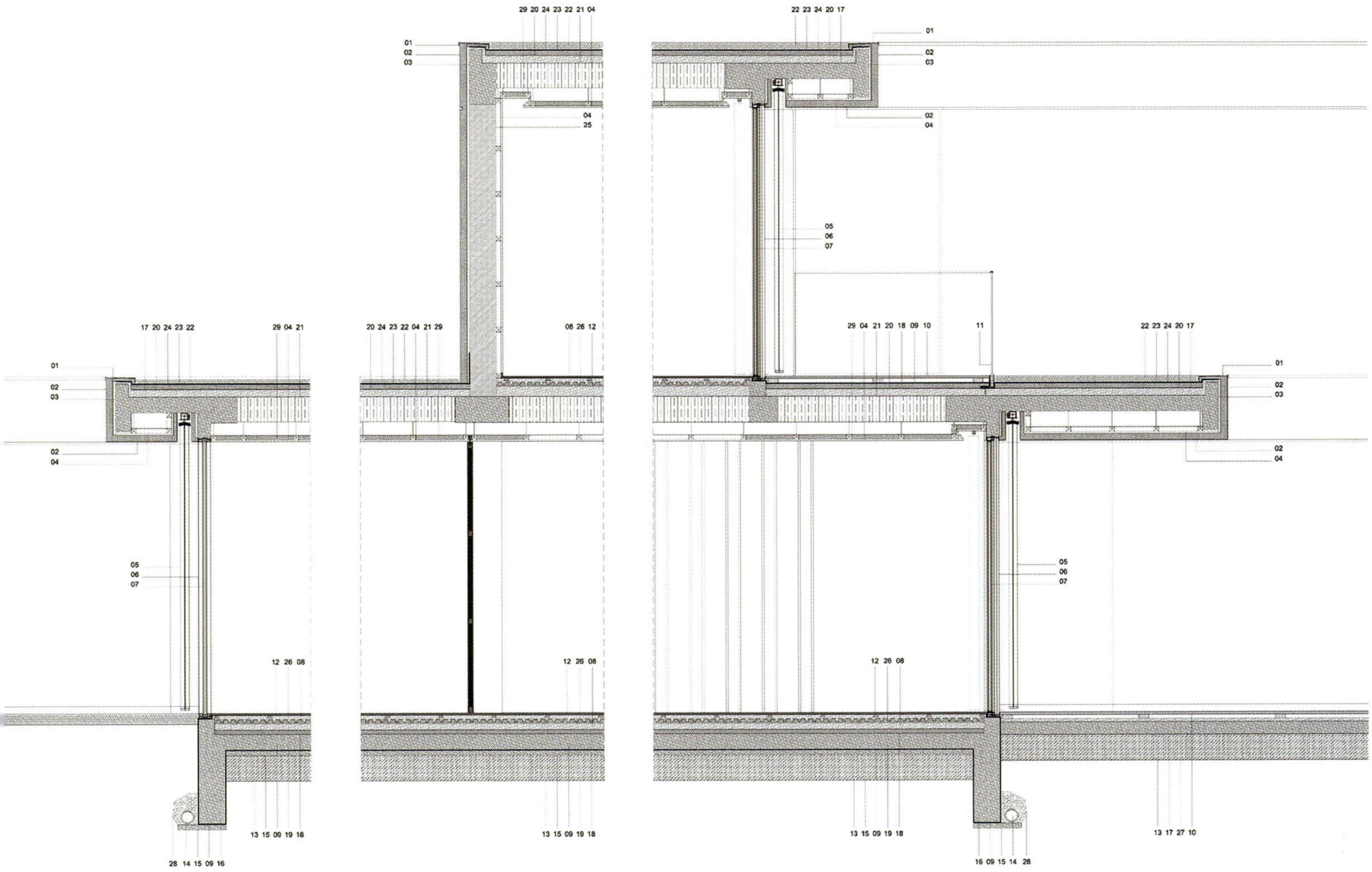

01 Zinc capping | 02 Etics system | 03 Reinforced concrete beam | 04 Plasterboard | 05 Adjustable shutters | 06 Aluminium window frame | 07 Clear double glazing | 08 Lightweight concrete layer | 09 Waterproofing membrane | 10 Stone flooring | 11 Glass railing | 12 Linkfloor vinyl flooring | 13 Tout-venant | 14 Drain pipe | 15 Concrete blinding | 16 Concrete lintel | 17 Reinforced concrete slab | 18 Insulation | 19 Reinforced concrete slab with steel grid | 20 Shape layer (lightweight concrete) | 21 Lightened slab | 22 Pebbles | 23 Waterproofing membrane + textile | 24 Thermal insulation (pir system) | 25 Thermal block | 26 Water underfloor heating system | 27 Draining concrete | 28 Gravels/chippings wrapped in geotextile felt | 29 Stone wool

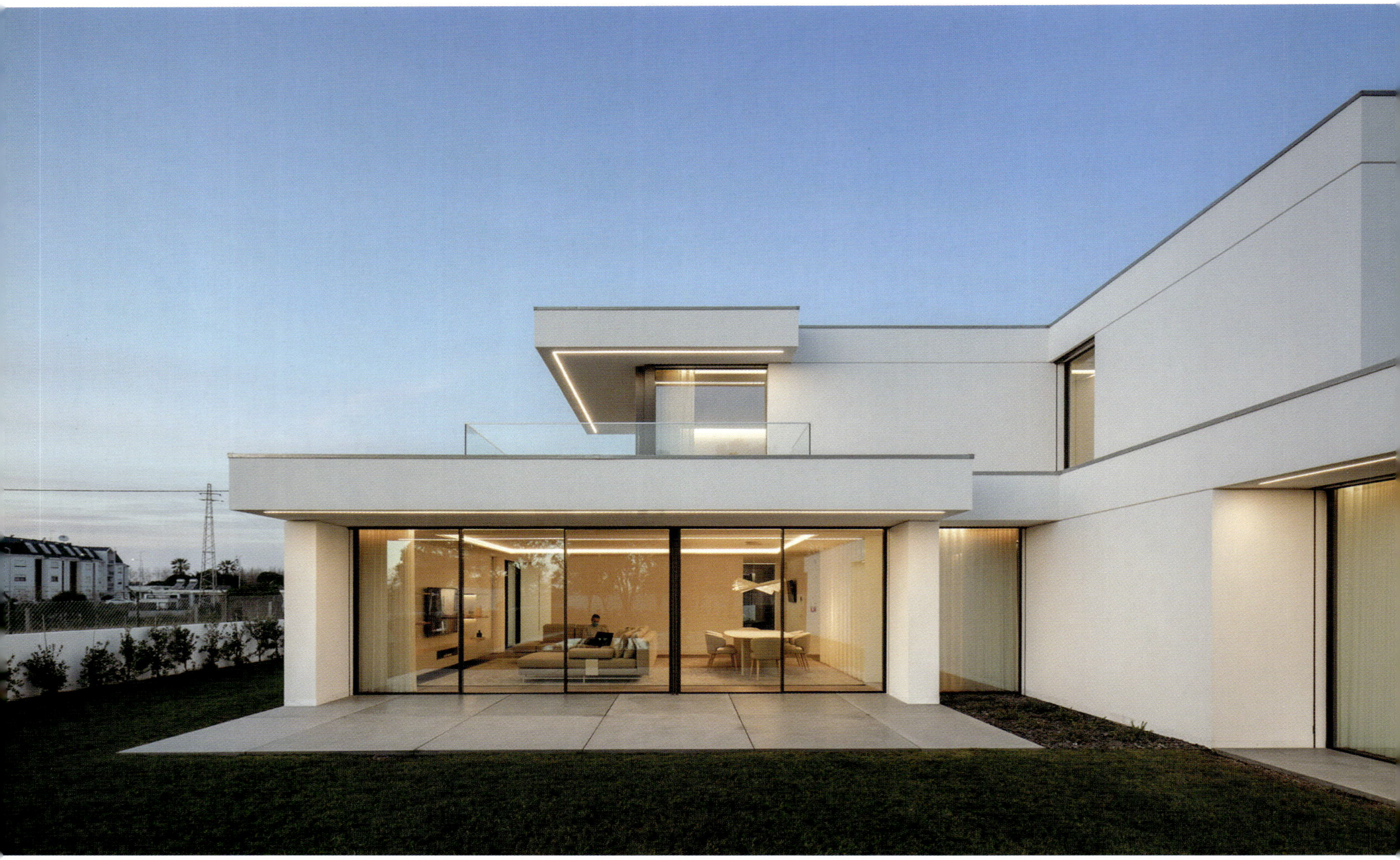

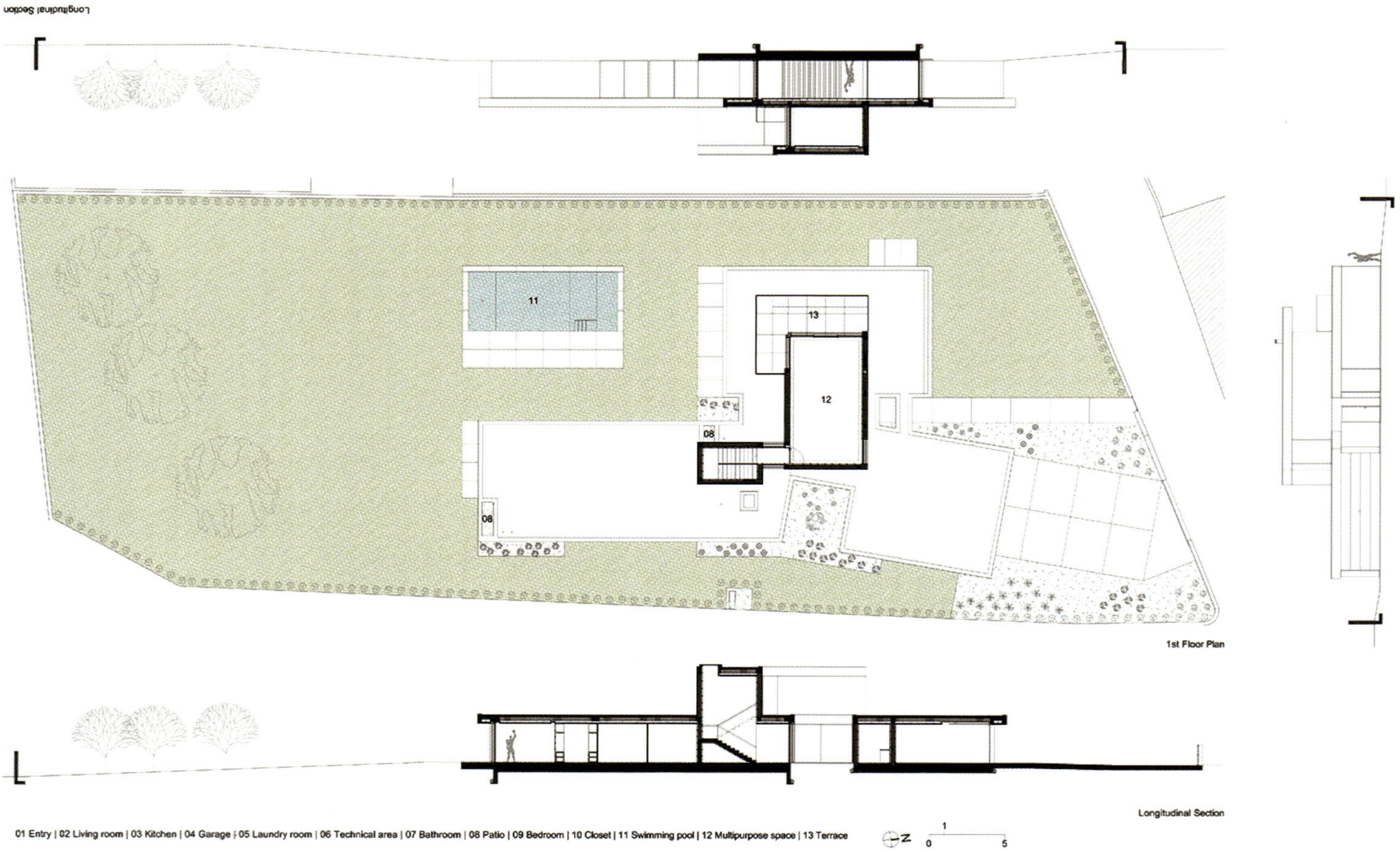
Longitudinal Section
11
13
12
08
08
1st Floor Plan
North Elevation
Longitudinal Section
N
1
0
5

01 Entry | 02 Living room | 03 Kitchen | 04 Garage | 05 Laundry room | 06 Technical area | 07 Bathroom | 08 Patio | 09 Bedroom | 10 Closet | 11 Swimming pool | 12 Multipurpose space | 13 Terrace

1st Floor Plan

2nd Floor Plan

Longitudinal Section

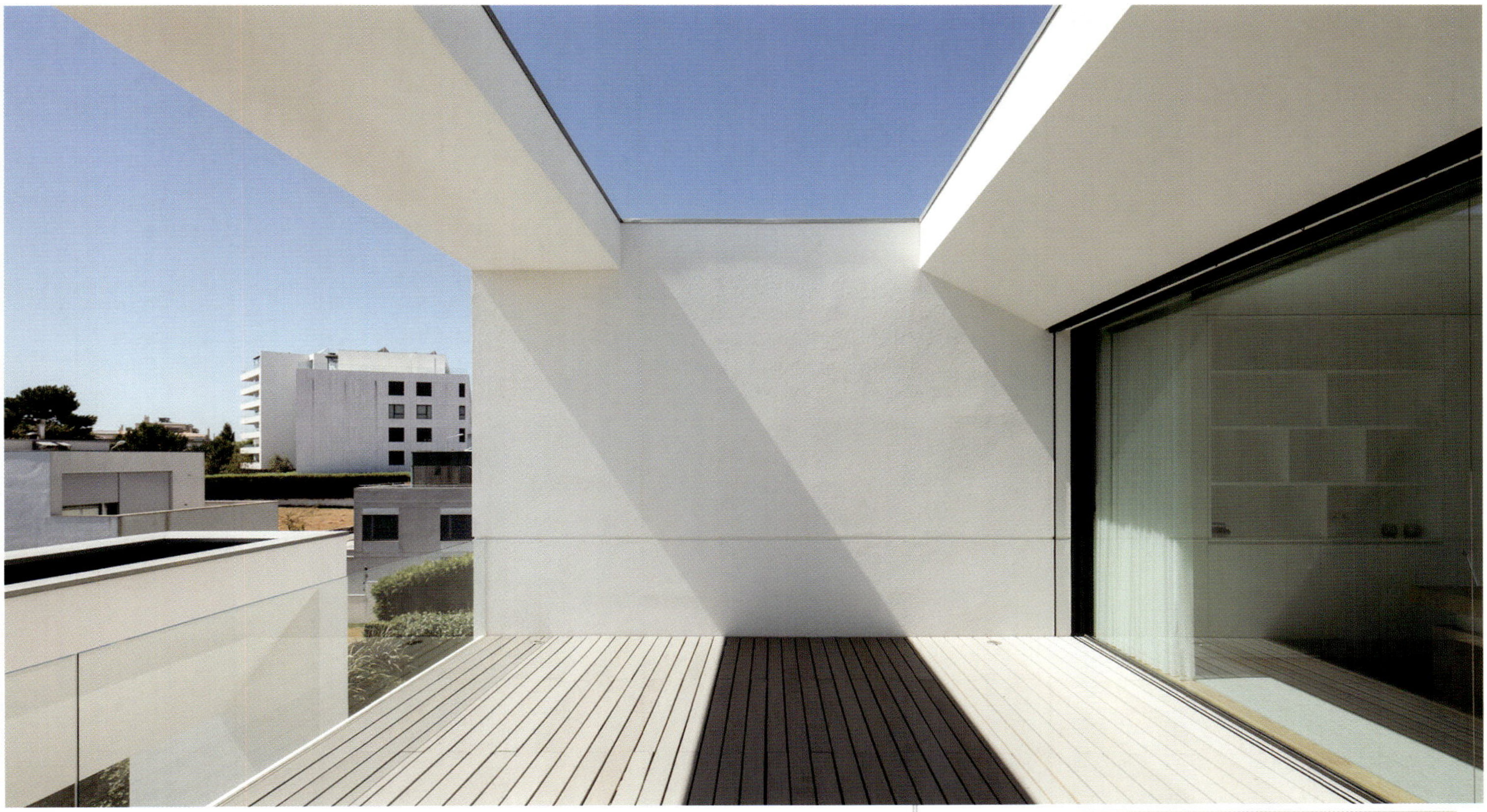

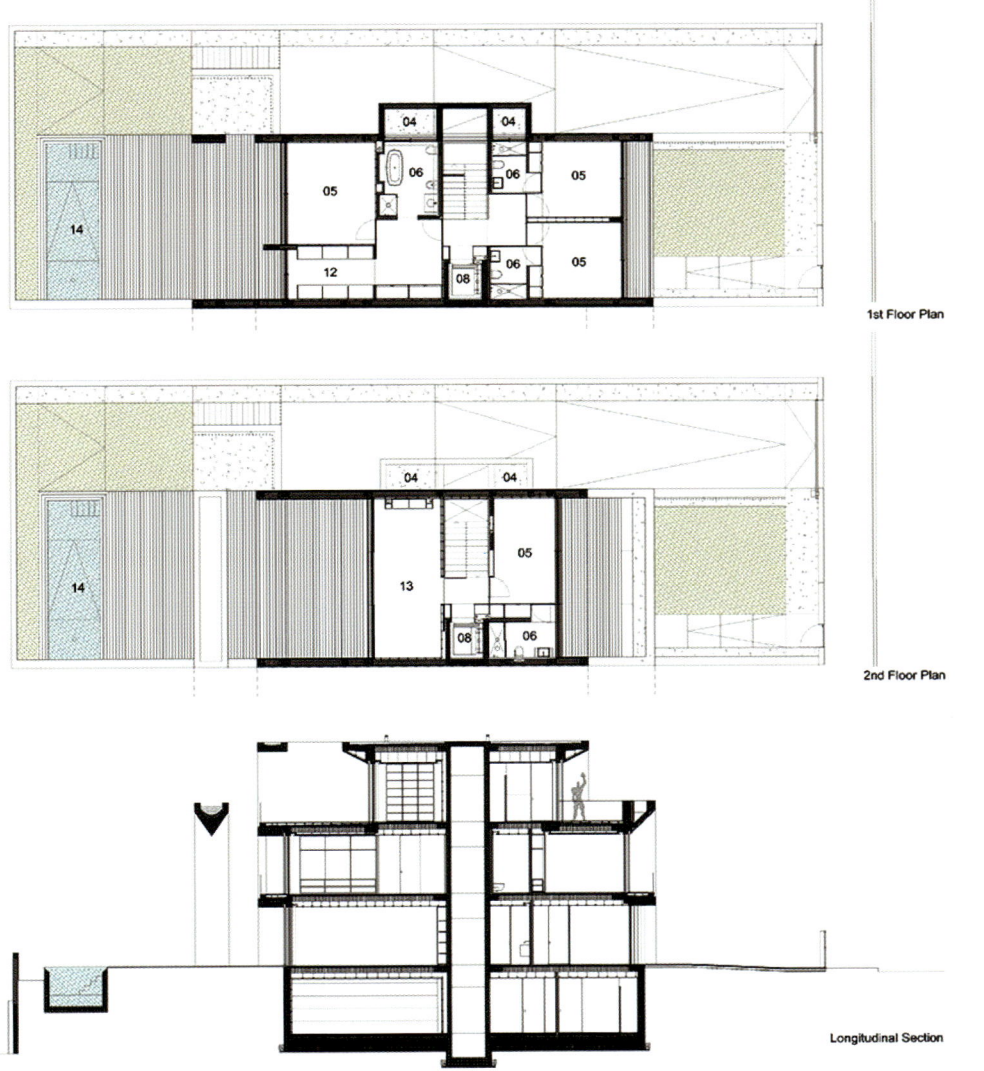

The extreme transparency between inside and outside allows the villa to transform everyday experiences. The living room and the dining room are in fact oriented towards the garden and the swimming pool. A white lacquered mdf slat separates the living area from the main atrium.

Transversal Section

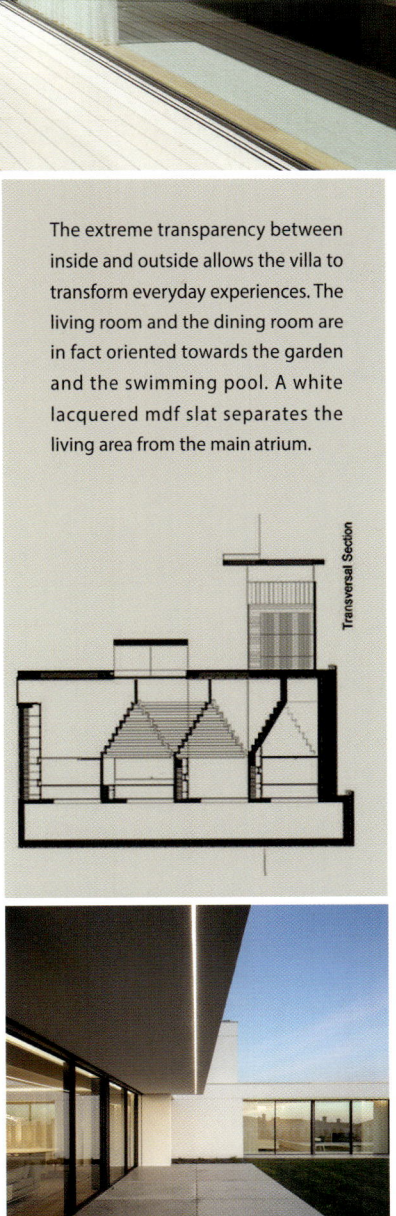

01 Entry | 02 Living room | 03 Kitchen | 04 Garage | 05 Laundry room | 06 Technical area | 07 Bathroom | 08 Patio | 09 Bedroom | 10 Closet | 11 Swimming pool | 12 Multipurpose space | 13 Terrace

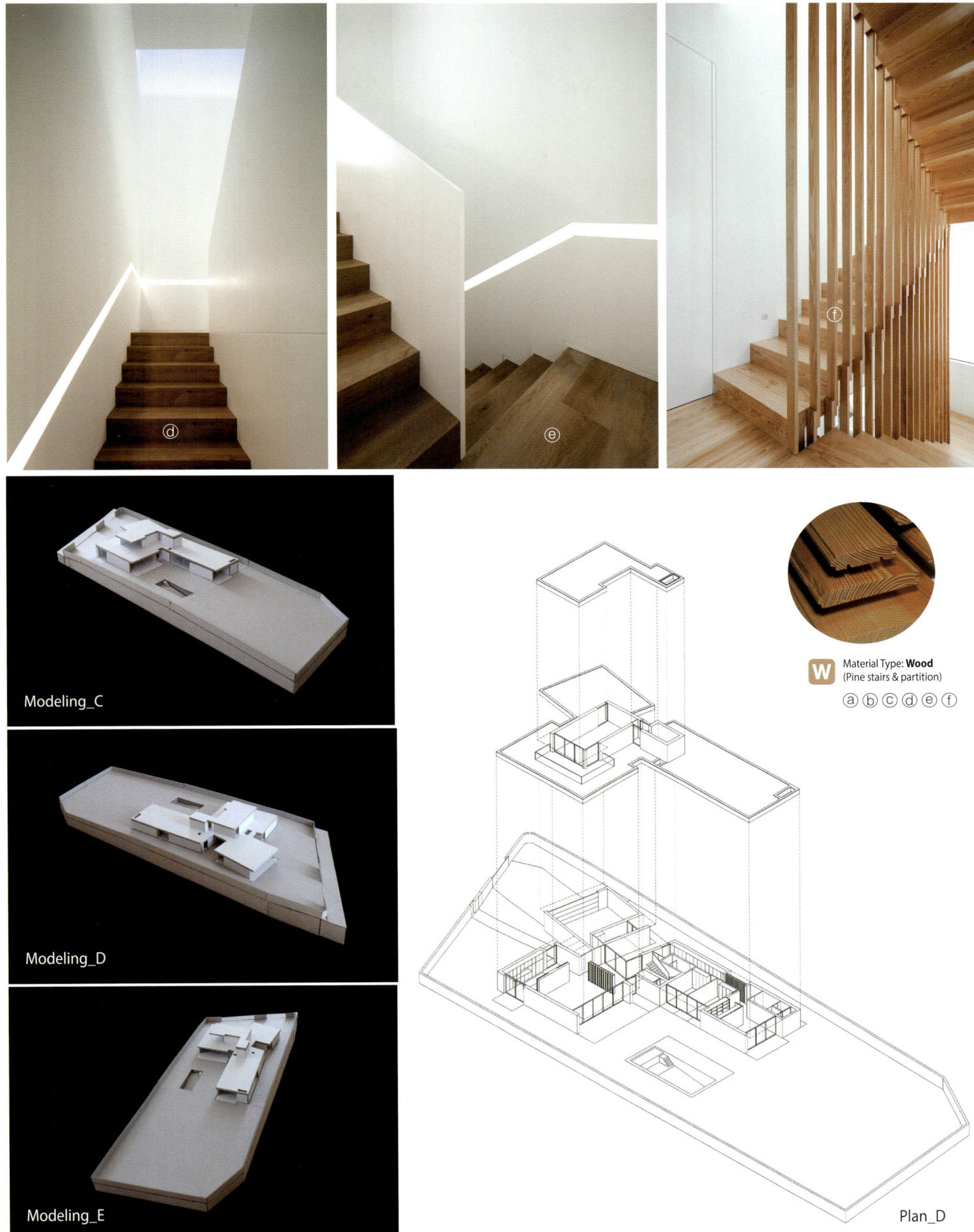

Modeling_C

Modeling_D

Modeling_E

Material Type: **Wood**
(Pine stairs & partition)
ⓐ ⓑ ⓒ ⓓ ⓔ ⓕ

Plan_D

The bedrooms are also connected to the garden. In the body that extends to the ground on the east side is the sleeping area, with three bedrooms, of which the main one is a suite with anteroom. The top floor is intended for the area for relaxation and recreation, here there is a multipurpose room suitable for playing or watching TV. This space is connected to the terrace, which is located on the roof of the ground floor, overlooking the Atlantic Ocean.

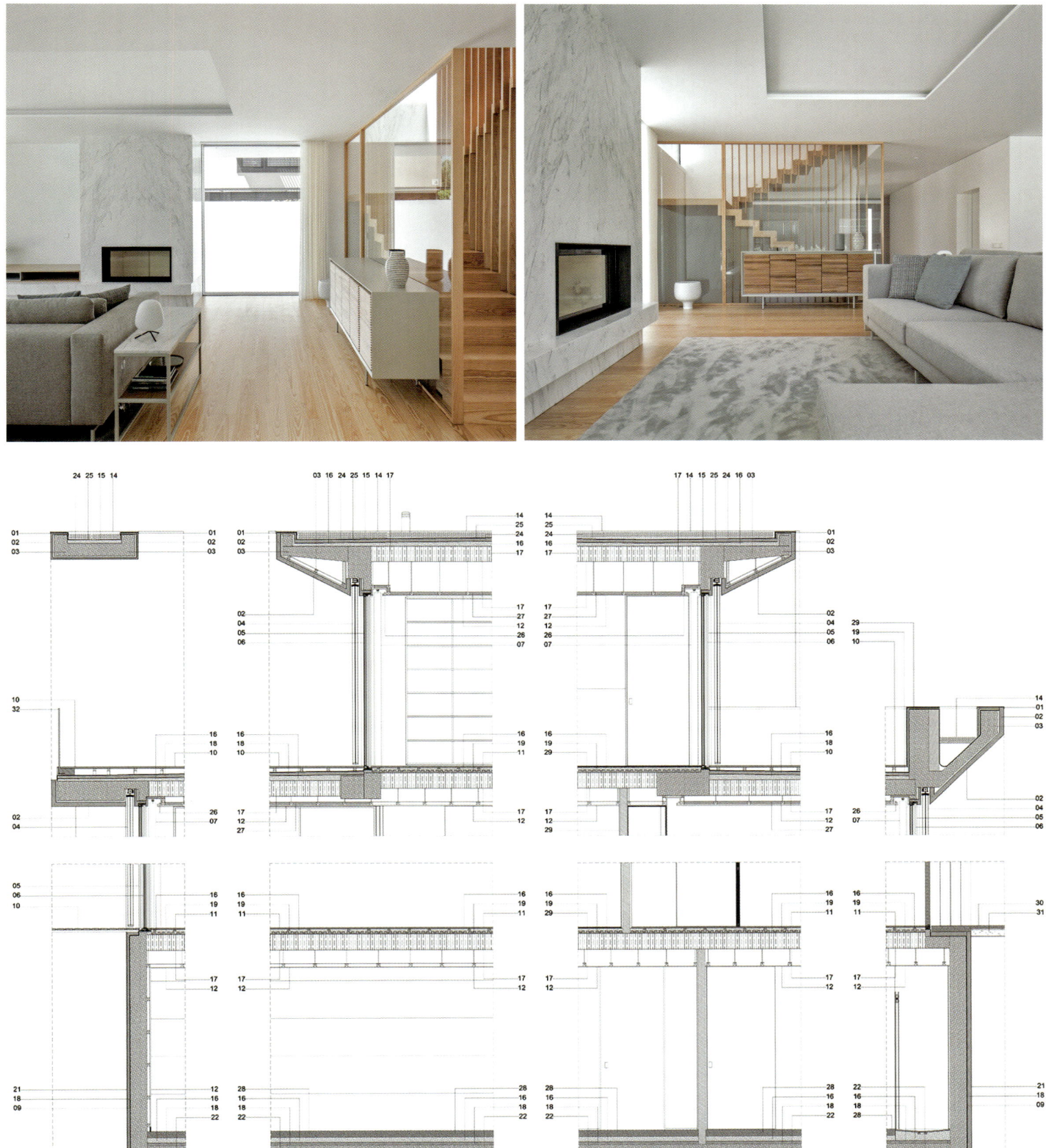

01 Zinc capping | 02 ETICS system | 03 Reinforced concrete beam | 04 Rolling shutters | 05 Aluminium window frame | 06 Clear double glazing | 07 Curtain | 08 Lightweight concrete layer | 09 Waterproofing membrane | 10 Wooden deck | 11 Pine wood floor | 12 Plasterboard
13 Tout-venant | 14 Pebbles | 15 Vapor barrier | 16 Shape layer (lightweight concrete) | 17 Lightened slab | 18 Insulation | 19 Water underfloor heating system | 20 Drain pipe | 21 Studded drainage membrane | 22 Concrete blinding | 23 Gravels/chippings wrapped in geotextile felt
24 Thermal insulation (PIR system) | 25 Waterproofing membrane + textile | 26 Led lighting | 27 Stone wool | 28 Black concrete flooring | 29 Marble | 30 Aquastone aggregate | 31 Geotextile filter | 32 Balcony glass railing

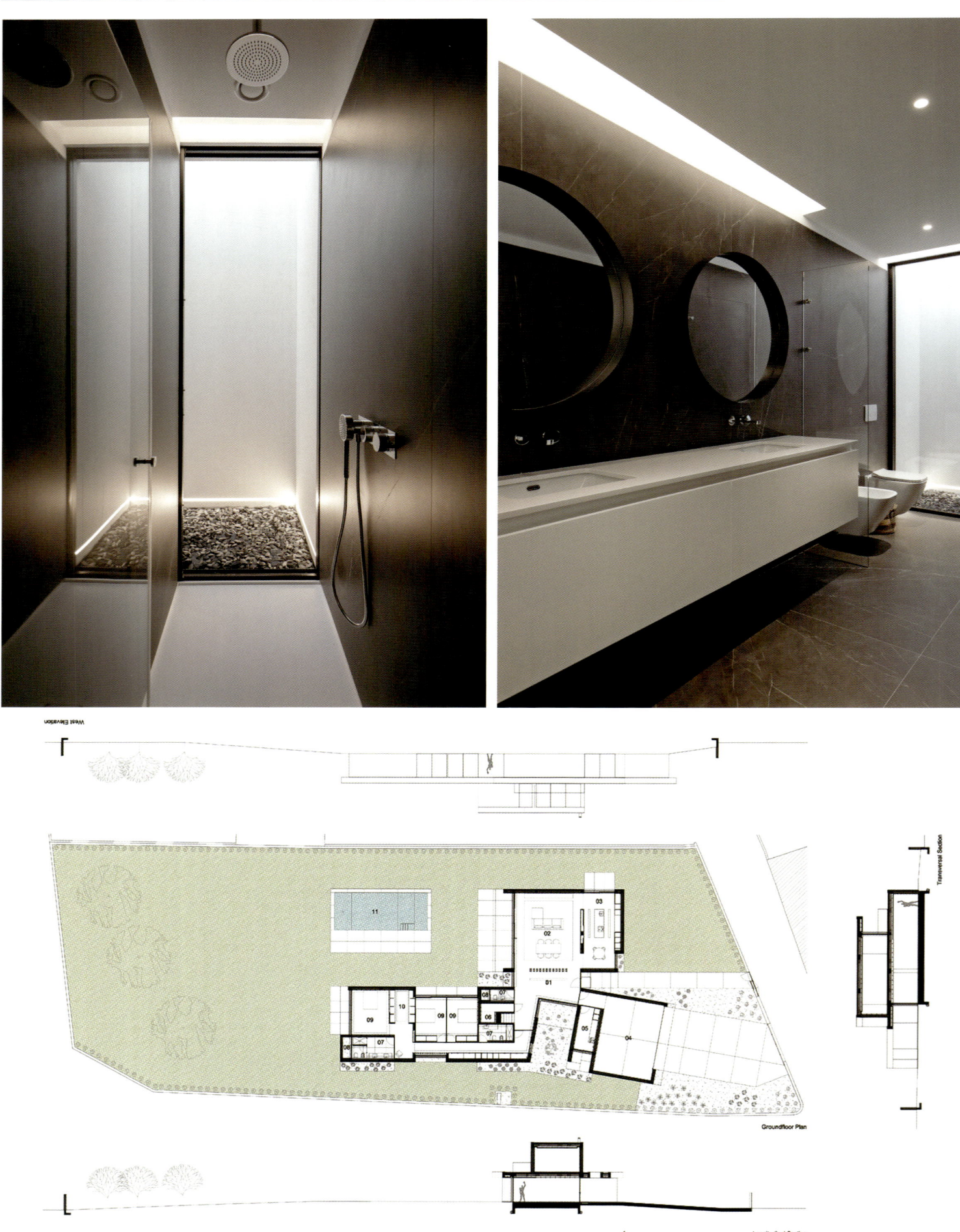

01 Entry | 02 Living room | 03 Kitchen | 04 Garage | 05 Laundry room | 06 Technical area | 07 Bathroom | 08 Patio | 09 Bedroom | 10 Closet | 11 Swimming pool | 12 Multipurpose space | 13 Terrace

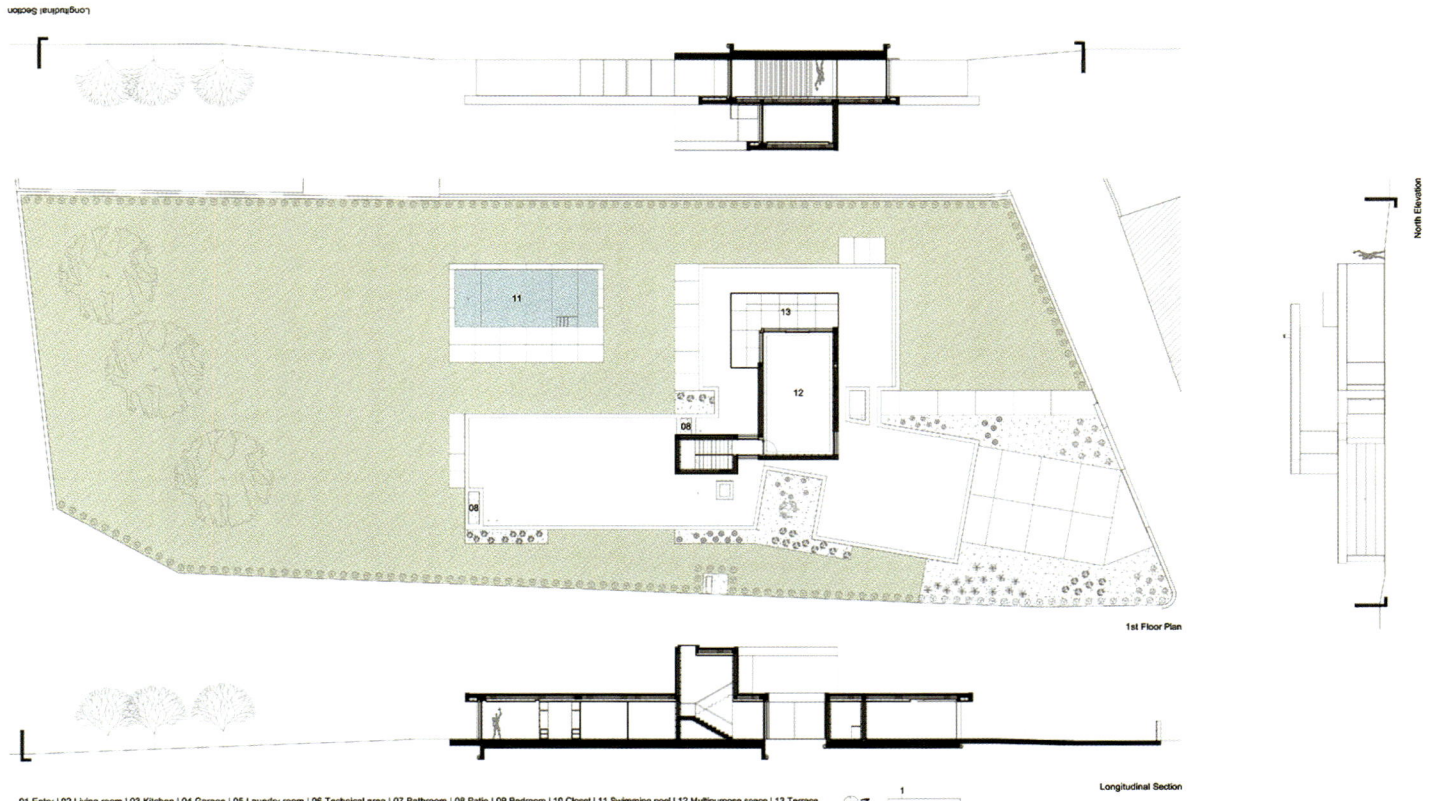

01 Entry | 02 Living room | 03 Kitchen | 04 Garage | 05 Laundry room | 06 Technical area | 07 Bathroom | 08 Patio | 09 Bedroom | 10 Closet | 11 Swimming pool | 12 Multipurpose space | 13 Terrace

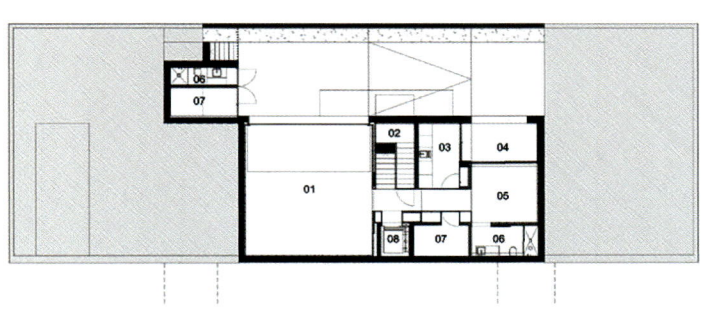

Basement Plan

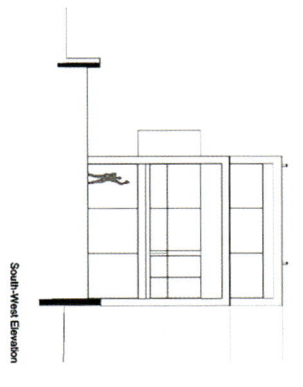

South-West Elevation

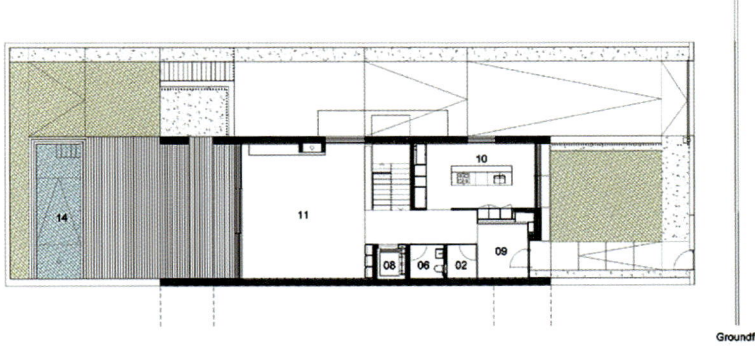

Groundfloor Plan

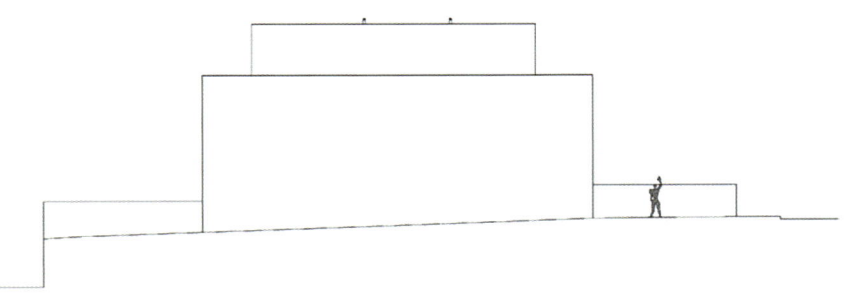

South-East Elevation

The House No.6

Architects: Sara Kalantary + Reza Sayadiyan
Location: Tehran, Iran
Area: 1600 m²
Photographs: Deed studio, Parham Taghiof, Reza Sayadian, Ryan Sayadian
Main Material: Concrete + Moscato Beige is a light beige marble

House No 6, critically looks at the bygone urban area as well as contemporary disoriented human; which in turns, caused a new space to be created. City is a bed where homes lie on. Home calms the heart and soul, a place where life flows. City is composed of numerous layers; and although these are the most important areas where the human interaction occurs and social behavior emerges, privacy and security are established as well. In the city of Tehran, man's privacy has sustained severe damages due to destruction of urban outer layers, as well as loss of inner ones. Buildings are immediately adjacent to the streets, and urban open-air spaces are disappearing.

Human's presence in the city is in jeopardy and group activities are gradually fading. Generally speaking, we had "sub-neighborhoods" in the past, where neighbors interacted with one another and the social interaction occurred. This "sub-neighborhood" was located in between the city and home, acting as a connecting bridge. For instance, natives of the past along with their fellow citizens, living in a human condition and an environment compatible with nature, could easily go through such urban layers to get connected to the city.

Material Type: **Stone & Tiles**
(Moscato Beige; light beige marble)

Moscato Beige is a light beige marble. It is calcitic, and it has a homogeneous background. A combination of soft toasted colors compose its base. The veins partially cover the surface and are pearly and brighter compared to the base. This light beige marble presents the chromatic subtlety and softness of natural suede. Moscato Beige represents the perfect stone for warm spaces that also want to maximize the luminosity of the environment. Moscato Beige fosters elegant and radiant while welcoming spaces thanks to its light creamy tones. It fits perfectly in interiors, bathrooms, kitchens, countertops, and showers. It is selected by architects and interior designers for large residential projects as well as luxury hotel projects. The surface finishing standards of the TINO collection are: Polished, Honed, Domus, and Sandblasted. TINO offers Moscato Beige marble in any format and finishes desired by the client. Check the maximum size per finish. For other finishes, check availability and sizes.

But once lifestyles changed, and home turned into a machine of life, all the afore-mentioned layers in between the city and home disappeared as well. Although we might not be able to revive the "sub-neighborhoods" of the past, through our project, we decided to get creative and design a space with the same use, yet appropriate for today's living condition. That is how an area came into existence: An area between earth and the sky from one side, amid home and the city from another. This common area is a recess where every single resident can easily have access to. A place for a dialogue, socialization, an occasional spree, or simply a getaway from life. By creating this, we refreshed the memories of the lost sub-neighborhoods, and by donating this to the city and neighbors, we took a further step toward a beauteous life.

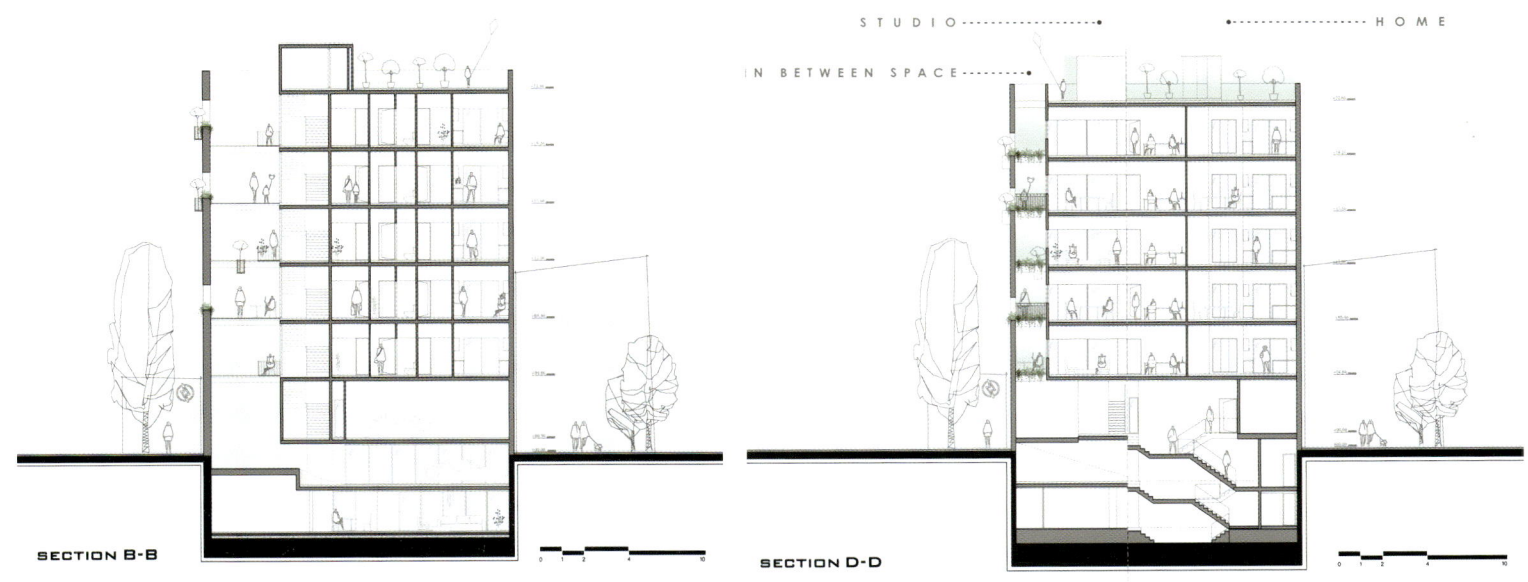

STUDIO · · HOME

IN BETWEEN SPACE

SECTION B-B

SECTION D-D

House No 6, has a unique perspective toward the essence of life, and renders a new living standard as well as modern man's correlation with the city. Through this project, we managed to present a new typology for typical construction paradigms in the country. In this building, home and work space are placed together to solve some of the difficulties of today's life. In fact, we altered massing and consequently, brought dead spaces back to life, vivified the third façade, and tied this inner space to the city. As Milan Kundera states: "You don't need to look for someone special to fall in love with, but to love someone in a special way." We categorically believe that in architecture, a structure doesn't need to be prodigious. It has to be appropriately built.

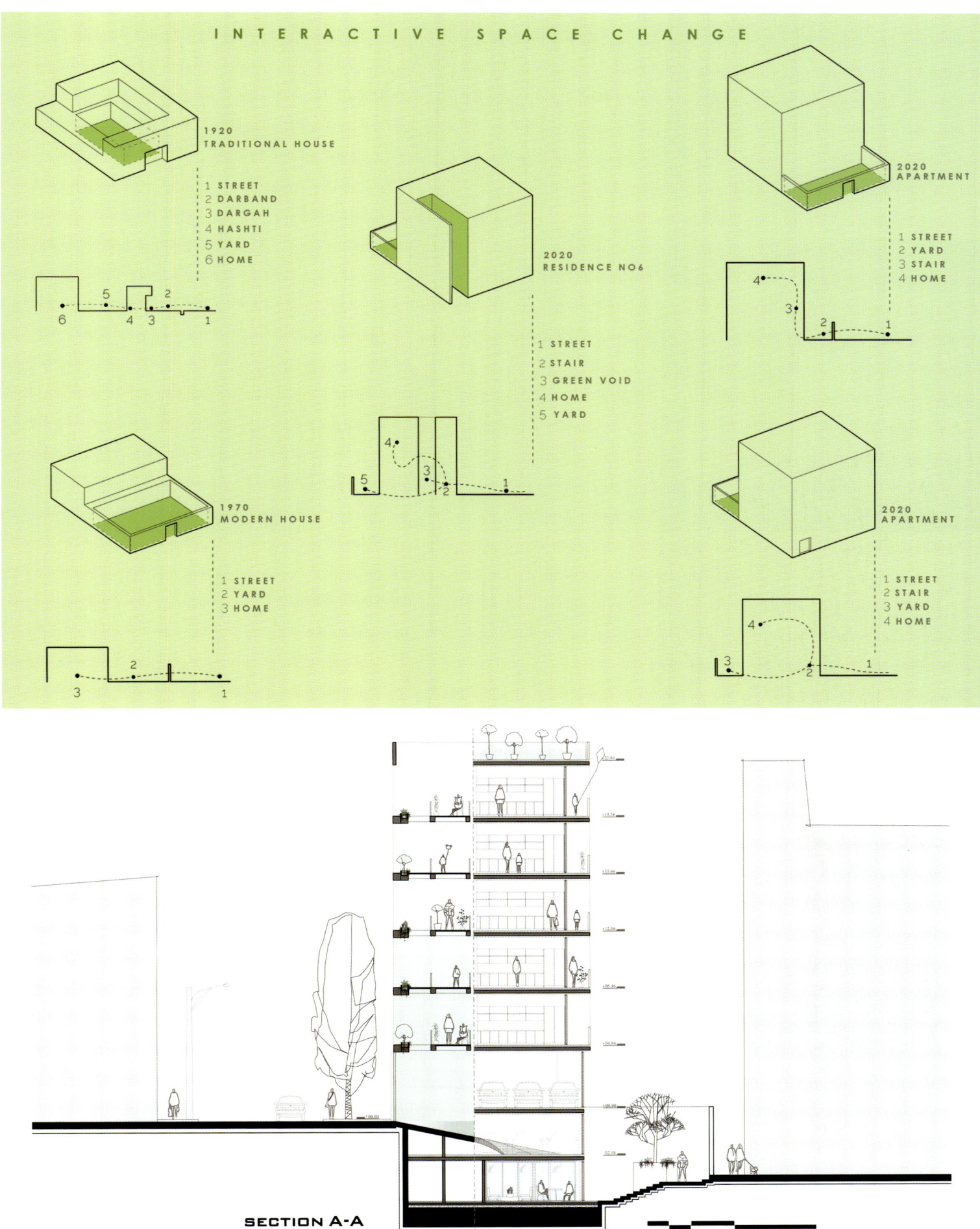

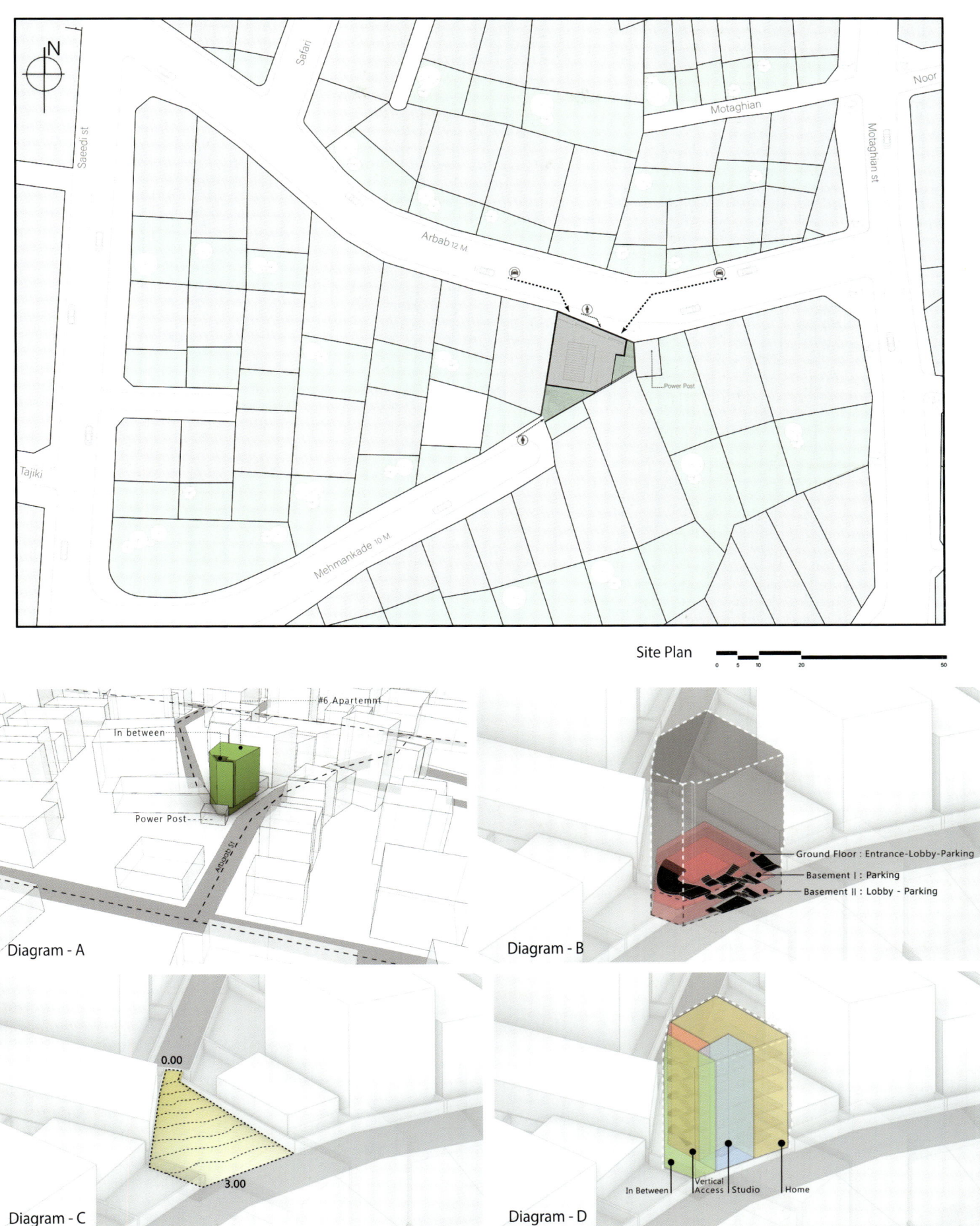

Site Plan

Diagram - A

Diagram - B

Diagram - C

Diagram - D

219

G Material Type: **Glass**
(Curtain Wall Facade Systems | Island Exterior Fabricators)

Island's unitized curtainwall assemblies are fully customizable aluminum-framed enclosure solutions, based upon a series of pre-tested curtainwall systems. Each curtainwall application is tailored to meet the specific needs of a project – from engineering considerations to thermal performance and nuances of the architectural design intent. The unitized system is developed through shop drawings and mockup testing before fabrication begins at Island's Calverton facility. The factory-glazed assemblies are shipped to the site and installed in-sequence to enclose each floor – allowing the greater project team to minimize on-site efforts and maximize quality-control.

IN BETWEEN SPACE CHANGE

1920

(ANDAROONI) (ANDAROONI) (ANDAROONI)
(BIROONI) (BIROONI) (BIROONI)
(STREET) (STREET) (SQUARE/PLAZA) (STREET)

IN BETWEEN SPACE CHANGE

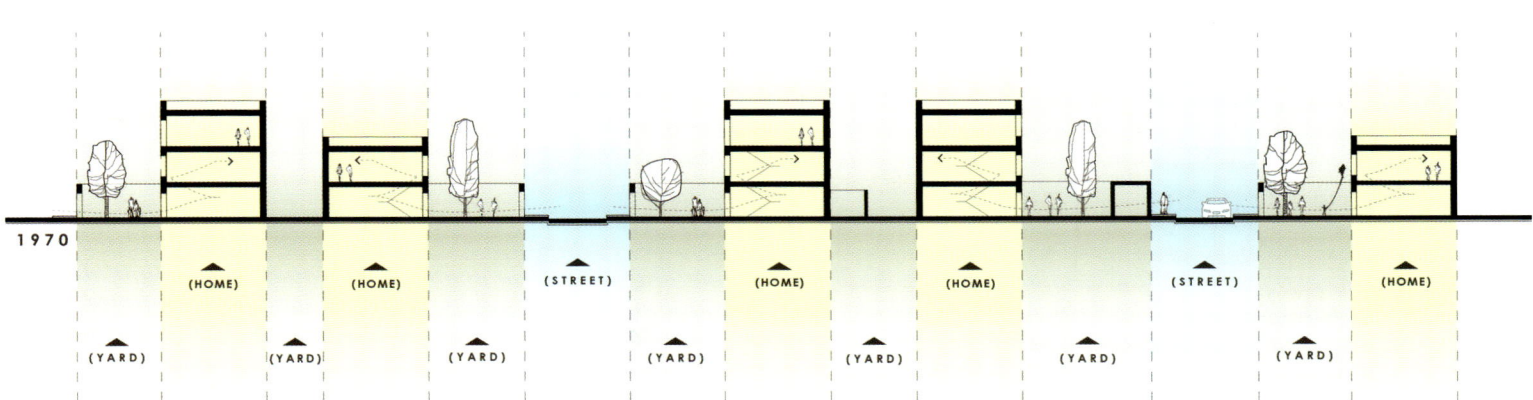

1970

(HOME) (HOME) (STREET) (HOME) (HOME) (STREET) (HOME)
(YARD) (YARD) (YARD) (YARD) (YARD) (YARD) (YARD)

IN BETWEEN SPACE CHANGE

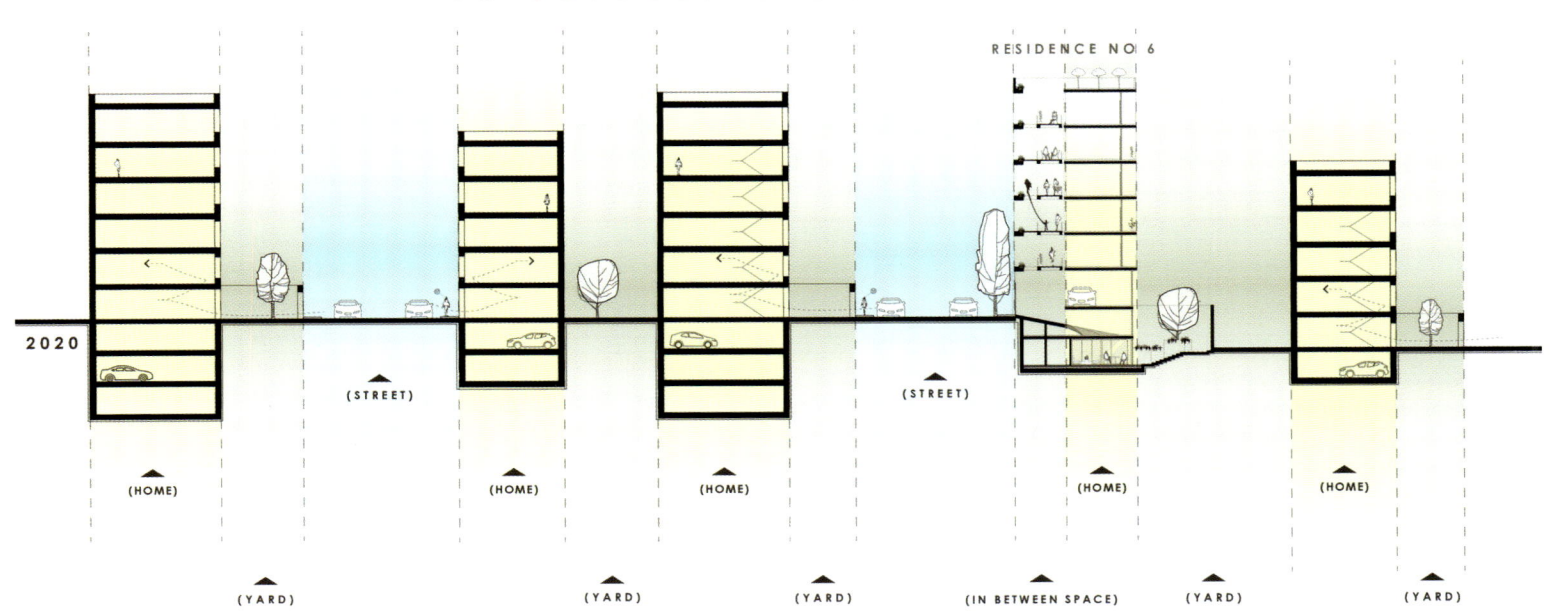

RESIDENCE NO. 6

2020

(STREET) (STREET)
(HOME) (HOME) (HOME) (HOME) (HOME)
(YARD) (YARD) (YARD) (IN BETWEEN SPACE) (YARD) (YARD)

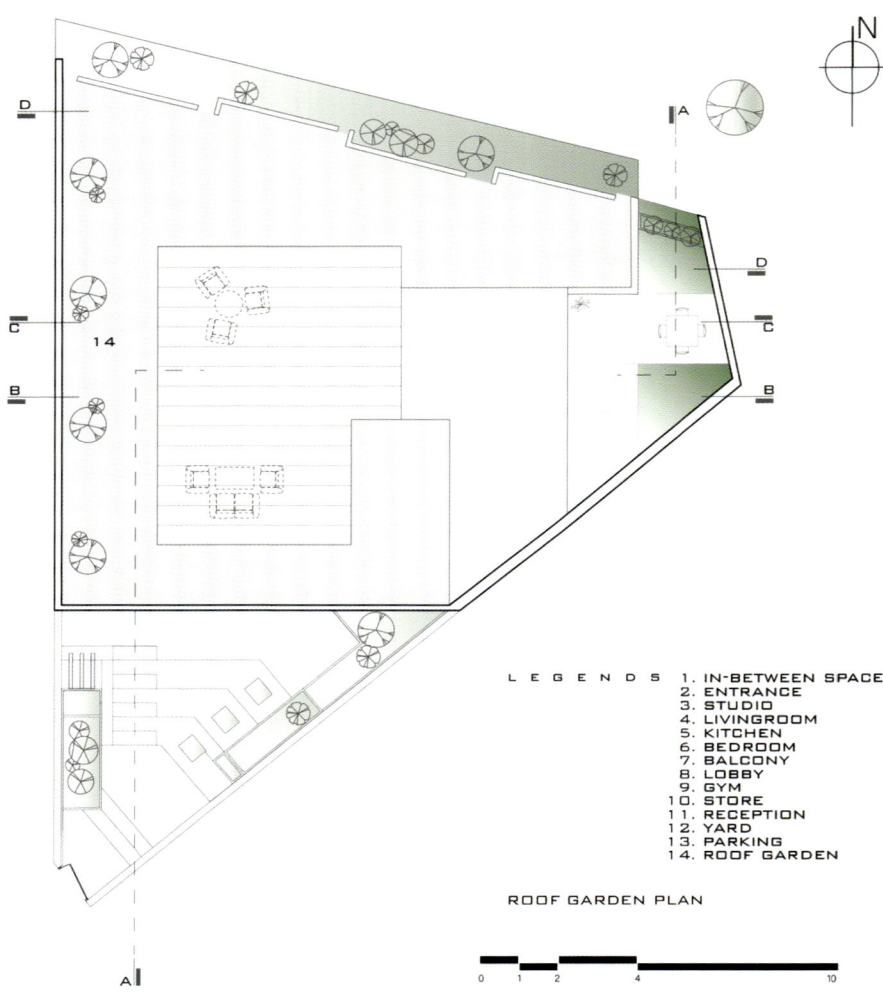

LEGENDS
1. IN-BETWEEN SPACE
2. ENTRANCE
3. STUDIO
4. LIVINGROOM
5. KITCHEN
6. BEDROOM
7. BALCONY
8. LOBBY
9. GYM
10. STORE
11. RECEPTION
12. YARD
13. PARKING
14. ROOF GARDEN

ROOF GARDEN PLAN

0 1 2   4   10

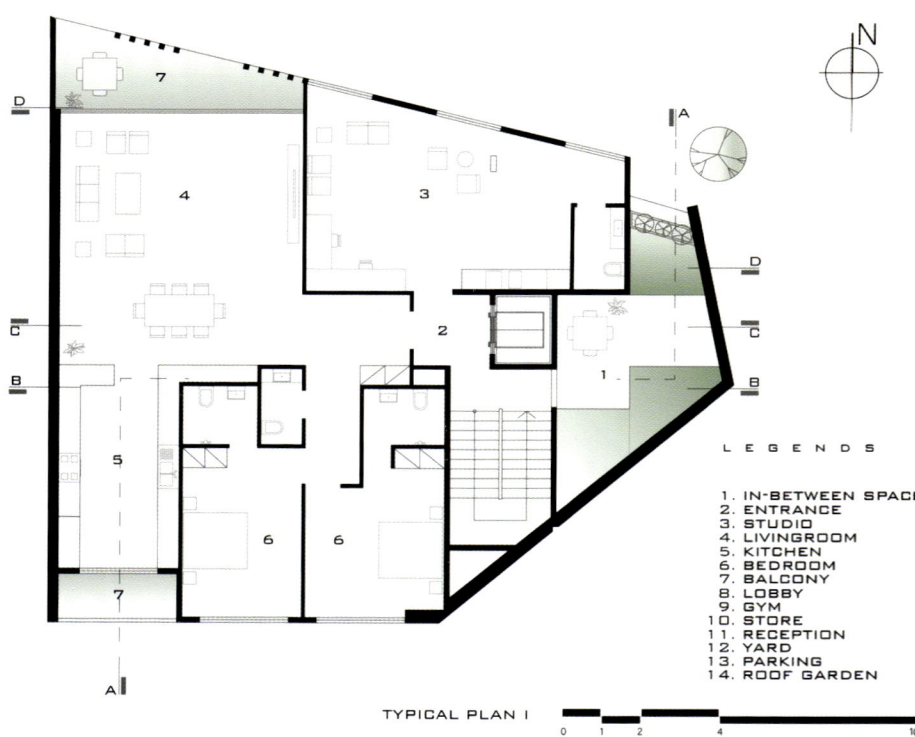

LEGENDS
1. IN-BETWEEN SPACE
2. ENTRANCE
3. STUDIO
4. LIVINGROOM
5. KITCHEN
6. BEDROOM
7. BALCONY
8. LOBBY
9. GYM
10. STORE
11. RECEPTION
12. YARD
13. PARKING
14. ROOF GARDEN

TYPICAL PLAN I

0 1 2   4   10

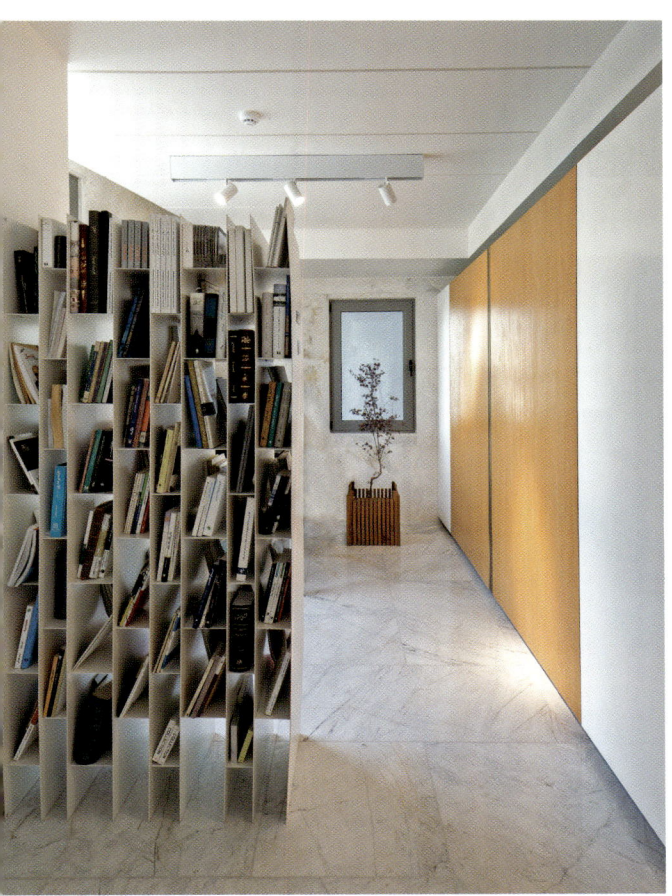

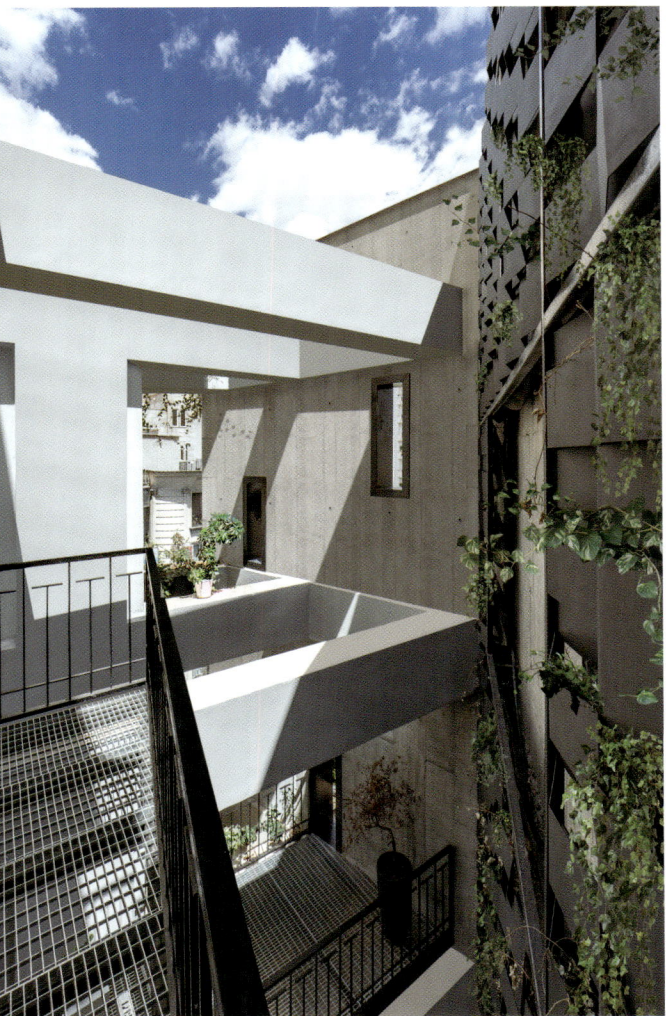

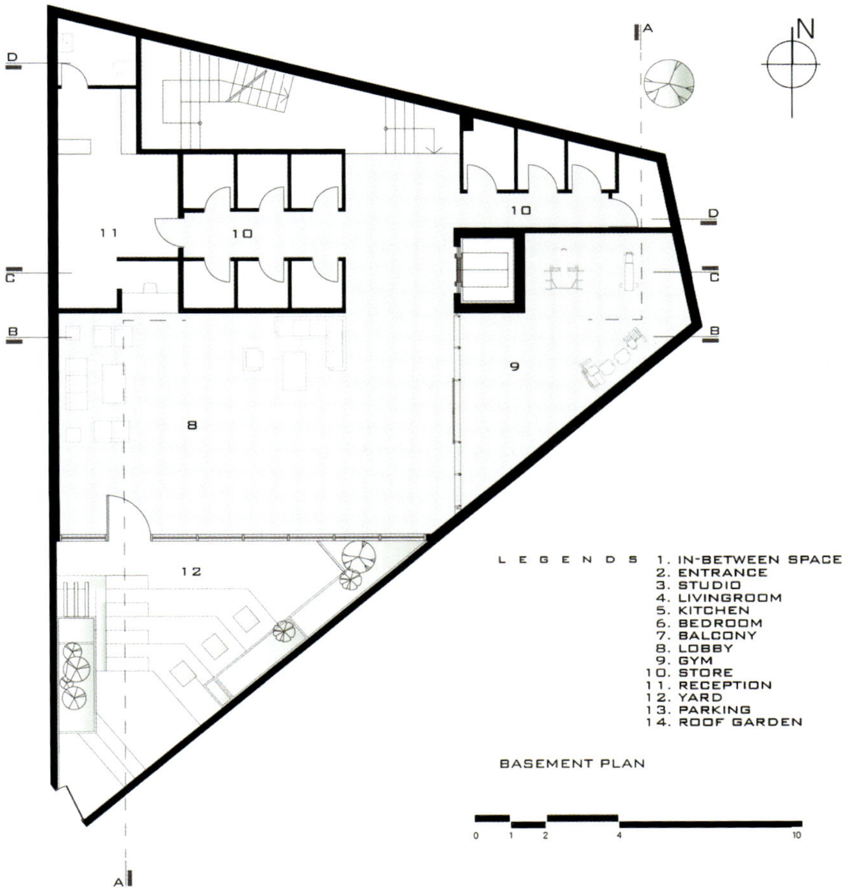
N
A
D
C
B
11
10
9
8
12
LEGENDS 1. IN-BETWEEN SPACE
2. ENTRANCE
3. STUDIO
4. LIVINGROOM
5. KITCHEN
6. BEDROOM
7. BALCONY
8. LOBBY
9. GYM
10. STORE
11. RECEPTION
12. YARD
13. PARKING
14. ROOF GARDEN
BASEMENT PLAN
0 1 2 4 10

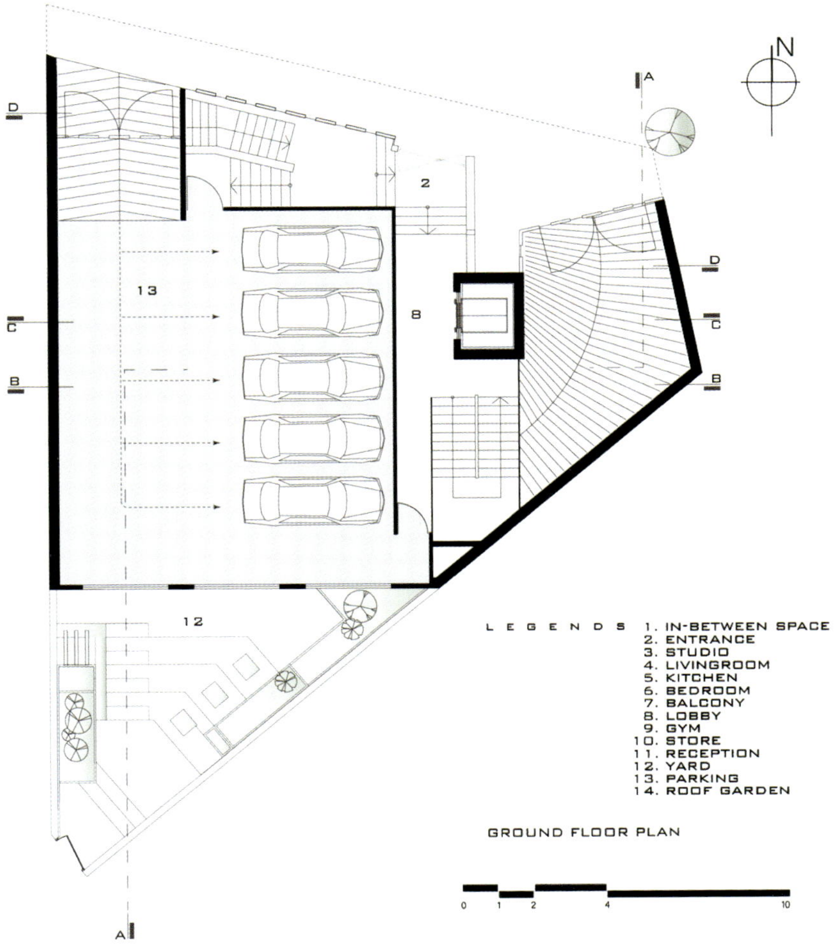
N
A
D
C
B
2
13
8
12
LEGENDS 1. IN-BETWEEN SPACE
2. ENTRANCE
3. STUDIO
4. LIVINGROOM
5. KITCHEN
6. BEDROOM
7. BALCONY
8. LOBBY
9. GYM
10. STORE
11. RECEPTION
12. YARD
13. PARKING
14. ROOF GARDEN
GROUND FLOOR PLAN
0 1 2 4 10

The Casa da Mole

Architects: Marchetti Bonetti
Location: Pond of Conceicao, Brazil
Area: 565 m²
Photographs: Ronaldo Azambuja

Responsible Architects: Giovani Bonetti
Project Team: Giovani Bonetti, José Darós, Cristhine Digiacomo
Main Material: Concrete(The textured acrylic coating with ECO Terracor finish) + Black ceramic cover finishing.

Casa da Mole is located in the condominium facing Praia Mole, so the focus of the project was to take advantage of the great look of this beach. For this, the living room was designed with double height and the floor to ceiling frames appear to give amplitude and ample luminosity to the environment, seeking to make the most of the view that this house close to the beach provides. The volume of the house was also designed to combine privacy for users, and to frame the view from the living room to the external environment. The straight lines bring modernity to the house, with emphasis on the volume of the upper floor, in which concrete finishing was used as the predominant material in the facades, contrasting the black ceramic covering of the ground floor. To meet the demands of customers, the house was divided into social and service environments on the ground floor, intimate environments on the upper floor and environments for integration with the landscaping on the lower floor. Also on this floor is the swimming pool that is connected to the sauna, thus making it a space of comfort and relaxation for its users.

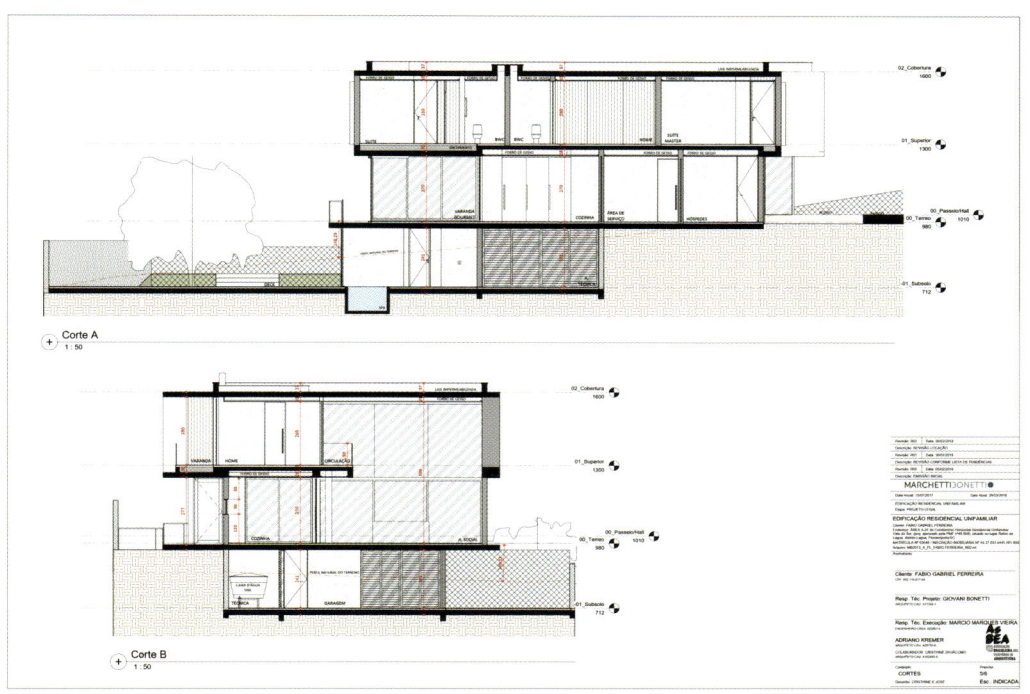

The textured acrylic coating with ECO Terracor finish, winner of the Planeta Casa Award, from Casa Claudia magazine, and certified by the Green Building Council Brazil. It is a textured acrylic coating, with exclusive aesthetic effect, which offers a very high level of sustainability when using as raw material 80% of recycled / renewable materials and water, in addition to presenting a very low VOC index.

Material Type: **Concrete**
(The textured acrylic coating with ECO Terracor finish)

Material Type: Other
(Black ceramic cover)

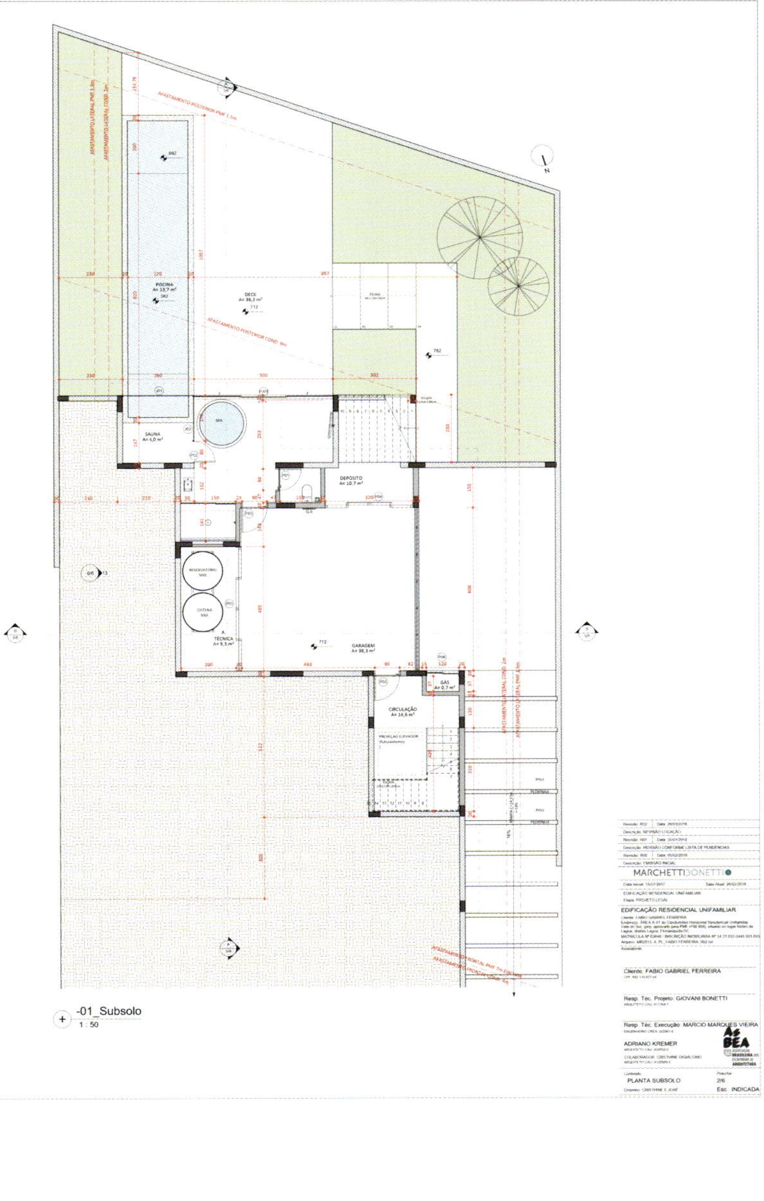

-01_Subsolo
1 : 50
MARCHETTIBONETTI
EDIFICAÇÃO RESIDENCIAL UNIFAMILIAR
Cliente: FABIO GABRIEL FERREIRA
Resp. Téc. Projeto: GIOVANI BONETTI
Resp. Téc. Execução: MARCIO MARQUES VIEIRA
ADRIANO KREMER
PLANTA SUBSOLO
Esc.: INDICADA

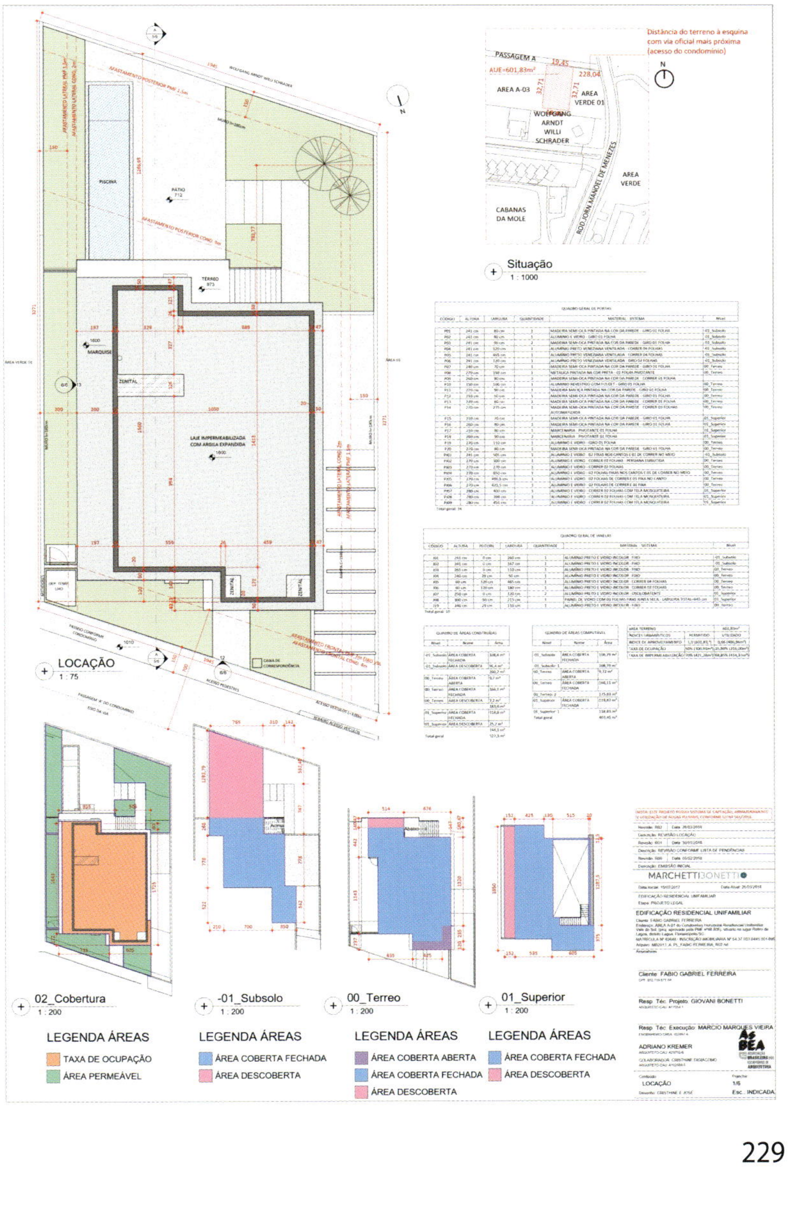

Situação
1 : 1000
LOCAÇÃO
1 : 75
LAJE IMPERMEABILIZADA COM ARGILA EXPANDIDA
02_Cobertura
1 : 200
-01_Subsolo
1 : 200
00_Terreo
1 : 200
01_Superior
1 : 200
LEGENDA ÁREAS
TAXA DE OCUPAÇÃO
ÁREA PERMEÁVEL
LEGENDA ÁREAS
ÁREA COBERTA FECHADA
ÁREA DESCOBERTA
LEGENDA ÁREAS
ÁREA COBERTA ABERTA
ÁREA COBERTA FECHADA
ÁREA DESCOBERTA
LEGENDA ÁREAS
ÁREA COBERTA FECHADA
ÁREA DESCOBERTA
MARCHETTIBONETTI
EDIFICAÇÃO RESIDENCIAL UNIFAMILIAR
Cliente: FABIO GABRIEL FERREIRA
Resp. Téc. Projeto: GIOVANI BONETTI
Resp. Téc. Execução: MARCIO MARQUES VIEIRA
ADRIANO KREMER
LOCAÇÃO
Esc.: INDICADA

Another point that stands out in the project is the concept of using natural resources, such as reusing rainwater. As well as, all internal and external lighting in the house with LED lamps. The Casa da Mole project sought a differentiator in meeting the wishes of customers, due to the care and respect for the location in which it is located.

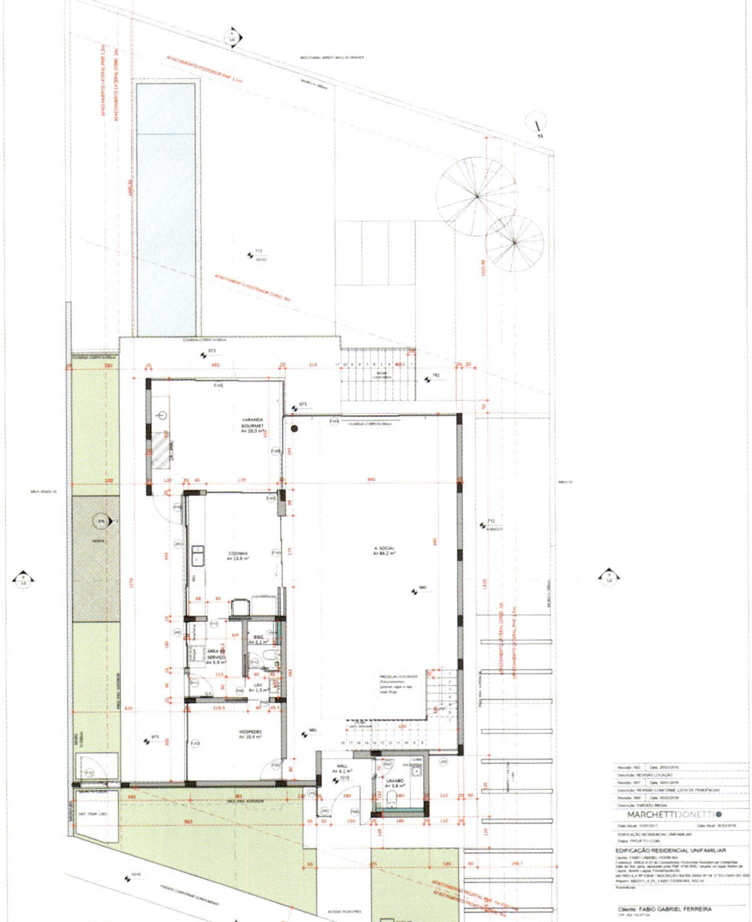

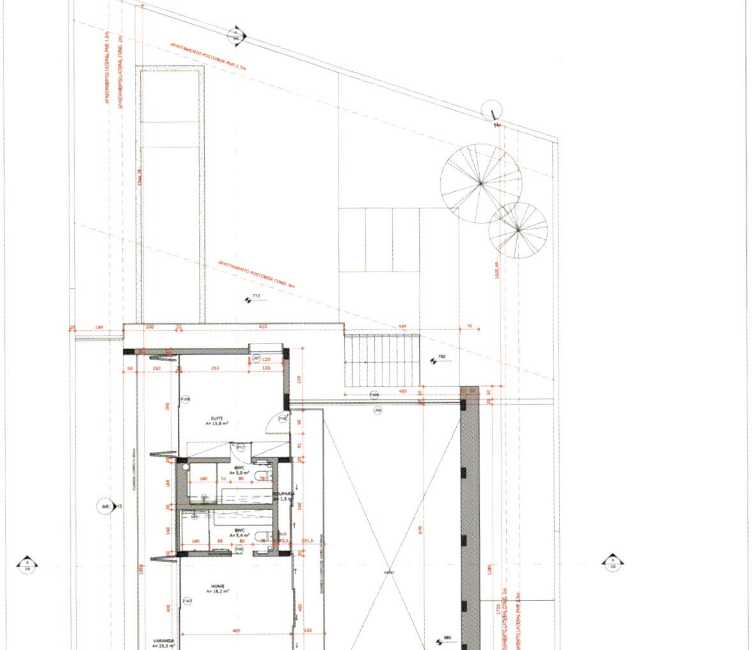

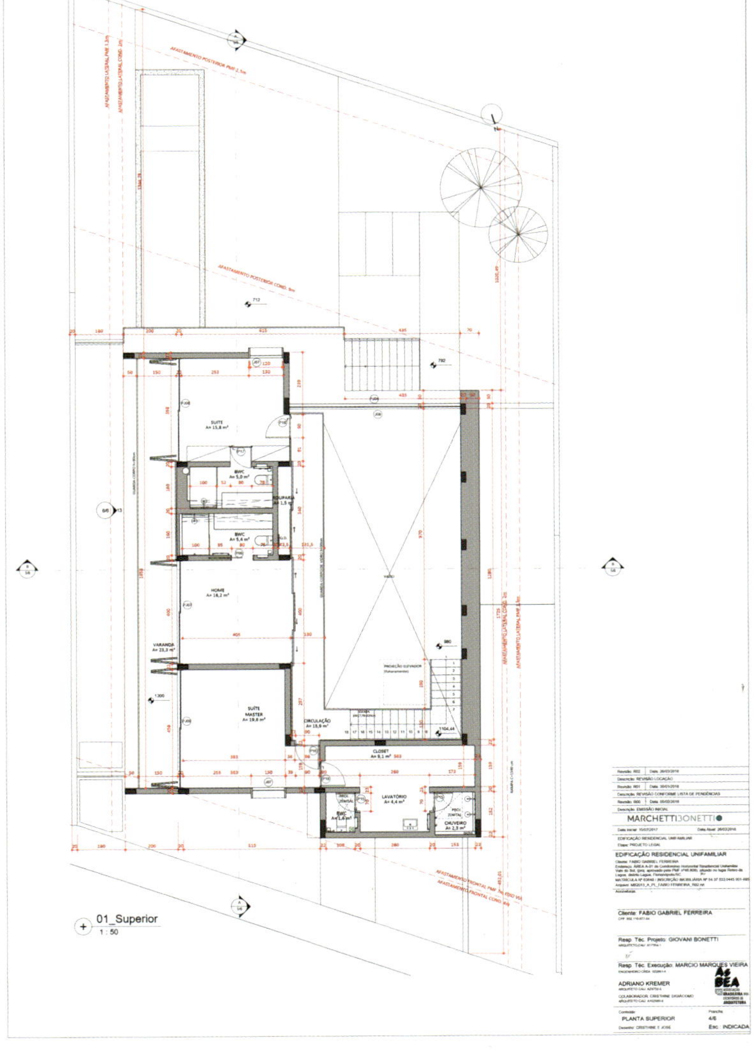

The Antoine de Ruffi School

Architects: TAUTEM Architecture + bmc2 architectes
Location: Marseille, France
Area: 4150 m²
Photographs: Luc Boegly

Lead Architects: TAUTEM Architecture
Associate Architects: bmc2
Main Material: Concrete(Mineral Concrete Stain - Concretal®-Lasur) + Wood; bio-sourced larch

Housing 22 classrooms and common areas, this mineral monolith features strict geometry and spectacular volumes. Its light-colored concrete façades are sculpted, and the openwork of this thickness forms a colonnade on the port side and a grand staircase on the city side, creating the interplay of light and shadow in its embrasures. In contrast to the building's envelope, the interiors are warm and comfortable thanks to the use of color and wood. A remarkable site. The Antoine de Ruffi school group occupies a strategic spot between the entrance to the new Méditerranée district, and its "inhabited park" coordinated by the urbanist Yves Lion. Its situation offers, on one hand, a view over the developing suburban fabric, with scattered warehouses, silos, soap factories, large-scale housing estates from the 1970s, and in the distance, the Massif de l'Etoile. In the reverse view, towards the west, one sees the port and its huge ships, the towers by Zaha Hadid and Jean Nouvel, as well as the continuous sweep of the highway viaduct.

Sculpted monolith. At first glance, this monolith combines massiveness and minerality. The monumentality is the condition guaranteeing its existence in this dense district where high rise apartment buildings (up to 17 stories) are slated for construction. The architects have voluntarily limited the number of architectural and technical components to guarantee simplicity and longevity and to ensure easy maintenance. Built with "low carbon," light-colored concrete, between the pearl white blanc and beige of the coquina sand (dear to Pouillon), the building was poured in place and without joints. The painstaking work of the "skin" has produced alternating parts of coquina and smooth, mat and shiny surfaces and an interplay of light and shadow in the embrasures.

Material Type: Concrete
(Mineral Concrete Stain - Concretal®-Lasur)

Built with "low carbon," light-colored concrete, between the pearl white blanc and beige of the coquina sand (dear to Pouillon), the building was poured in place and without joints. The painstaking work of the "skin" has produced alternating parts of coquina and smooth, mat and shiny surfaces and an interplay of light and shadow in the embrasures.

The façades play a protective role. With a thickness of 100 cm, they are the result of a "double wall", a process of simultaneous pouring of two veils of concrete between which a rigid form of insulation is inserted.

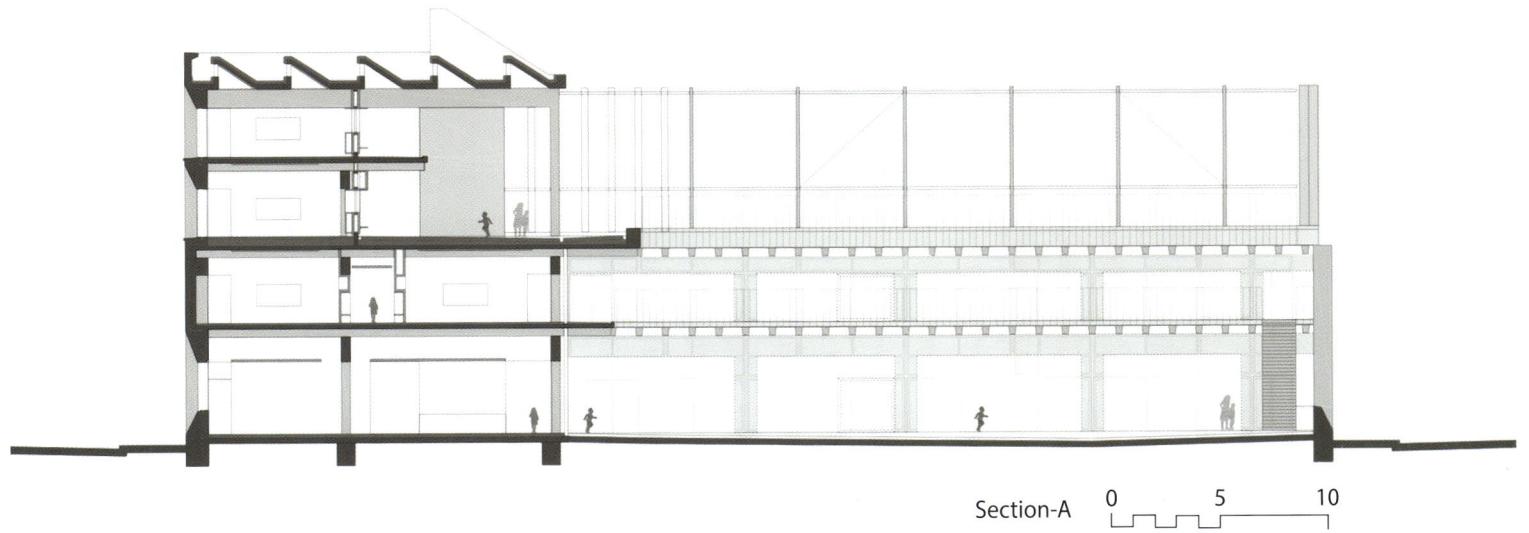

Section-A 0 5 10

Bio-climatic design: façades adapted to their exposure. The façades play a protective role. With a thickness of 100 cm, they are the result of a "double wall", a process of simultaneous pouring of two veils of concrete between which a rigid form of insulation is inserted. They combine thermal performance and massiveness to the two mineral façades. In their thickness, the deep embrasures placed here provide the interior with useful voids for installing storage, work stations and fluid circulations.

Detail Plan-A

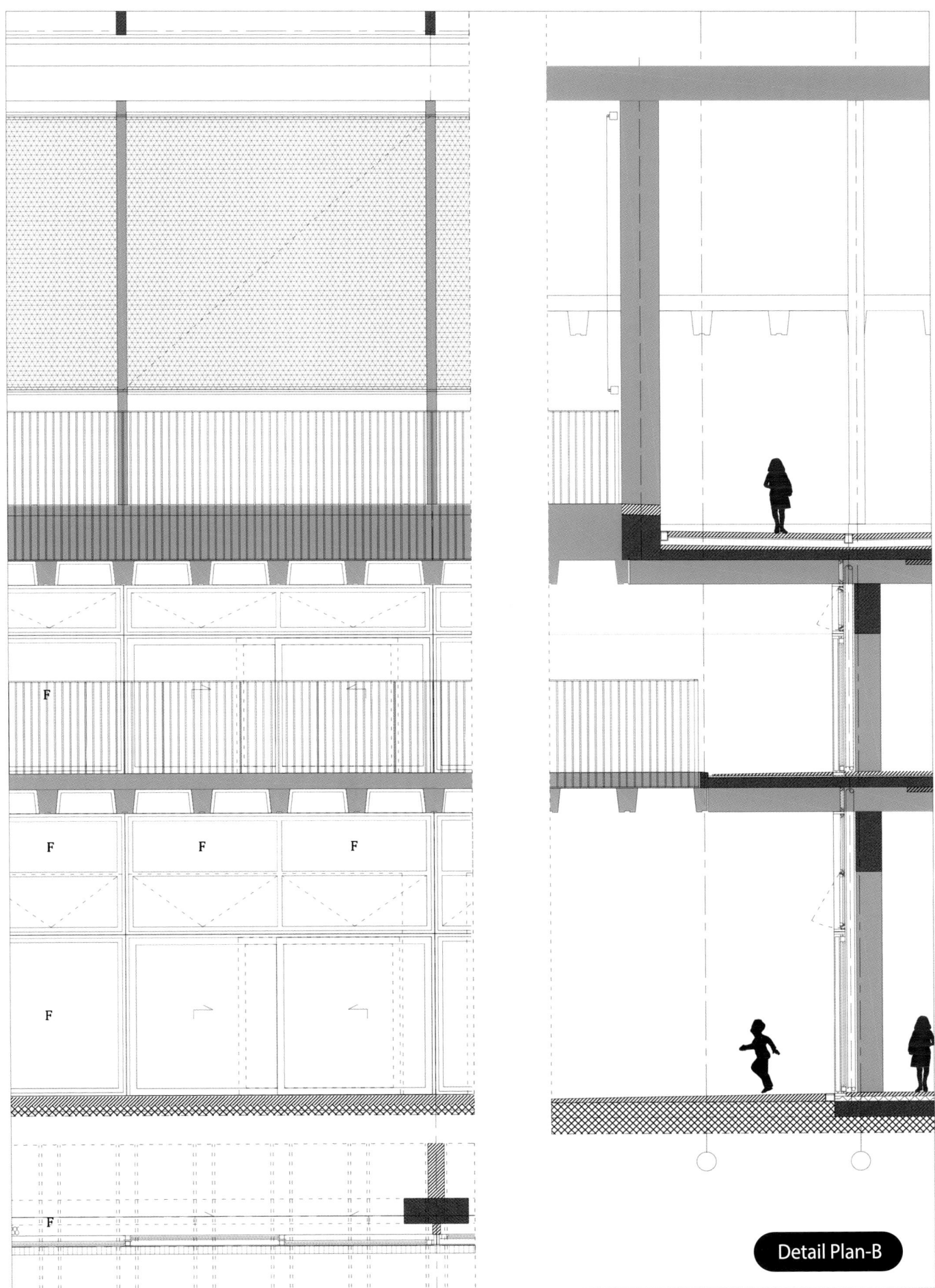

Detail Plan-B

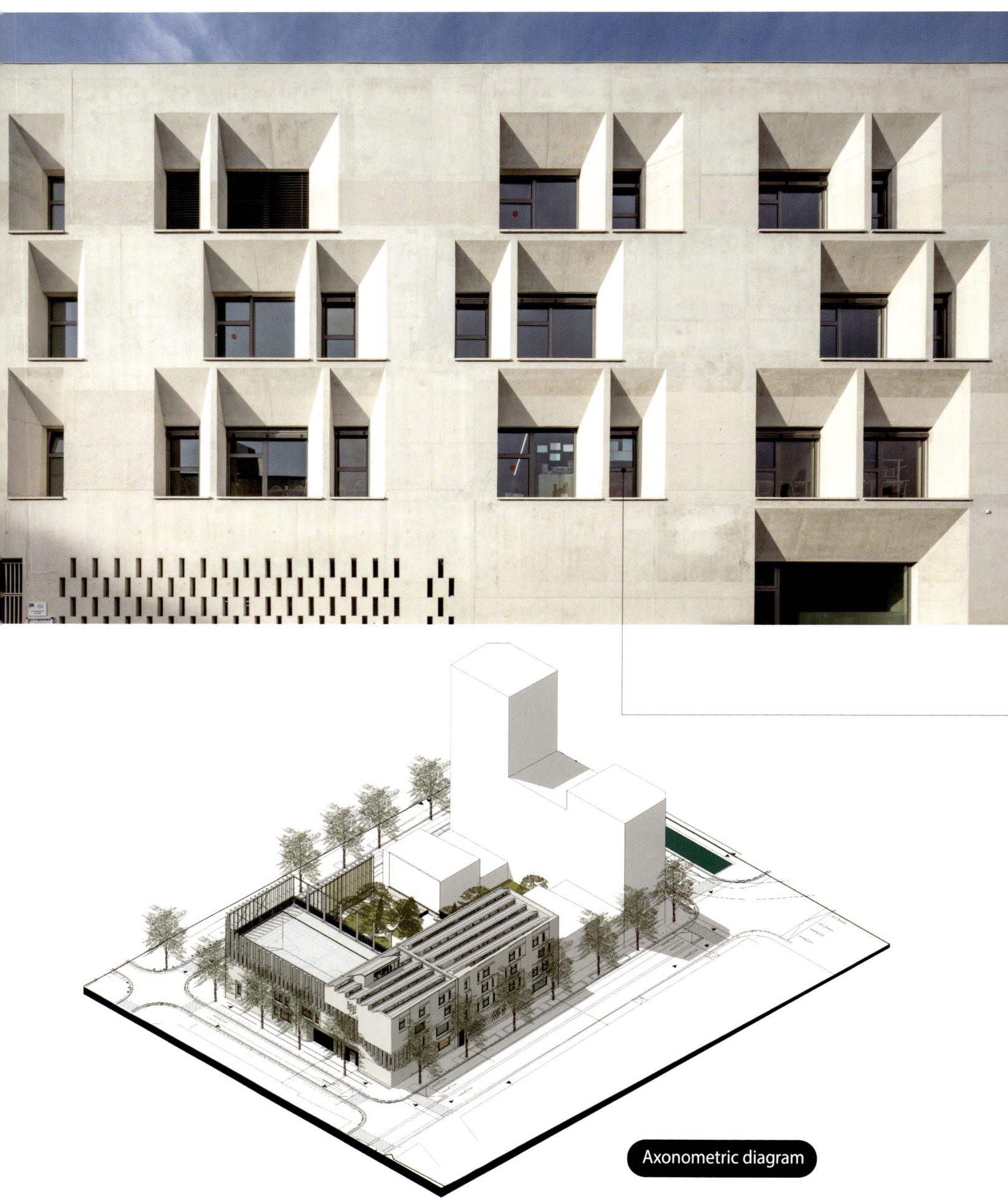

Axonometric diagram

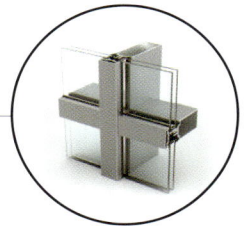

Material Type: **Glass**
(Glass Curtain Wall)

Joyful and luminous interiors. To create surprise and contrast with the minerality of the envelope and always in this Mediterranean style, the interior feels joyful and colorful. The softness of curves and the use of wood enabled this children's universe, warm and enveloping. This wood, bio-sourced larch from the Alpes, was used with restraint, for the major walls covered in wood paneling and glazed between the class rooms and circulations and for built in furnishings.

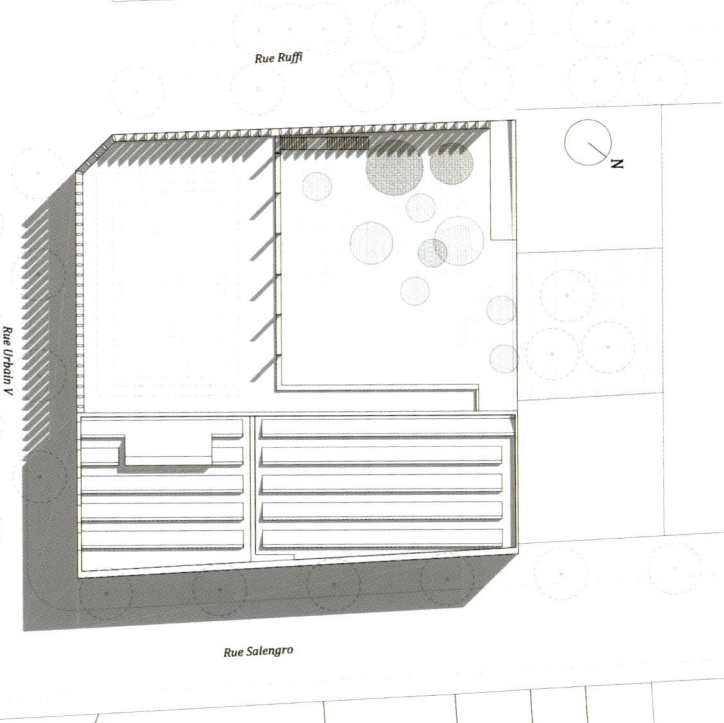

Site Plan

243

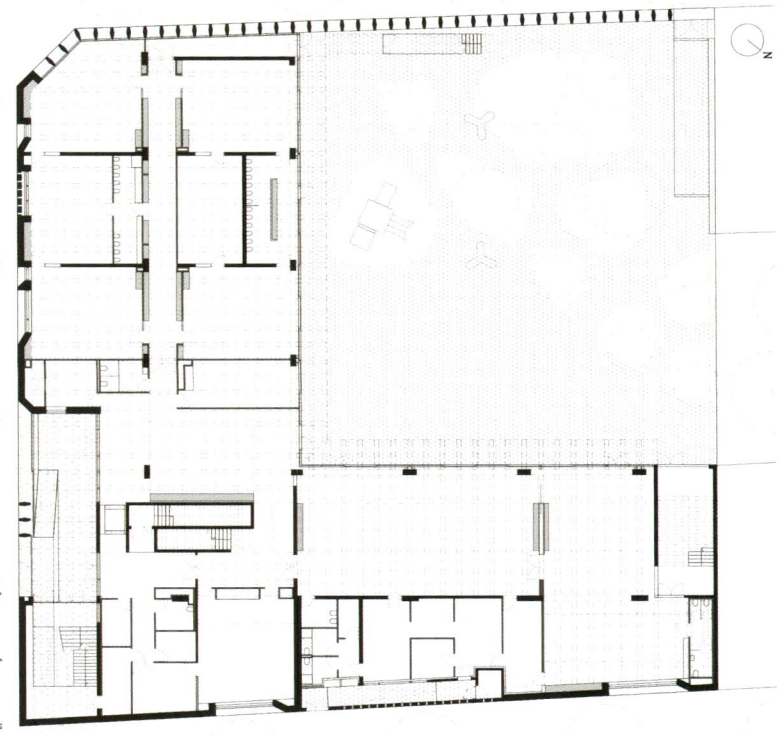

Ground Floor Plan

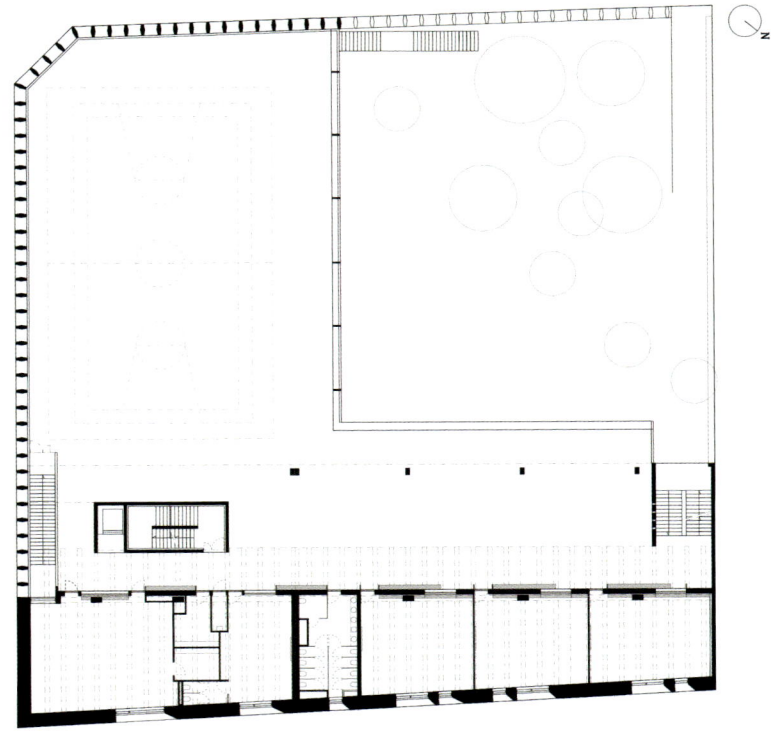

Second Floor Plan

245

The 55 Southbank Boulevard

Architects: Bates Smart
Location: Southbank, Australia
Area: 27000 m²
Photographs: Peter Clarke

Manufacturers: RC+D, KONE
Project Leader: Allan Lamb, Celine Herbiet
Main Material: Glass + Concrete + CLT (Cross Laminated Timber)

The new Adina Apartment Hotel Melbourne Southbank is the first cross-laminated timber (CLT) high-rise hotel in Australia. Utilising the latest in timber construction technology this project attains new levels in sustainable design. Approximately 5,300 tonnes of CLT were used to add 10 levels to an existing commercial building, offsetting nearly 4,200 of CO_2 from the atmosphere. CLT Pushes Urban Adaptive Reuse Even Further The existing commercial office building built in 1989, was able to accommodate an extension of six levels with the use of concrete framed construction. The design challenge Bates Smart faced was how to surpass this, in order to deliver a hotel with 220 rooms. The solution involved the use of Cross Laminated Timber (CLT) construction which means the existing building is able to support an additional 10 levels, thereby achieving the target room number spread across 13,000 square metres of new floorspace. CLT is approximately 20% the weight of concrete, essentially doubling the number of levels that could be built above the existing structure. Using CLT also allowed for components to be prefabricated off-site, resulting in increased construction efficiencies and decreased impact on surrounding buildings. CLT presents a more sustainable approach to increasing density within our cities. With a limited supply of developable sites, lightweight timber structures can increase yields, which would not be achievable using traditional concrete and steel construction.

Material Type: **Glass**
(Curved Glass Curtain Wall)

Material Type: **Steel & Matal**
(Kriskadecor's aluminum cladding system)

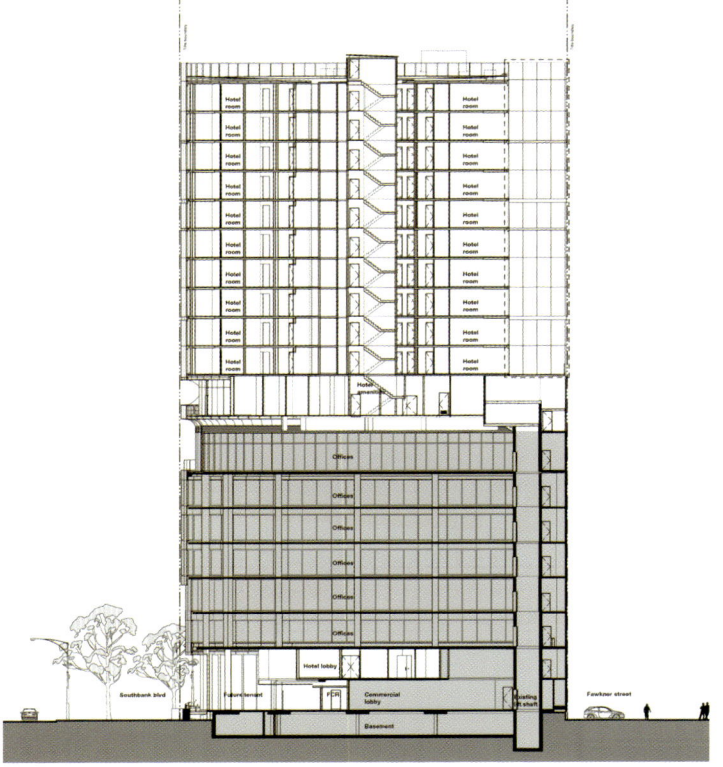
Section-A

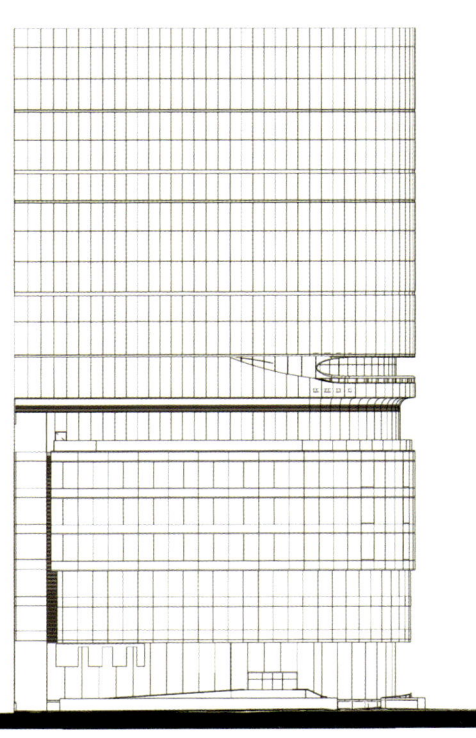
North Elevation

A Striking, Contemporary Design The design of the new CLT levels complements the curved architecture of the existing building without overtly expressing the timber, as is typical in many CLT buildings. A large recessed balcony helps celebrate the corner and the building's new height. It also gives guests a place to take in expansive Melbourne skyline views. Inside, the ground floor lobby is lined with timber to create a warm and welcoming environment that also nods to the building's innovative structure. Curved walls also distinguish the space, and work in accordance with the curved façade and sinuous lines of the new extension. Guest floors offer 70 studio apartments, 140 one-bedroom apartments, and 10 two-bedroom apartments, all complete with kitchens and lounge and dining areas. Guests also have access to a 20m lap pool and gym, which have been oriented to the north to take full advantage of natural light.

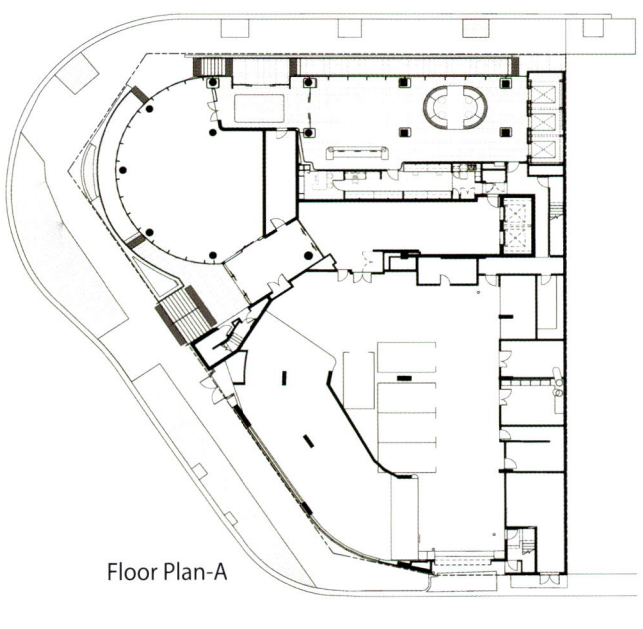

Floor Plan-A

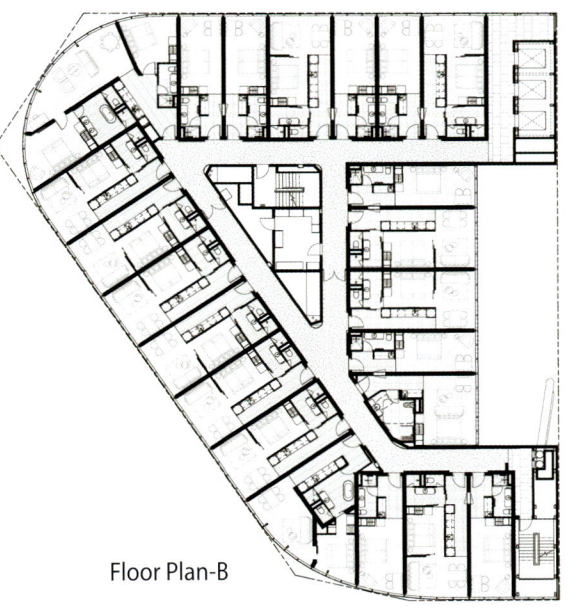

Floor Plan-B

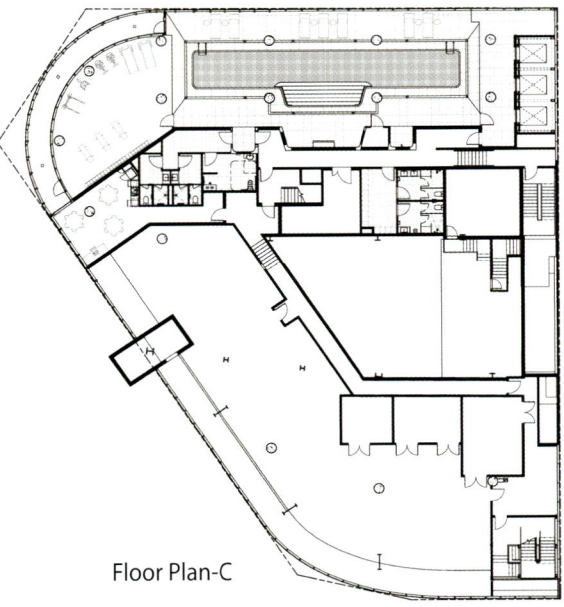

Floor Plan-C

A new paradigm in sustainable construction In addition to being significantly lighter than concrete, CLT use results in lower transport costs and therefore reduced carbon emissions. Due to the quantum of offsite fabrication, including prefabricated modular hotel bathrooms, the time spent on site is significantly reduced. Most importantly, sequestered within the timber itself is approximately 4,000 tonnes of CO2 emissions, the equivalent of the annual carbon emissions of 130 homes. The predominant use of this carbon-negative material represents an opportunity for the creation of a new paradigm in sustainable construction. Timber for the hotel was sourced from suppliers with Forest Stewardship Council certification—one of only two internationally recognised forest certification networks, reaffirming the commitment the Adina Southbank makes to achieving high sustainability standards.

White Canvas House

Architects: ACA Architects
Location: Bangkok, Thailand
Area: 300 m²
Photographs: DOF Sky|Ground

Manufacturers: AutoDesk, COTTO, TOA, Trimble Navigation
Architect In Charge: Anon Chitranukroh
Design Team: Anon Chitranukroh, Waranthorn Intuputi, Kirin Chaichana
Main Material: Precast Concrete + Steel mesh + Aluminum composite panels

This renovation project began when the clients wanted to expand the space for their newborn child and to give an improvement the exterior and interior design to suit the new lifestyle. Situated in an allocated land village in Bangkok, Thailand, the existing buildings are composed of a typical precast concrete house with a hip roof and an unfinished reinforce concrete extension building structure besides.

The main idea of the design is to create one single house for a family of 3 members by esthetically combining both buildings' structures into one beautiful design language. Since the existing house is made from precast concrete walls and panels, it will be unappropriated to modify and add more façade loads to the building structure. Thus, from the limitation, we created a new façade layer as a "Shell" covering both buildings by composing with the exterior and landscape wall and space.

Material Type: **Steel & Matal**
(aluminum composite panels)

Material Type: **Concrete**
(Precast concrete walls and panels + painting finish)

"White Canvas" is the designing concept we brought up as the modern minimalist white planes with a composition of textures which are steel mesh, aluminum composite panels, and simple paint finish wall. Steel mesh is the essential material we selected on behalf of the owner's family business is a steel mesh distributor, creating solid and void connecting to interior space for privacy and ventilation.

Material Type: **Steel & Metal**
(Steel mesh Wall)

Elevation-A

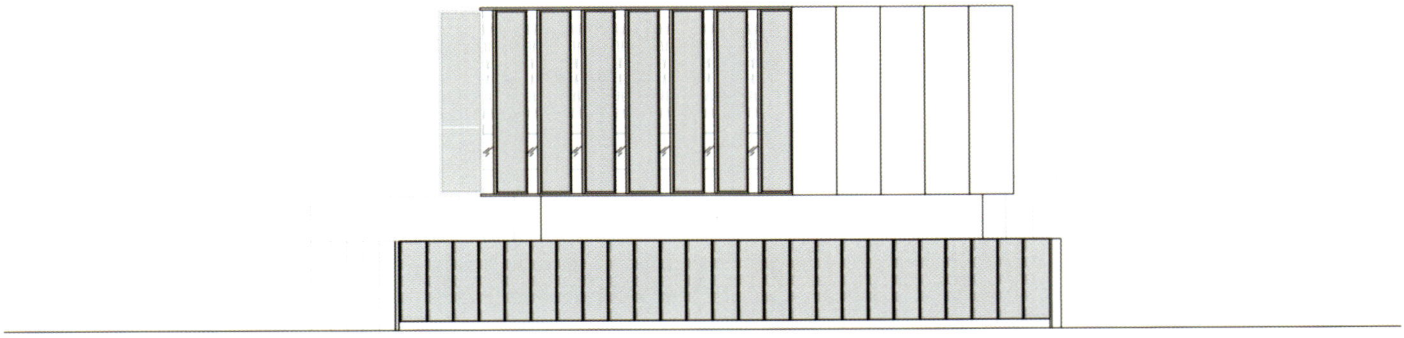

Elevation-B

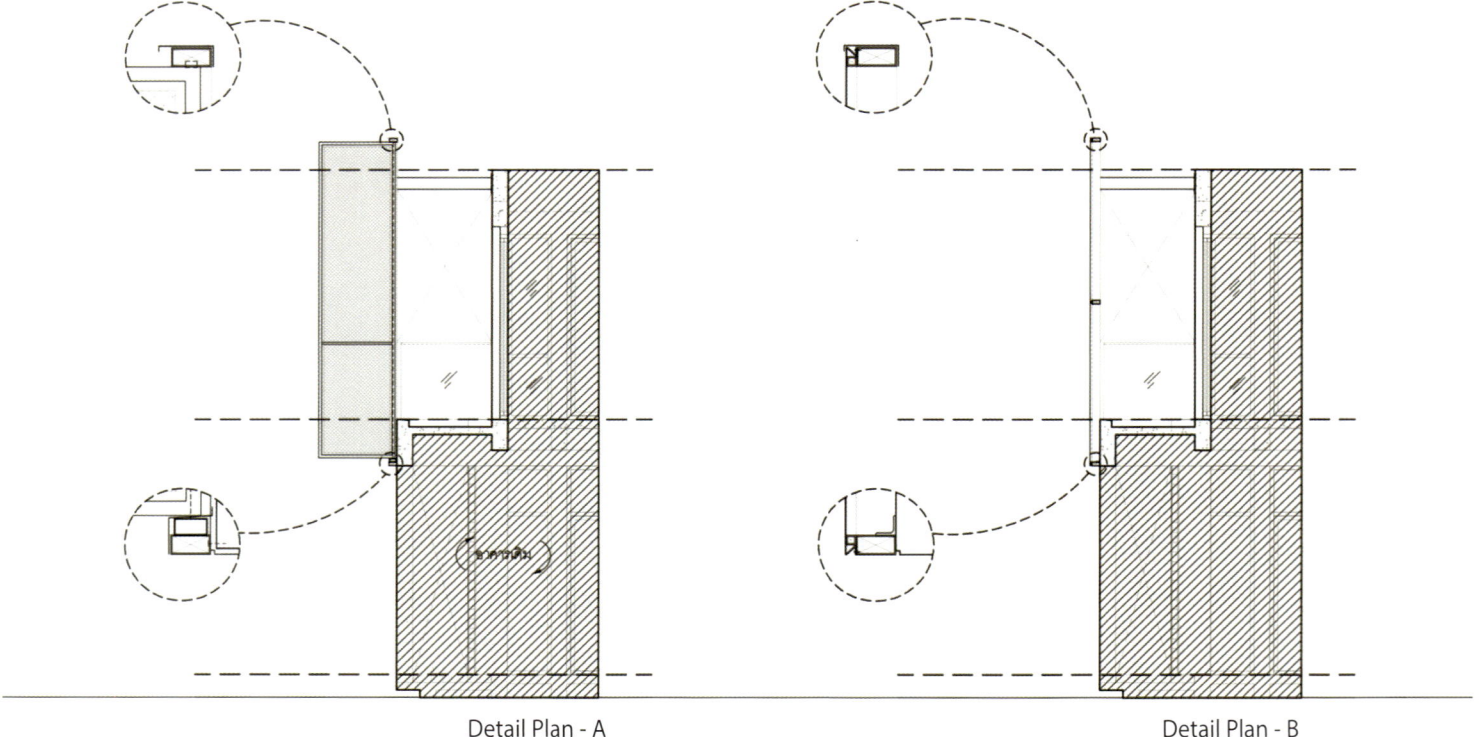

Detail Plan - A

Detail Plan - B

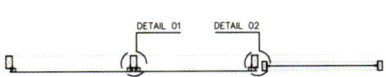
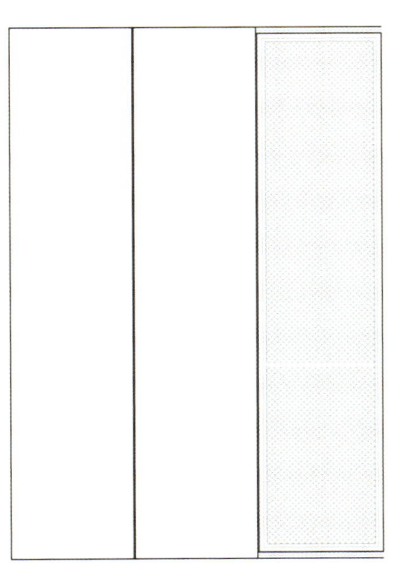
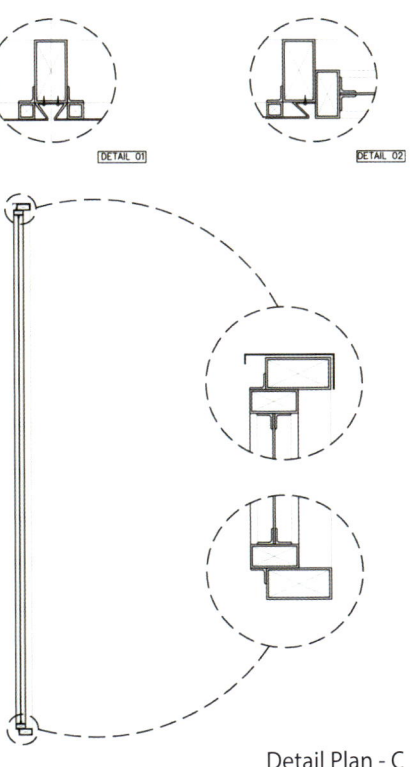

Detail Plan - C

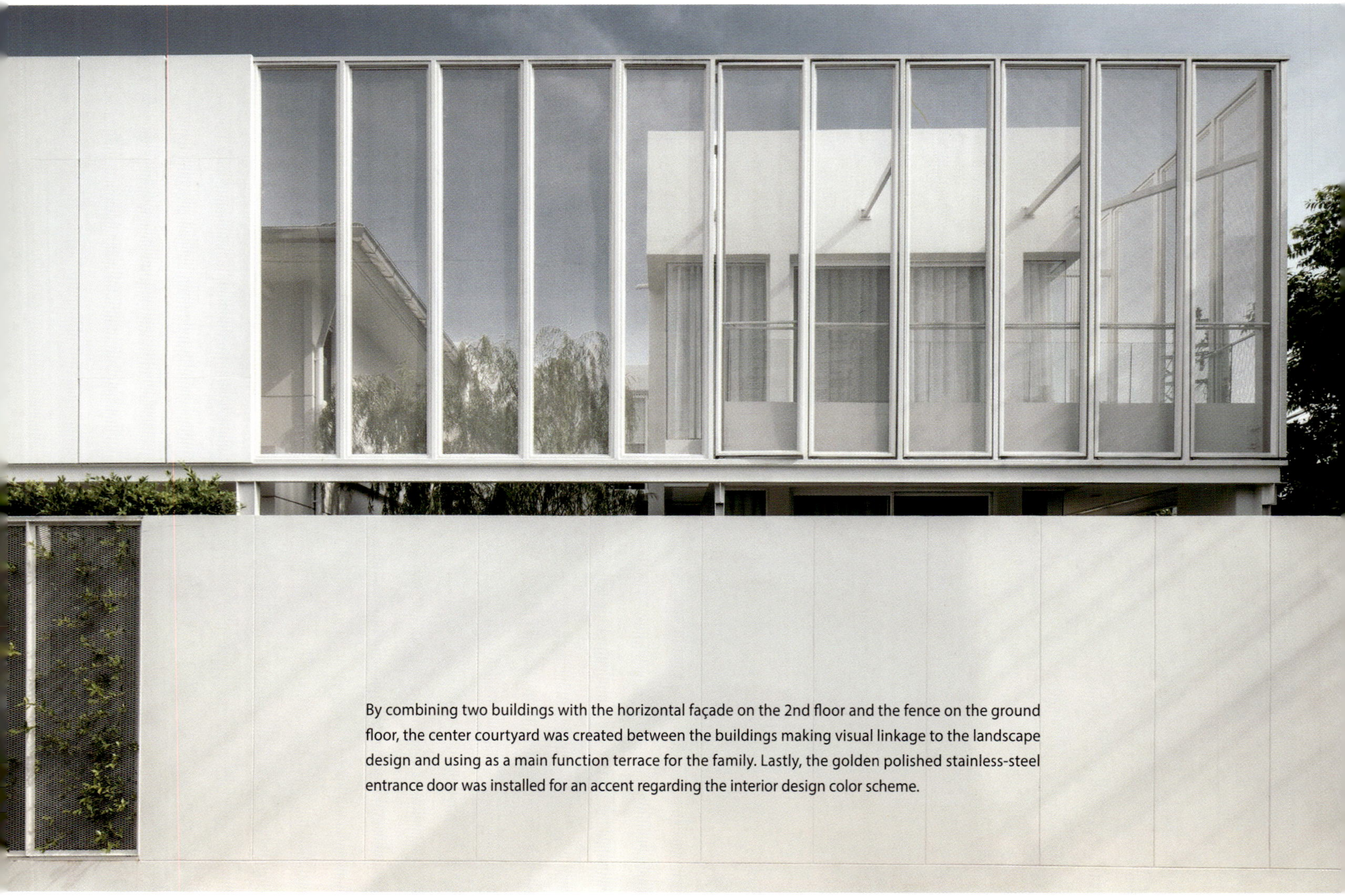

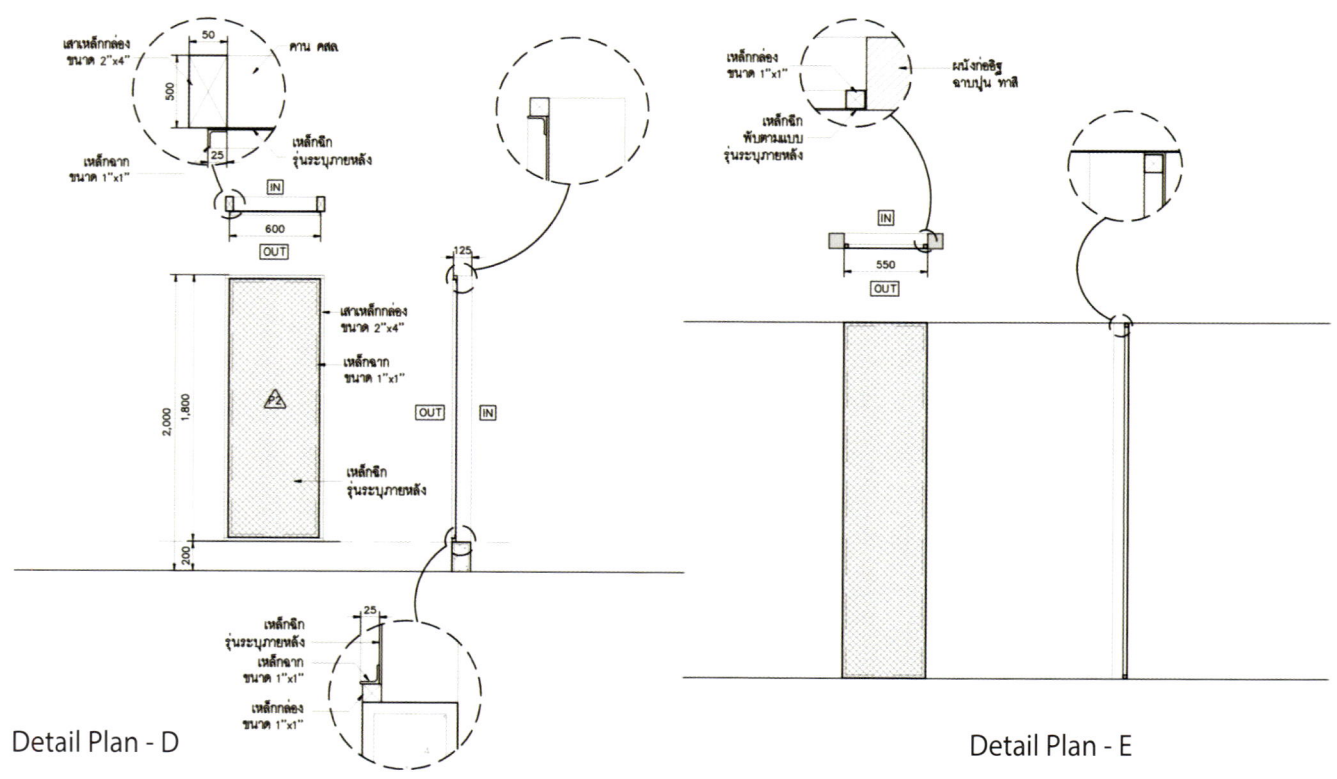

By combining two buildings with the horizontal façade on the 2nd floor and the fence on the ground floor, the center courtyard was created between the buildings making visual linkage to the landscape design and using as a main function terrace for the family. Lastly, the golden polished stainless-steel entrance door was installed for an accent regarding the interior design color scheme.

Detail Plan - D

Detail Plan - E

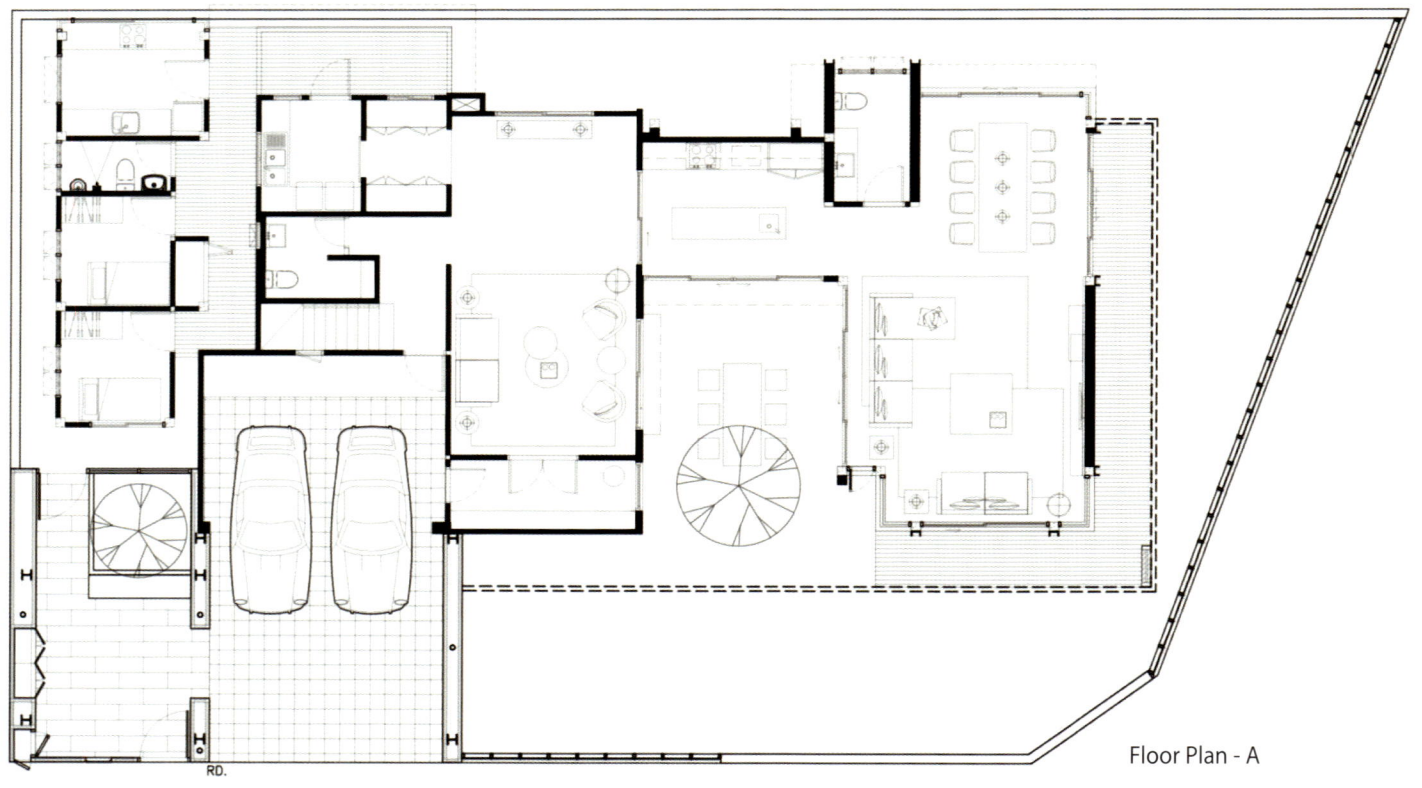

Floor Plan - A

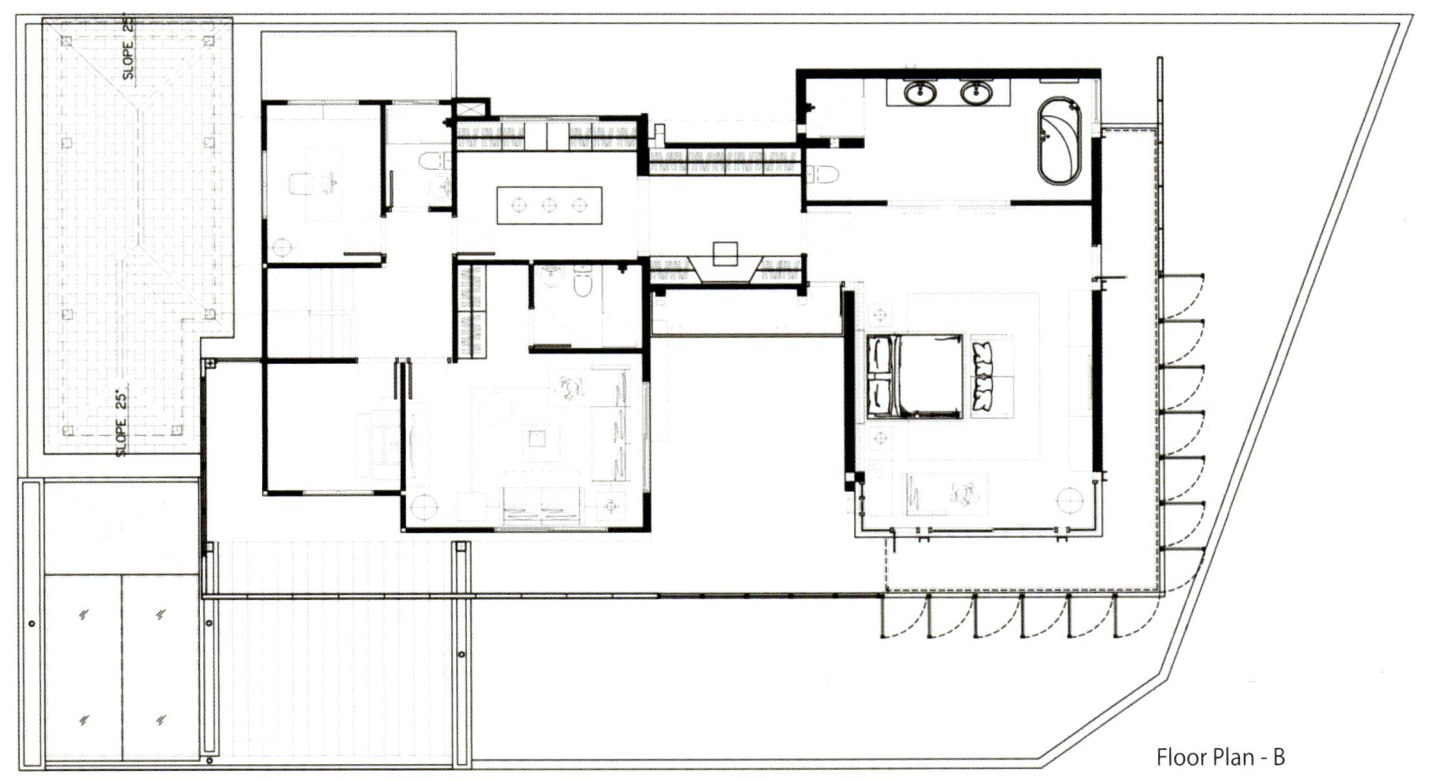

Floor Plan - B

SMOORE Liutang Industrial Park Shenzhen

Architects: CM Design
Location: Shenzhen, China
Area: 13658 m²
Photographs: Chao Zhang

Design Team: Jun Liao, Danping Chen, Heng Yang, Musen Li, Mingming Yao, Yingchuan Zhong, Jing Zhao, Shihua Long, Jinyao Liu
Main Material: Concrete(Ductal® Cladding Panels (EU) | Ductal®) + Rust steel + Aluminum meshes + Glass Curtain Wall

SMOORE Liutang Industrial Park locates in Xixiang Street, Baoan District, Shenzhen, at the northeast corner of the Liutang Park and adjacent to the Christian Baoan Church. The project covers an area of 5,017 square meters, and has a total floor area of 13,658 square meters, including 6,260 square meters in Building 2 and 7,398 square meters in Building 3. In the 1980-90s, a large number of small and medium-sized manufacturing plants came out in this area, which however, has been gradually surrounded by new residential communities, parks, offices, and related infrastructures due to the rapid process of urbanization. As a result, the area has changed from the outskirt of the city to the center, production workshops began to move away, and the old industrial buildings were facing urban renewal and function upgrade.

The design needs to form a new relationship between the buildings. The industrial park consists of two separate factories, walls between them were removed to form a pedestrian loop in the park. A roadway through the bottom of Building 1 connects the two sites closely, creating an interconnected and convenient freight network. The characteristics and materials of the original buildings were carefully preserved. New design elements were applying to the ground floor, passage facade, and windows to contrast and resonate with the preserved building, which was also a response and respect to the original texture. The first floor is the equipment storage and mechanical loading area, rust steel plate was used on the wall as a skin for protection. A trumpet-shaped entrance was built into Building 1 as a public area for trucks to pass and load goods, connecting the two sites of the park.

Material Type: **Concrete**
(Ductal® Cladding Panels (EU) | Ductal®)

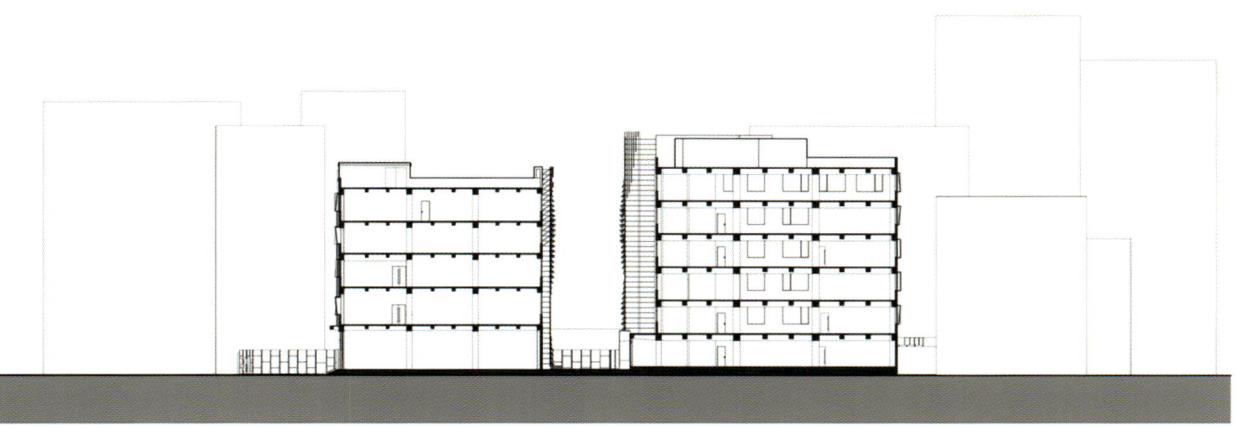

剖面图 / Sections

The most featured colorful mosaic facade on the second floor and above remains in use after cleaning and restoration. As an important component of the facade, renovation of the windows had to simultaneously maintain their original dimensions and show a new image. Original windows were replaced by different angled windows, with which dynamic scenes of the city could be portrayed and reflected from various directions, presenting a wide variety of facade forms. The design did not conform to the usual stereotype of an industrial park, giving the buildings vitality and more expression.

Material Type: **Steel & Metal**
(Rust steel mosaic facade)

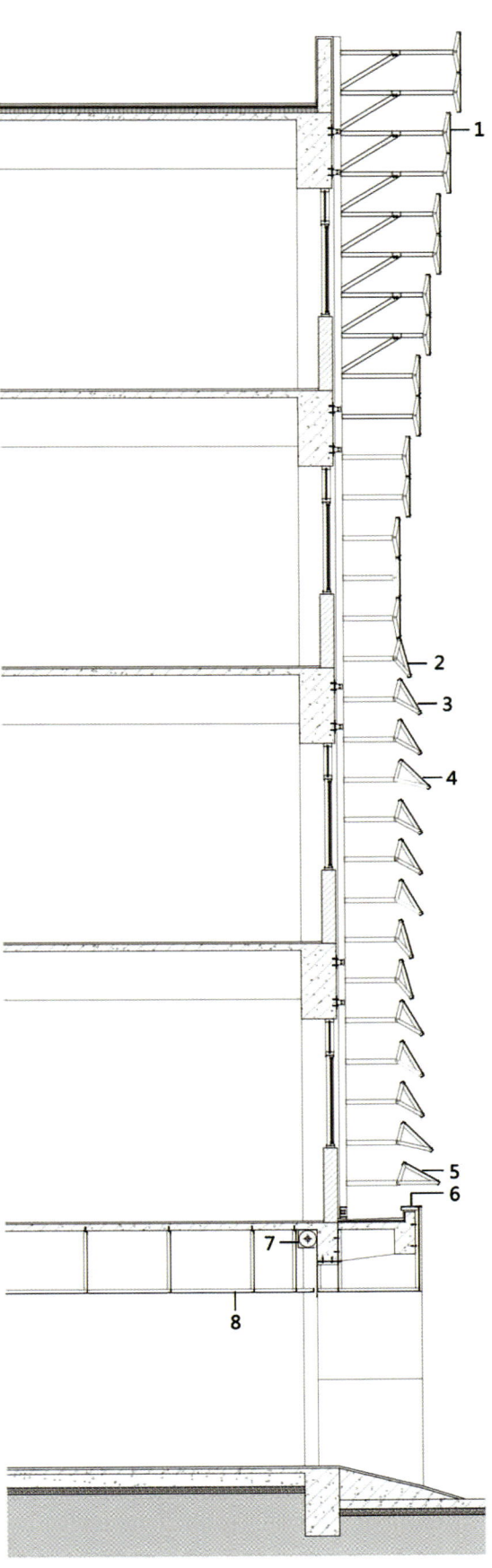

剖面图
比例 1:50

1. 1200/600mm 铝板网幕墙单元：
   60/28mm 铝板网
   40/40mm 角钢
   10mm 扁钢Y型支撑
   80mm 槽钢
   150mm 槽钢立柱
   原有建筑结构梁
2. 15°旋转幕墙单元
3. 30°旋转幕墙单元
4. 45°旋转幕墙单元
5. 60°旋转幕墙单元
6. 3mm 锈蚀钢板，3%找坡
   50/50mm 角钢支撑结构
   20mm 水泥砂浆保护层
   沥青防水卷材
   原有建筑雨板
7. 金属卷帘
8. 60/28mm 铝板网天花
   40/40mm 角钢

Section
scale 1:50

1. 1200/600mm aluminium expanded sheets facade module :
   60/28mm luminium expanded sheets
   40/40mm steel angle
   10mm Y-shape steel flat supporting structure
   80mm U-steel
   150mm U-steel column
   existing beam
2. 15° rotated facade module
3. 30° rotated facade module
4. 45° rotated facade module
5. 60° rotated facade module
6. 3mm sheet steel to 3% fall
   50/50mm steel angle supporting structure
   existing canopy
7. metal rolling gate
8. 60/28mm luminium expanded sheets ceiling
   40/40mm steel angle

Detail Plan

As a symbol of SMOORE's atomization technology, the "Fog Valley" was put into the narrow strip between the two buildings, which were both covered by a curtain wall composed of aluminum meshes, became the landmark of the park. Six kinds of mesh panel units with different angles were customized due to indoor lighting and sight view requirements, greatly reduced the construction difficulty and cut down the cost.

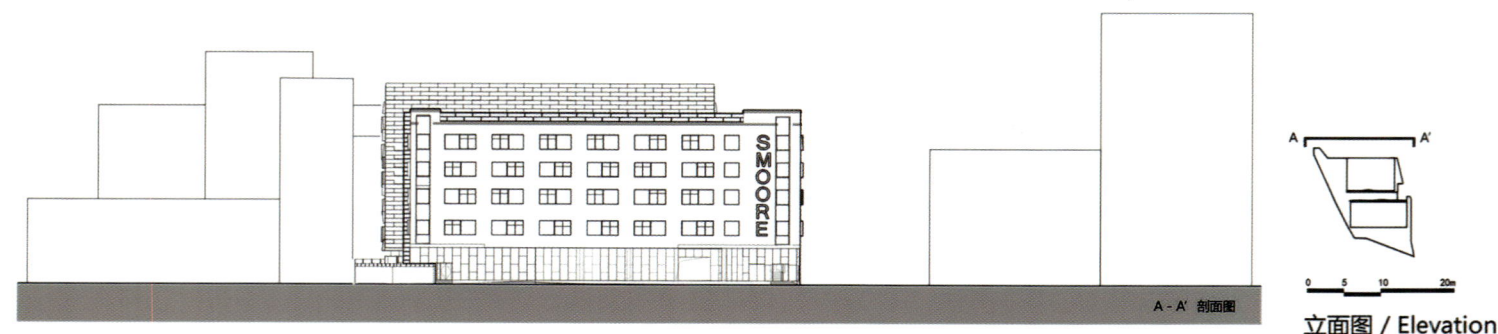

立面图 / Elevation

The wall on the north side of the park was removed to form an outdoor square. The single-story dilapidated building at the edge was reinforced and converted into electric room and garbage station. A completely new boundary of the park was integrated with the facade of the first floor of the buildings. Badminton and basketball fields as well as rest stools were arranged in the site to provide a leisure and fitness place for the employees.

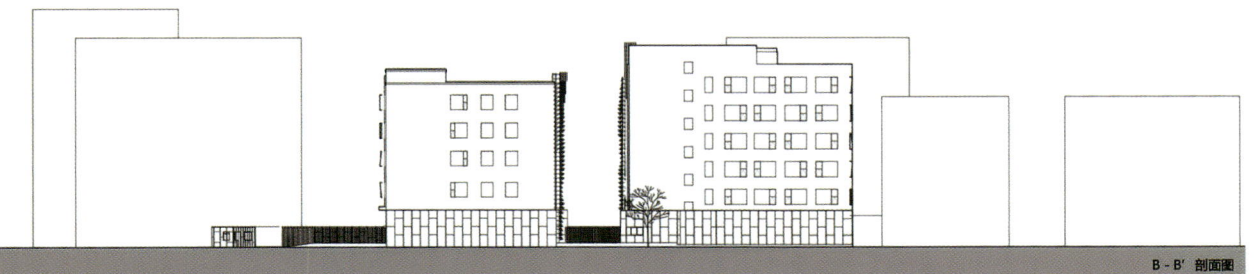

B - B′ 剖面图

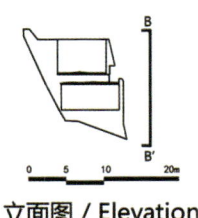

立面图 / Elevation

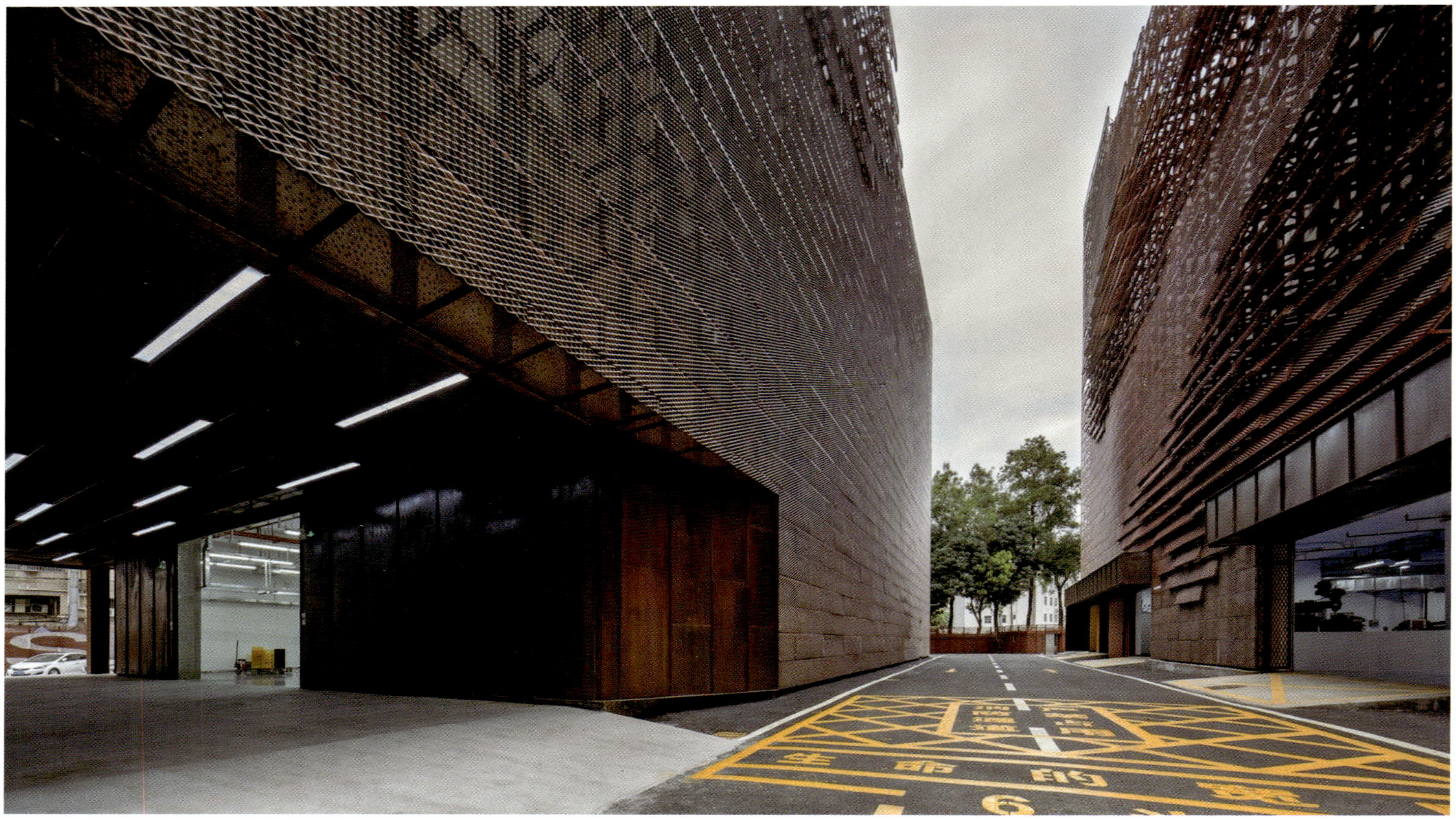

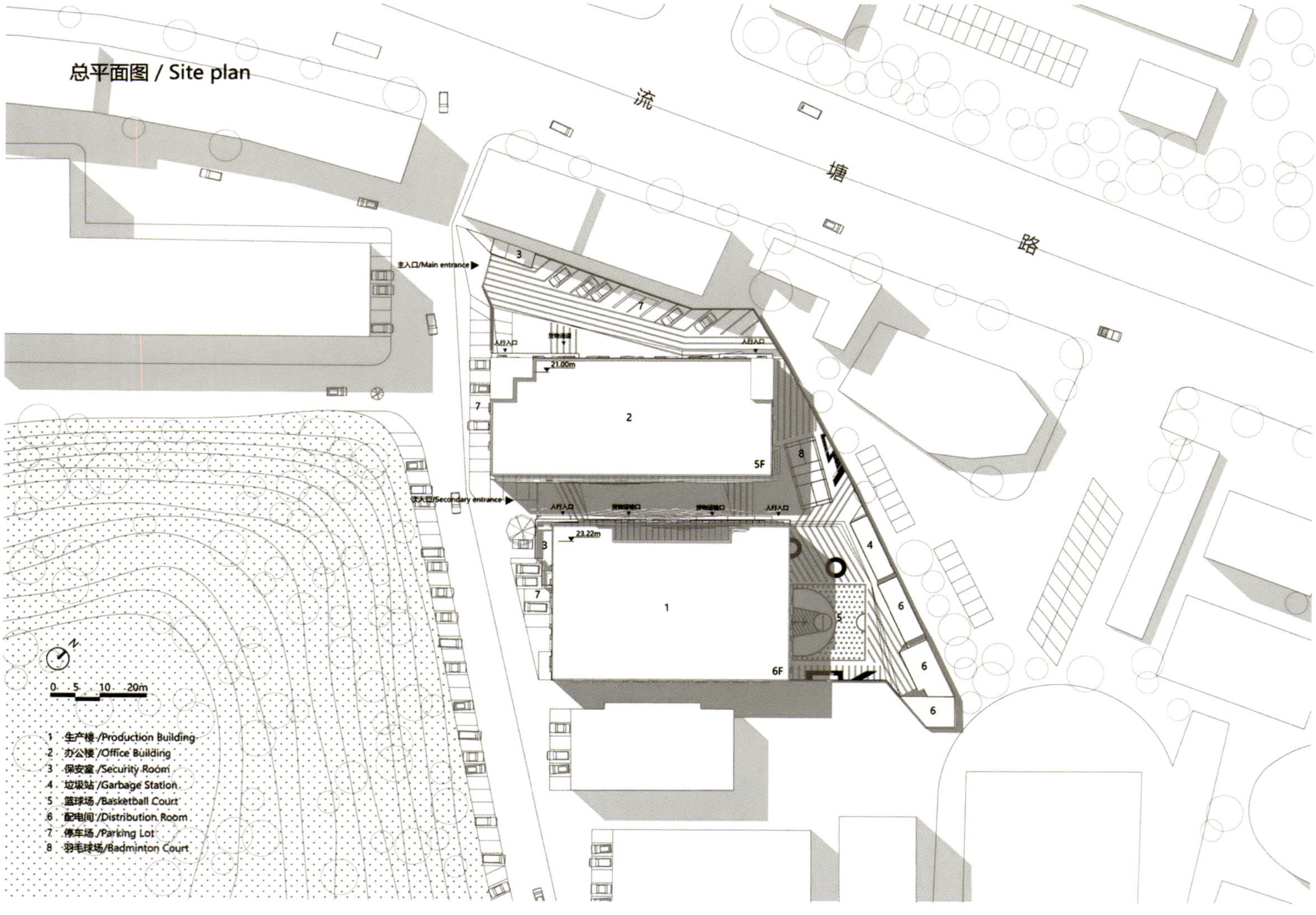

总平面图 / Site plan

1 生产楼 /Production Building
2 办公楼 /Office Building
3 保安室 /Security Room
4 垃圾站 /Garbage Station
5 篮球场 /Basketball Court
6 配电间 /Distribution Room
7 停车场 /Parking Lot
8 羽毛球场 /Badminton Court

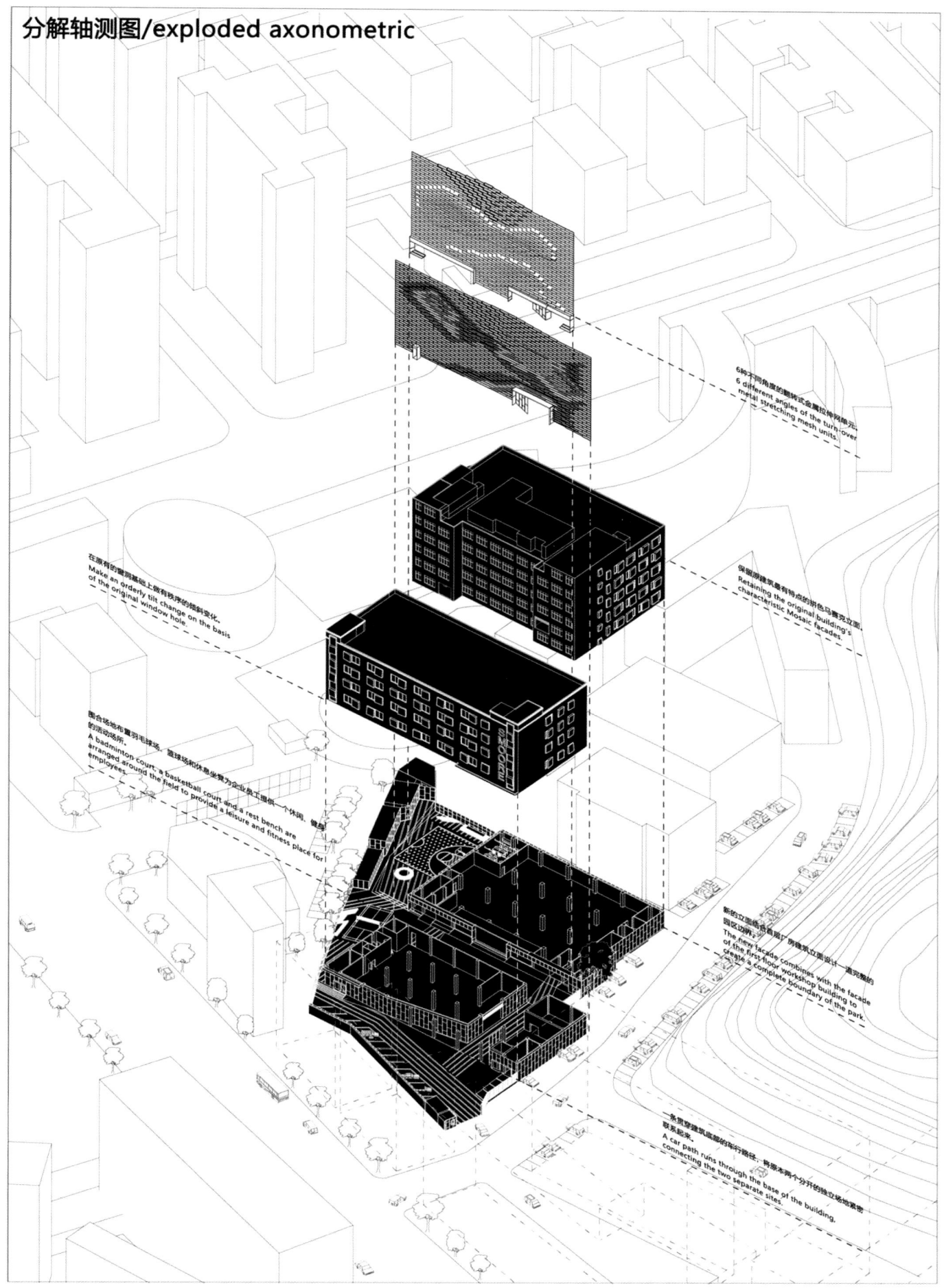

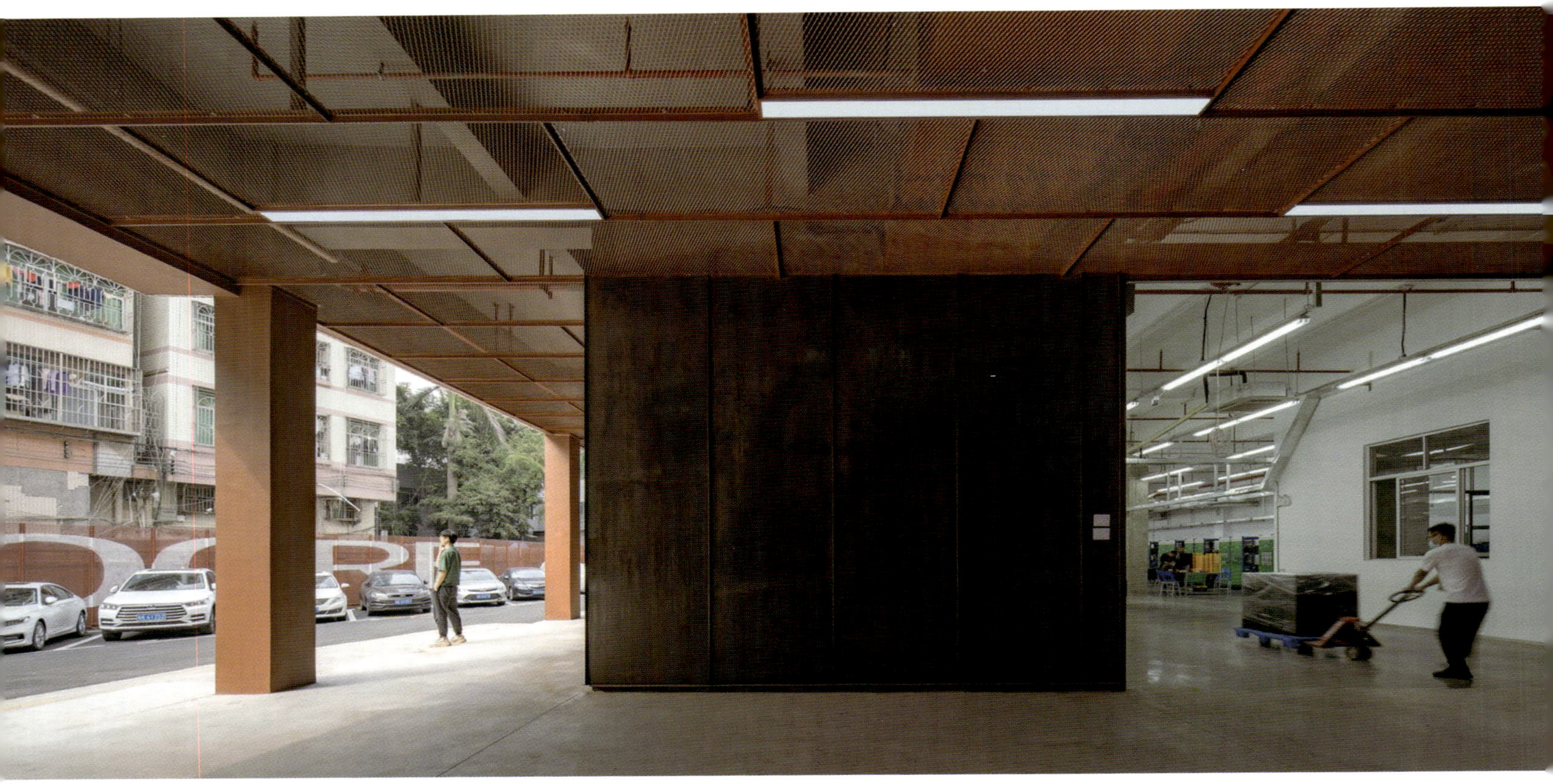

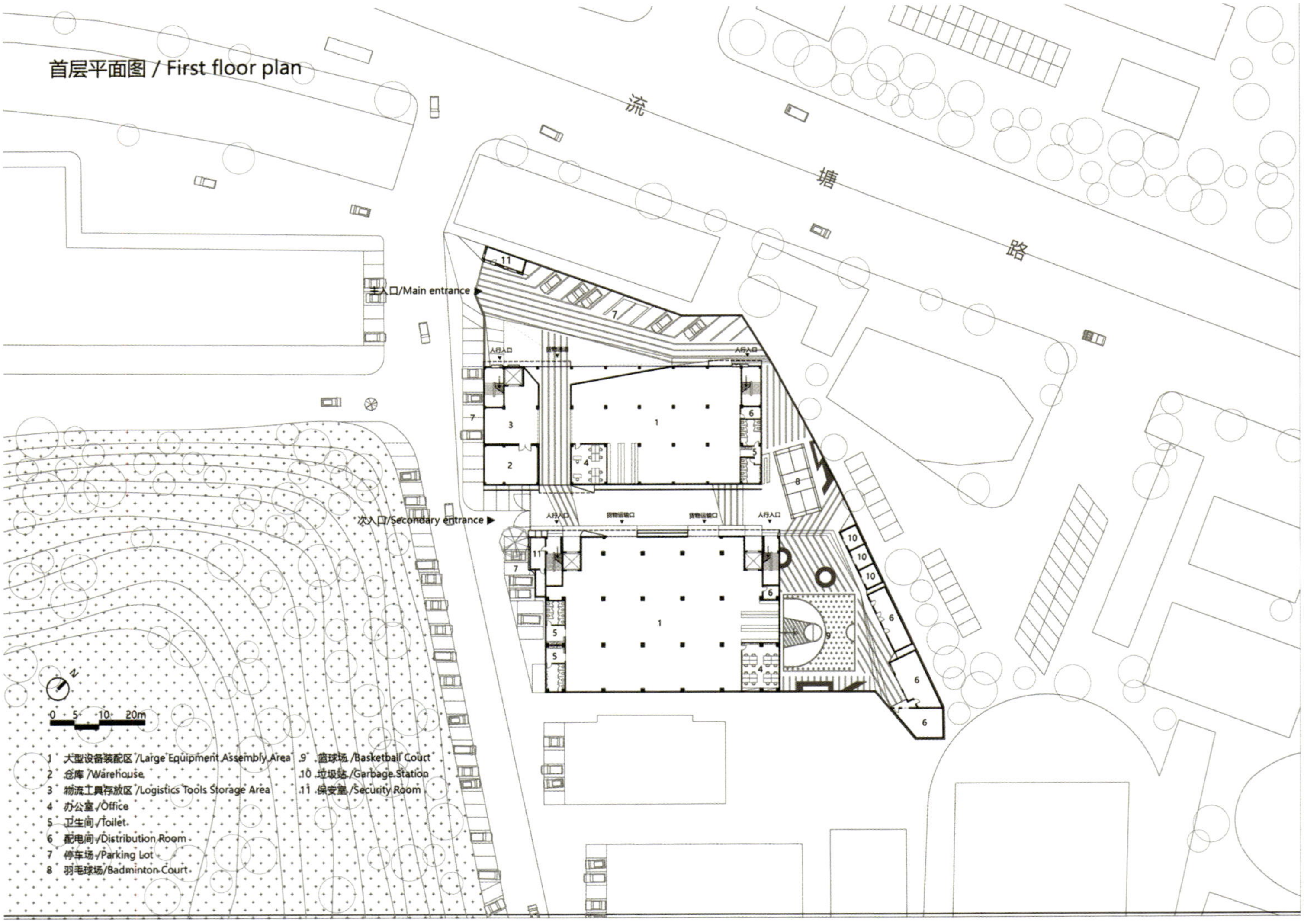
首层平面图 / First floor plan
流
塘
路
主入口/Main entrance
次入口/Secondary entrance
0 5 10 20m
1 大型设备装配区 /Large Equipment Assembly Area
2 仓库 /Warehouse
3 物流工具存放区 /Logistics Tools Storage Area
4 办公室 /Office
5 卫生间 /Toilet
6 配电间 /Distribution Room
7 停车场 /Parking Lot
8 羽毛球场/Badminton Court
9 篮球场 /Basketball Court
10 垃圾站 /Garbage Station
11 保安室 /Security Room

0 5 10 20m

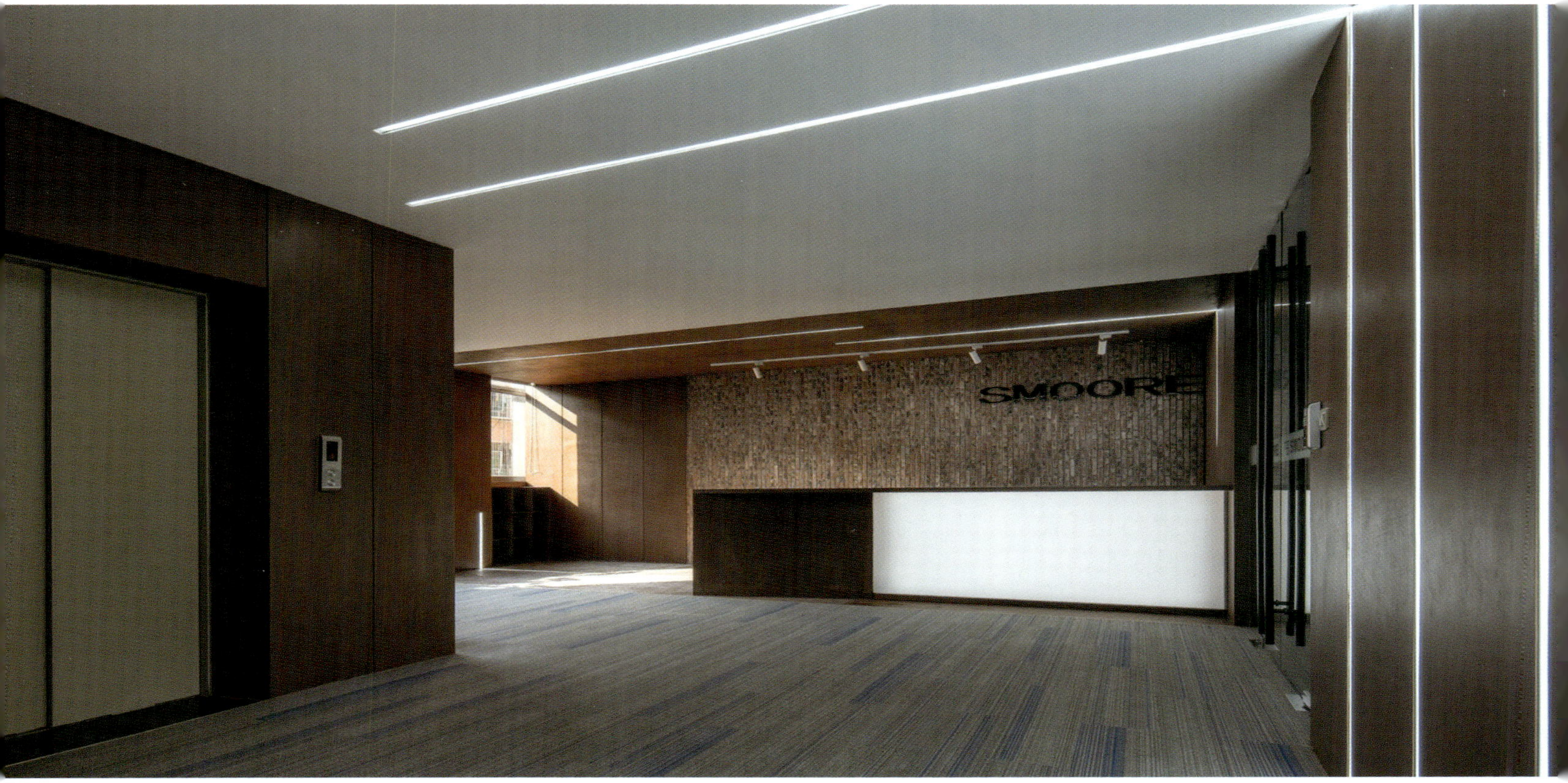

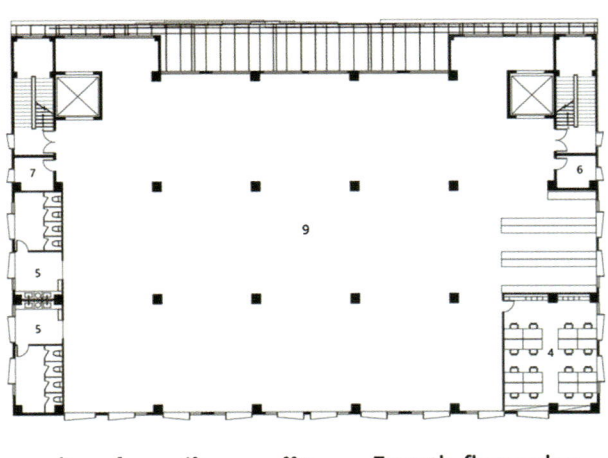

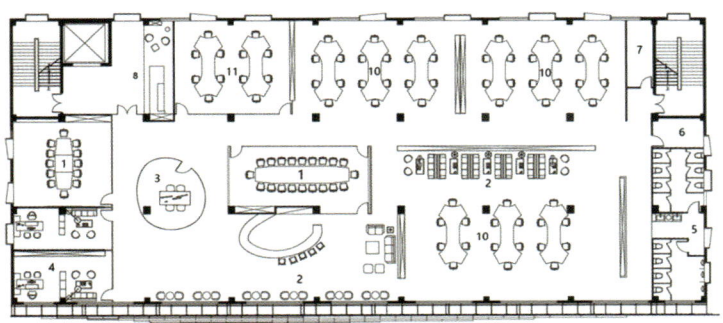

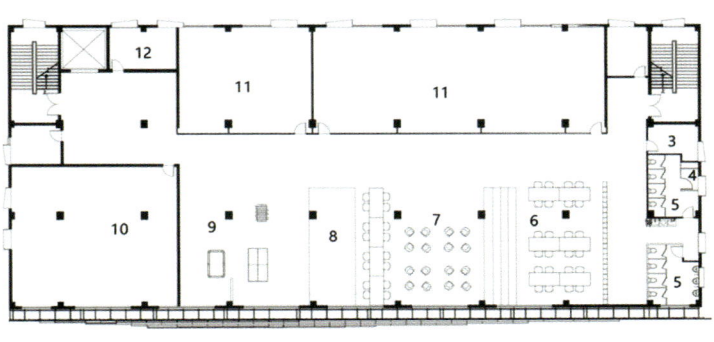

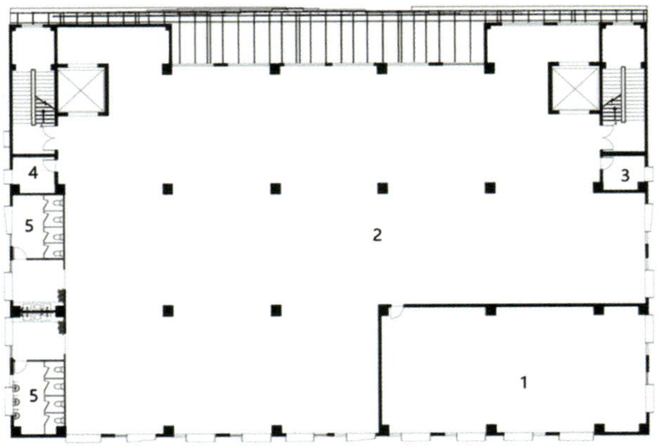

Fourth floor plan

平面图 / Fifth floor plan

0 5 10 20m

1 会议室 /Conference Room
2 休息区 /Rest Area
3 洽谈室 /Negotiation Room
4 办公室 /Office
5 卫生间 /Toilet
6 配电间 /Distribution Room
7 储藏间 /Storeroom
8 入口门厅/ Entrance Hallway
9 小型设备装配区/Small Equipment Assembly Area
10 开敞办公区 /Open Office Area
11 办公区 /Office Area

1 展示中心 /exhibition center
2 生产车间 /Production WorkShop
3 配电间 /Distribution Room
4 储藏间 /Storeroom
5 卫生间 /Toilet
6 办公区 /office area
7 会议室/conference room
8 水吧/water bar
9 娱乐区 /entertainment district
10 资料室 /resource center
11 多功能厅/hunction hall
12 导播间 /director room

Runxuan Textile Office Building

Architects: Masanori Design Studio
Location: Foshan, China
Area: 250 m²
Photographs: Yun Ouyang

Chief Designer: Terry Xu
Design Team: Gorry Huang, Gavin Peng
Main Material: Concrete(+ White aluminium bars

The project is located in Zhangcha, which is a famous textile town in Foshan, China. Most villages here have an industry park, and textile workshops are scattered throughout the town. Rationale International — Masanori Design Studio was entrusted to conceive a workspace for RUNXUAN TEXTILE, which hoped the new office could help enhance its brand image and competiveness in the market. The space occupies the first floor of a property. The facade of the entire building are available for the brand, but the property owner required that facade renovation should ensure ventilation of the second floor and respect the existing tile cladding.

Interweaving of Straight Lines and Curves; The curvilinear bottom of the building implicitly expresses the flexibility of cloth. The curve of the front facade is consistent with the logo of the brand, thereby creating a visual highlight and impressing visitors at first sight. Both the exterior and interior are dominated by white hue, as pure as cotton. The large French window reveals the pure interior, which is endowed with a sense of layering by spatial structures.

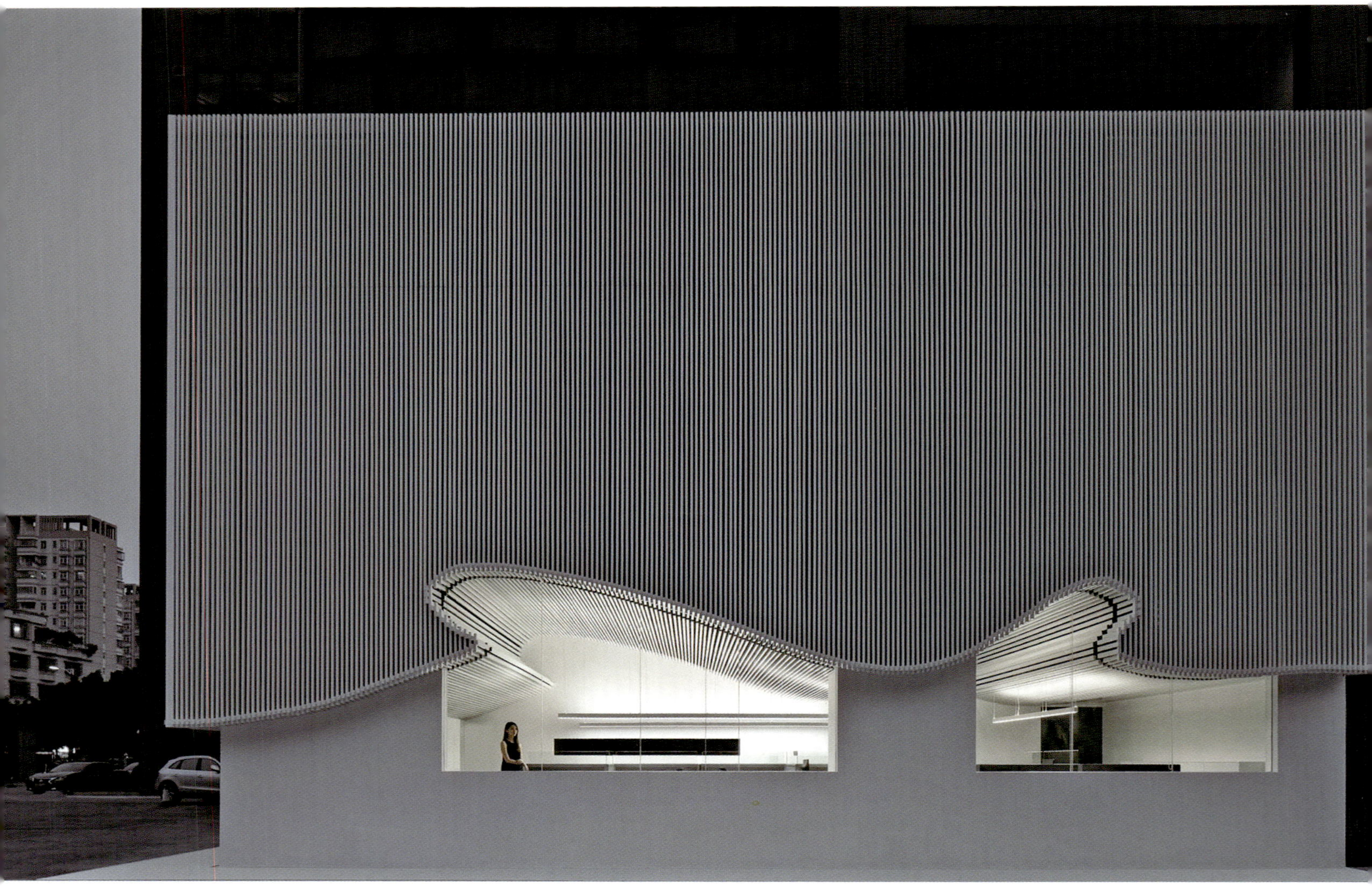

White aluminium bars extend from the top of 2F facade to the ceiling of the first floor. Looking like dense, orderly rows of yarn on looms, those neat lines envelop the entire building. The bottom ends of external aluminium bars form an undulating curve, which echoes with the curving form of the interior ceiling. In this way, aluminium bars integrate the top area of the building facade and the 1F ceiling into a whole. The design interprets cotton and yarn straightforwardly, while the imagery of cloth is expressed in a relatively euphemistic manner. The designers innovatively applied cloth elements to both the facade and indoor space via architectural languages, hence producing three-dimensional visual effects and realizing the unity of interior and exterior.

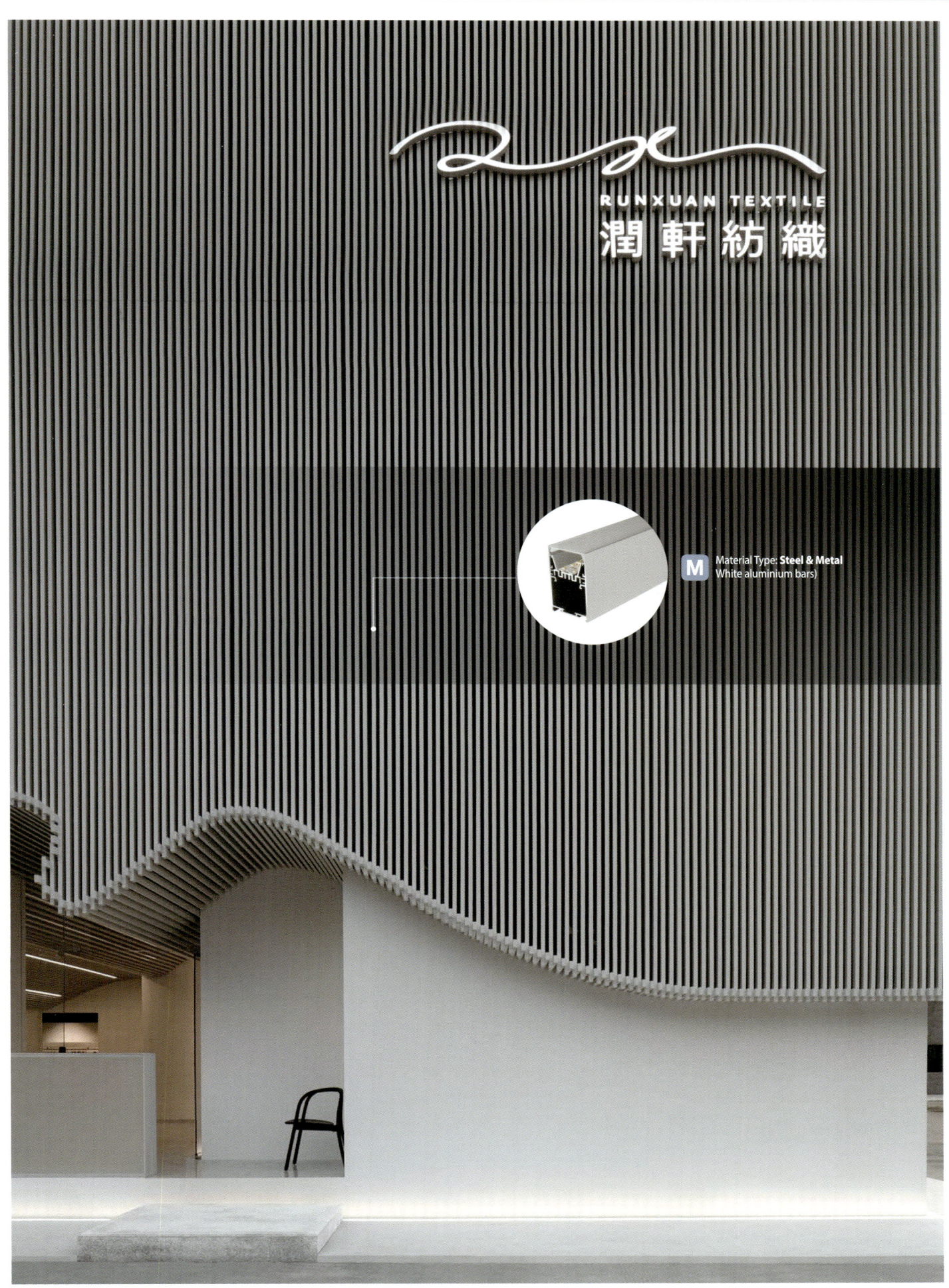

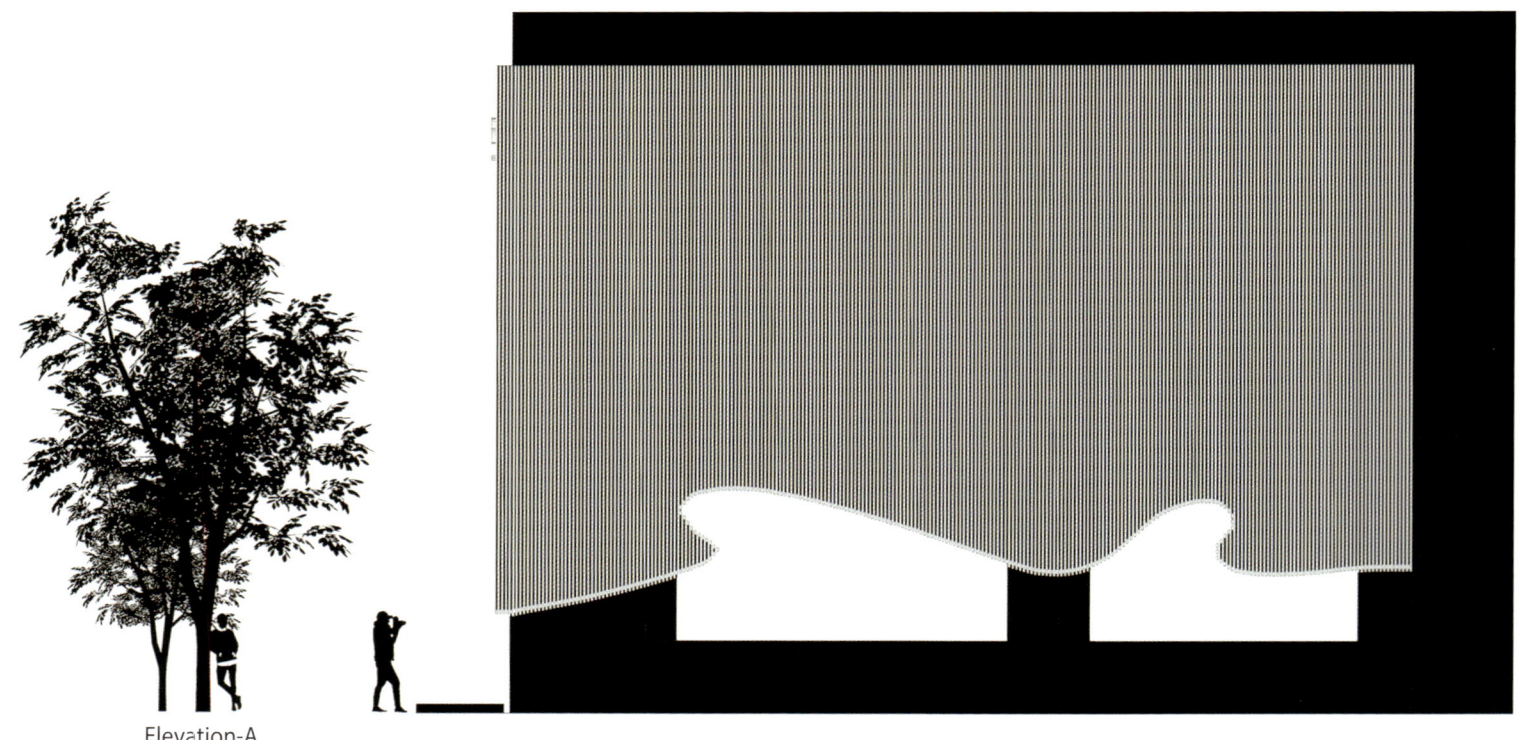

Elevation-A

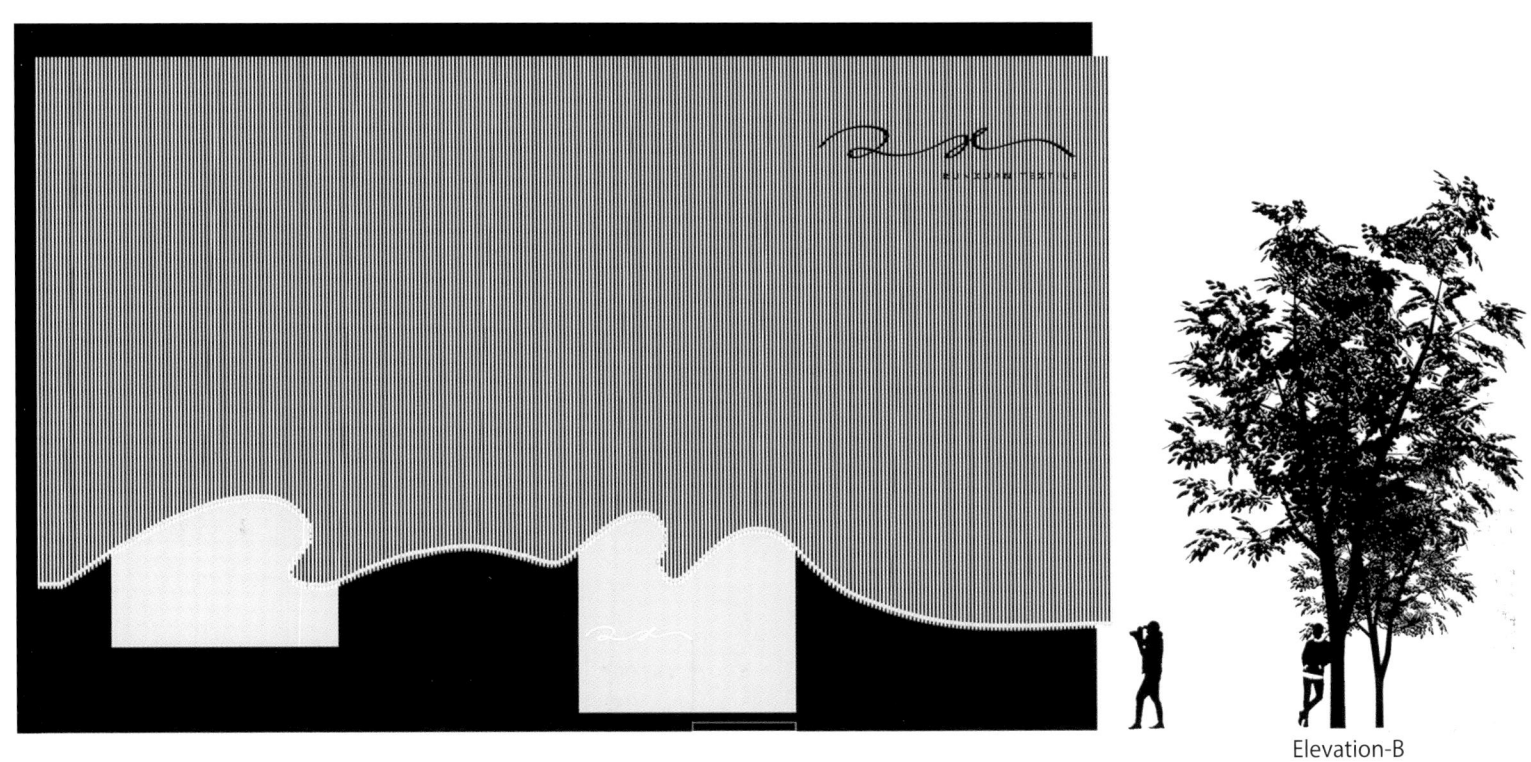

Elevation-B

Design Concept - A

Pure White Tone: The white tone that extends from the facade to interior becomes a visual highlight. Besides, the undulating ceiling gives the space a strong sense of architecture. Meticulous details, connection between ceiling and walls, columns, and play of light and shadows, bring various changes to the space. Minimalist white office furniture blends with floor and ceiling, and is complemented by green plants and black chairs dotted throughout the space. The functional layout is clear, which creates a sense of order in the free white-hued space.

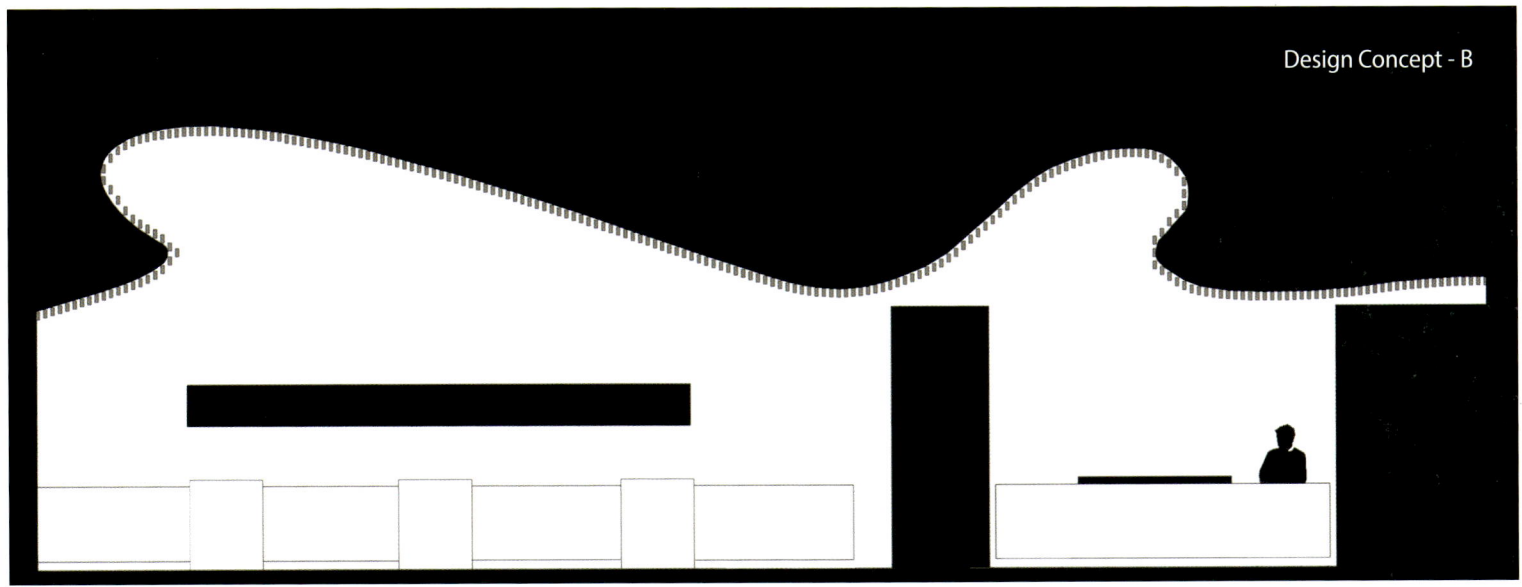

Design Concept - B

The aluminium structure is also used in the interior, forming an undulating three-dimensional ceiling that shapes like the mountain or wave, which gives people a strong sense of natural atmosphere and vitality. Such design adds interest inside and helps people to adjust the mood when working. In this space, one seems like standing under a loom. The "yarns" are interwoven with the light, and such black and white colours form artistic scenes from different angles.

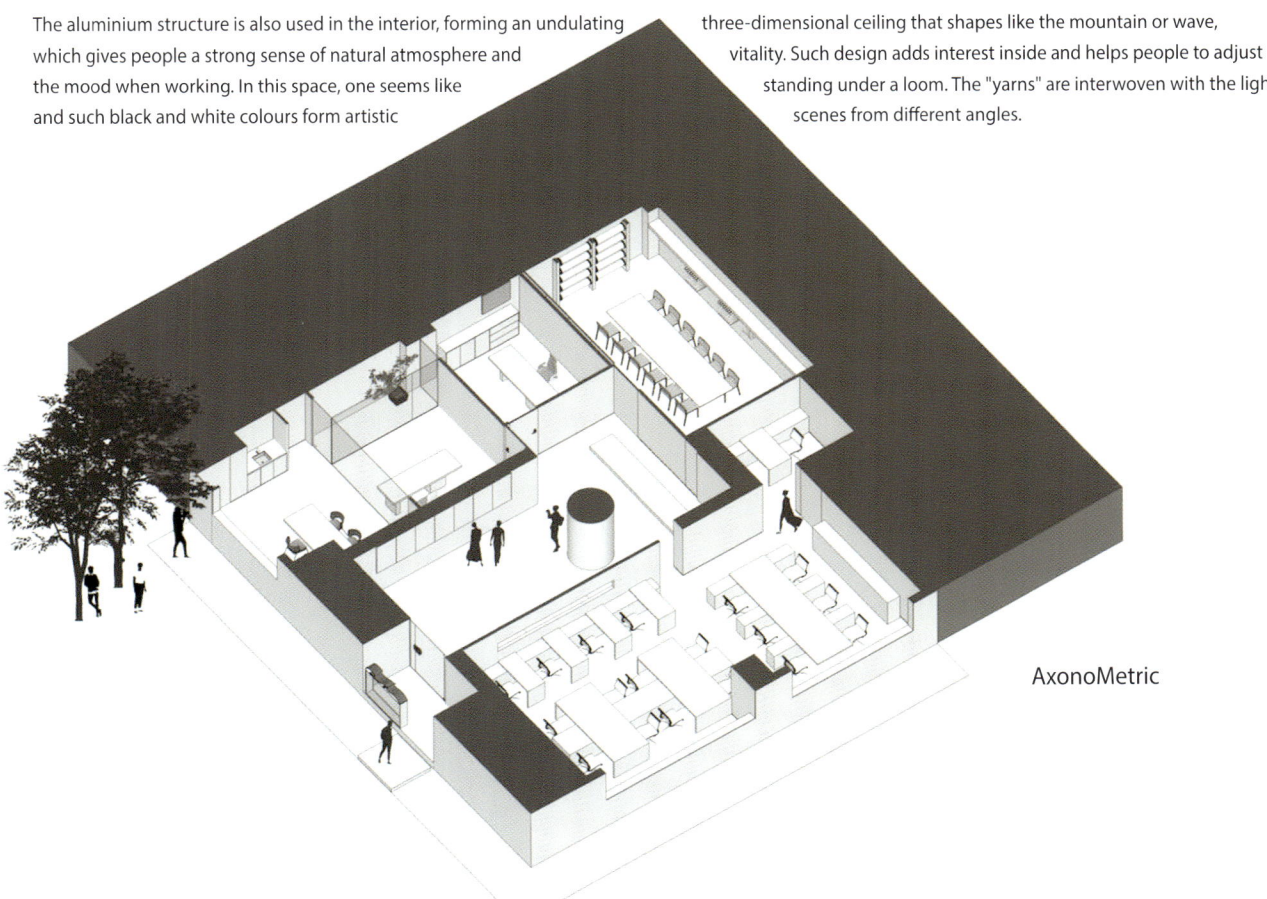

AxonoMetric

To meet the client's needs, the designers integrated the brand connotation into the working space. Moreover, by optimizing the indoor functional layout and solving its lighting and ventilation problems, they ingeniously created a healthier and more comfortable office, which became another feature of this project. A reception room and two independent offices were orderly set on the left section inside the entrance. The main problem was the lack of natural light and ventilation. To deal with it, the designers created a separate passageway at the side of the three rooms, and made an artificial "skylight" via a stretch ceiling, which seems to bring natural light in. As the passage connects the three rooms, when the back door of the reception room is opened, the wind can be directly introduced into the two office rooms.

RUNXUAN TEXTILE

Design Concept - C

Landscape House

Architects: FORM | Kouichi Kimura Architects
Location: Japan
Area: 182 m²
Photographs: Norihito Yamauchi
Chief Designer: Design:Kouichi Kimura
Main Material: Polymer Concrete + mortar finishing + Metal Pannels

Located in a relatively spacious residential area, the house has been built with the client-run hair salon attached. There's a park across the front road from the site. The area allows to closely view Mt. Ibuki, one of 100 famous Japanese mountains. In such conditions, I came up with the architecture that would bring out good living environment while laying out the volumes of the requested shop and of the residential part in an appropriate proportion. The building is composed of symmetrical volumes for which the proportion, openings, and materials have been carefully considered. The dynamic form creates a new façade in the streetscape. The first floor consists of the hair salon, residential entrance hall, and bathroom.

LDK (living, dining and kitchen area) is laid out as well as bed rooms and children's rooms on the second floor. The LDK has varied ceiling heights and has been planned to make the room including the stair hall as one continuous space. It can be loosely divided with curtains as necessary. To add visual depth, an opening has been provided to the wall in the back, reinforcing the sequential space which stretches in the longitudinal direction. To the volume which is projected from the independent wall that characterizes the façade, a high-side light has been provided, blocking eyes from the environment and clipping the magnificent view. In addition, the low wall below the window has been recessed, creating a niche which makes you feel huddled and cozy.

C Material Type: **Concrete**
(Polymer Concrete + mortar finishing)

M Material Type: **Steel & Metal**
(Zinc panelss)

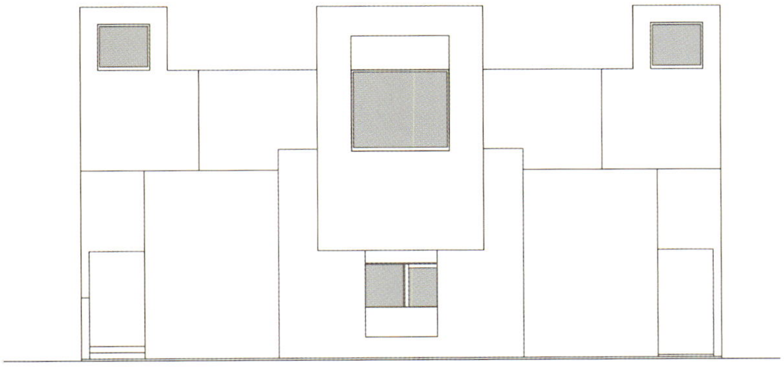

North Elevetion

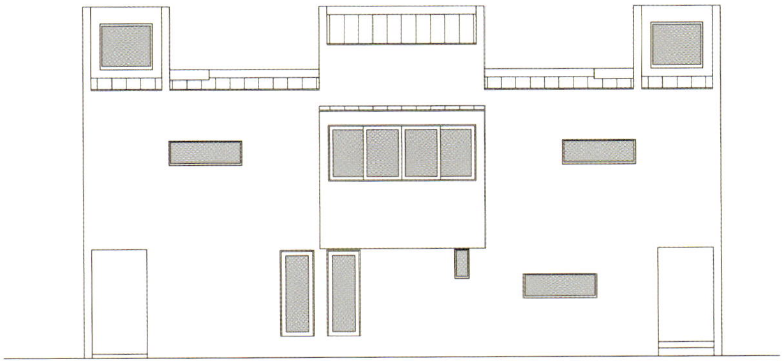

South Elevetion

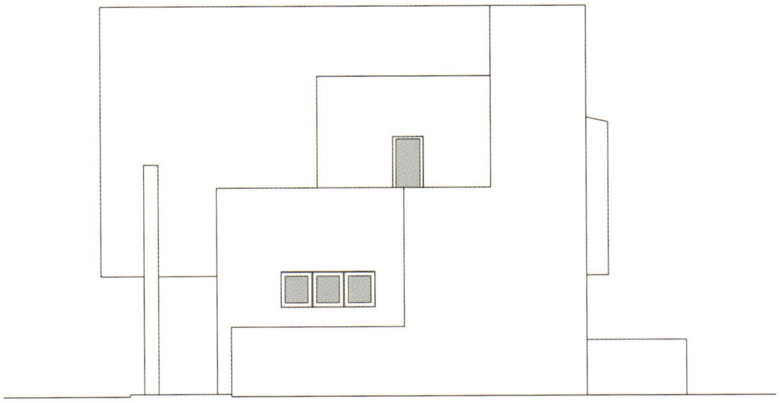

West Elevetion

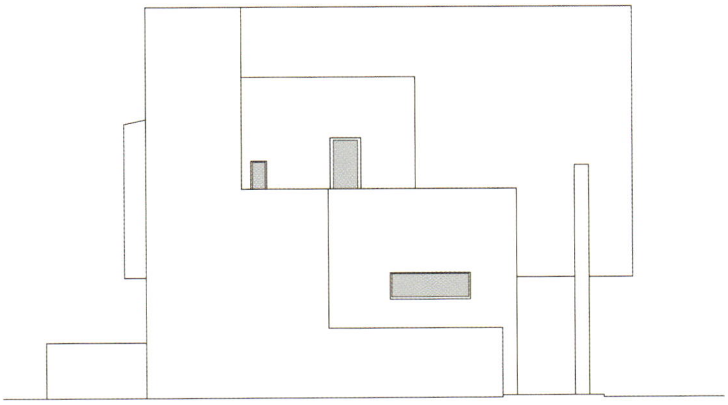

East Elevetion

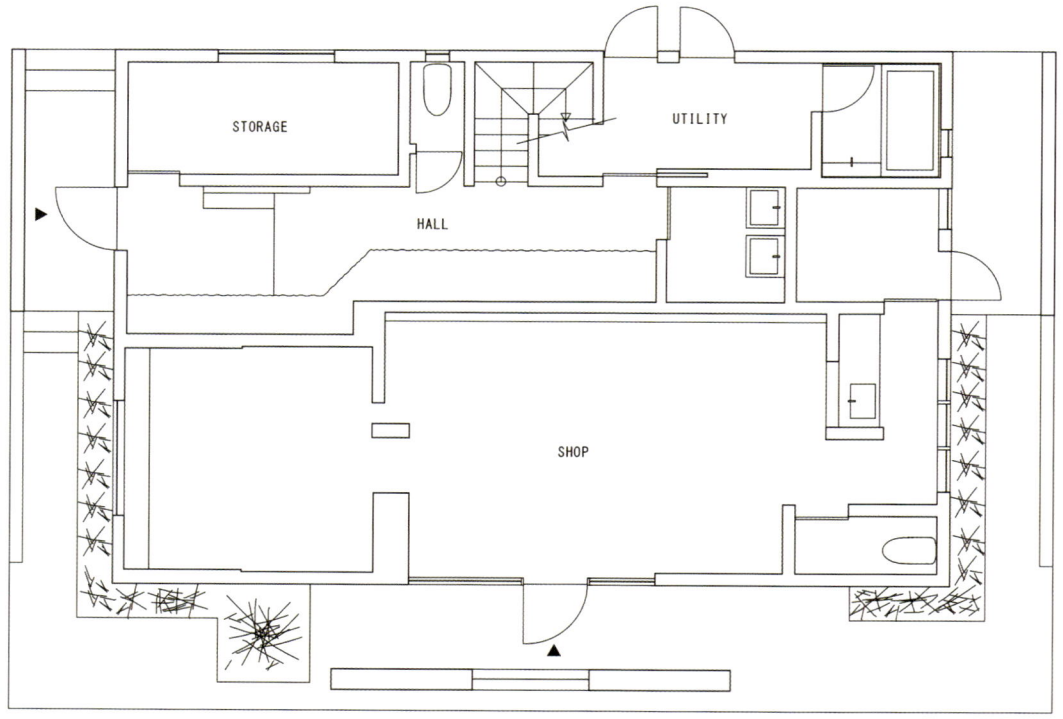

First Floor

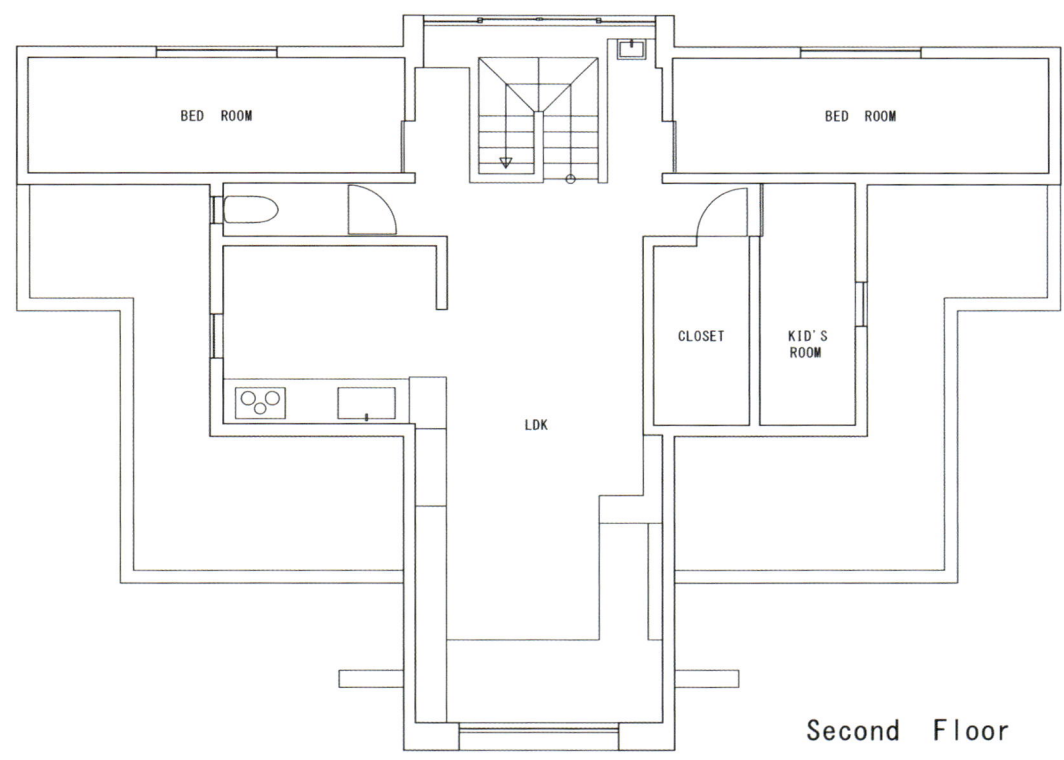

Second Floor

The modern-colored tiles on the wall along with the view from the high-side light produce a beautiful and expressive space. Also provided to the stair hall wall in the back is a large opening that allows to feel light and breeze near the window. The space trimmed with the mortar window frame can be flexibly used as a display or study. The view seen through the lace curtain and gently streaming light create emotional scenes. This house teaches us that it is the good relationship between external environment and internal space that gives openness or calmness to a residence with a good location.

Shenzhen Pingshan Art Museum

Architects: Vector Architects
Location: Shenzhen, China
Area: 47269 m²
Photographs: Shengliang Su, Chao Zhang

Design Principal: Gong Dong
Project Architects: Yue Han, Peng Zhang, Jinteng Li
Main Material: Polymer exposed concrete + Aluminum vertical bar

Shenzhen Pingshan Art Museum is situated on the boundary of an urban spatial transitioning — to the west, a high density neighborhood for typical urban living; to the east, a large scale urban park. The Art Museum is arranged along the north-south direction, parallel to the border on this long block In terms of the spatial arrangement on site, we fragmented the architectural volume to distribute the various functional spaces of the museum at different levels. The spaces are stacked vertically, allowing us to set up a multilevel public platform system that renders the architecture penetrable and porous. On one hand, this kind of spatial structure avoids the blockage that traditional centralized volume normally imposes on urban traffic.

On the other hand, the elevated ground floor also builds up a continuous spatial experience through the urban blocks into the museum. We placed the main entrance of the Art Museum on the ground floor, and introduced the daily commercial spaces into the periphery of the museum, with the intention to cultivate a more quotidian spatial atmosphere. People could walk through the museum from the residential area to the park at any given time during the day, and wander up the stairs. The promenade extends from the first floor to the second floor's platform, leading the urban dwellers to the entrance lobby on the second floor and other retail spaces.

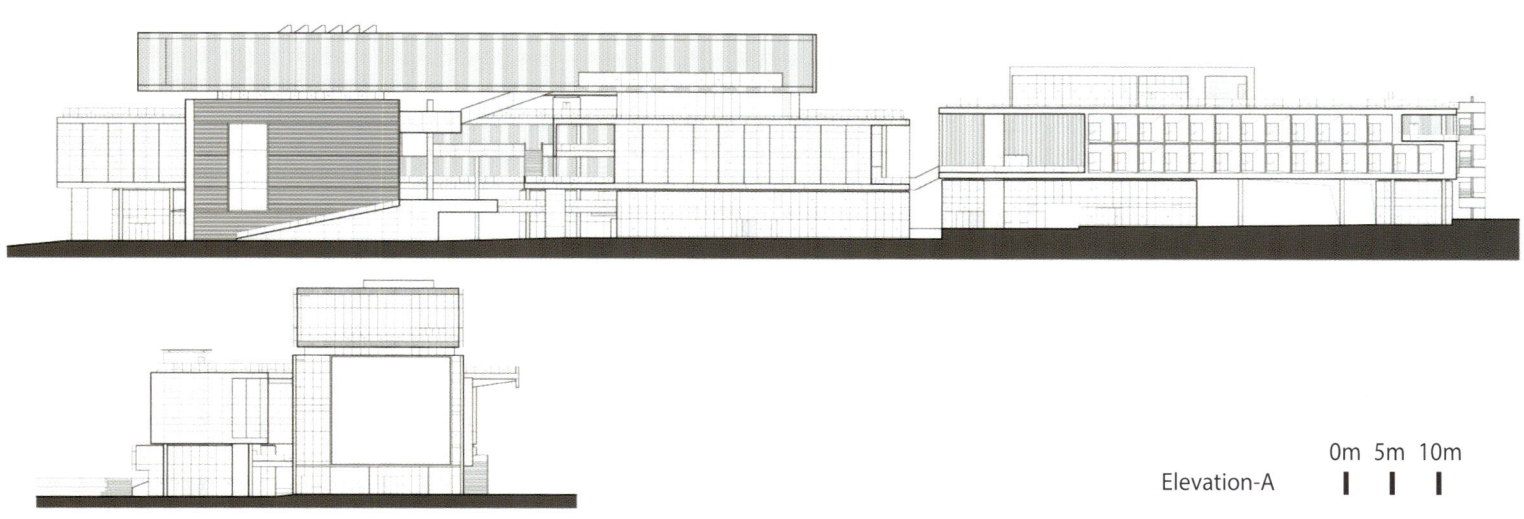

Elevation-A   0m 5m 10m

Material Type: Steel & Matal
(Aluminum vertical bar)

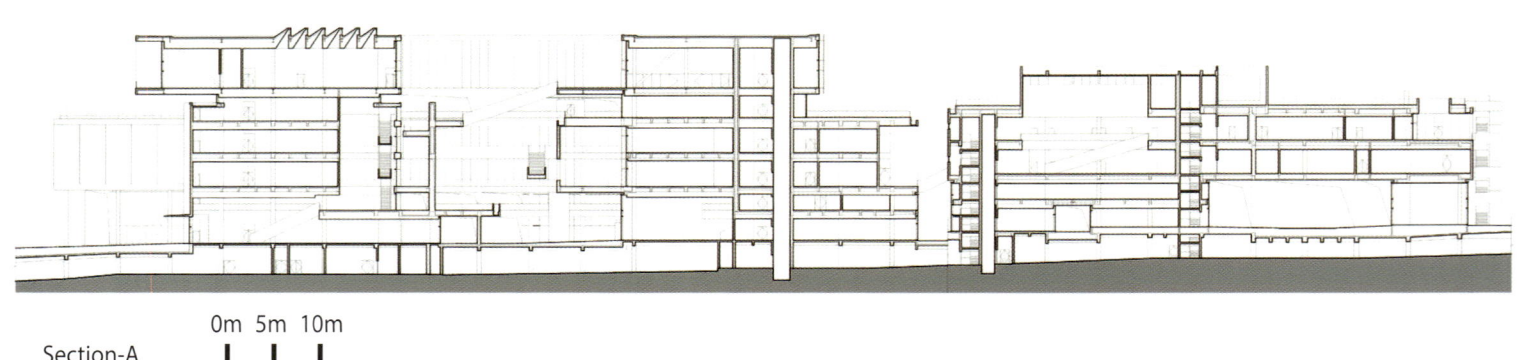

Section-A   0m 5m 10m

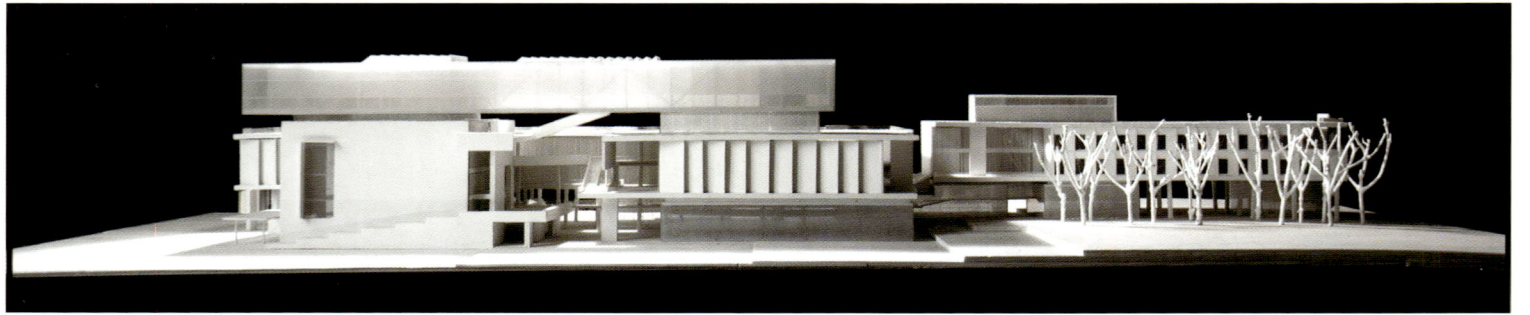

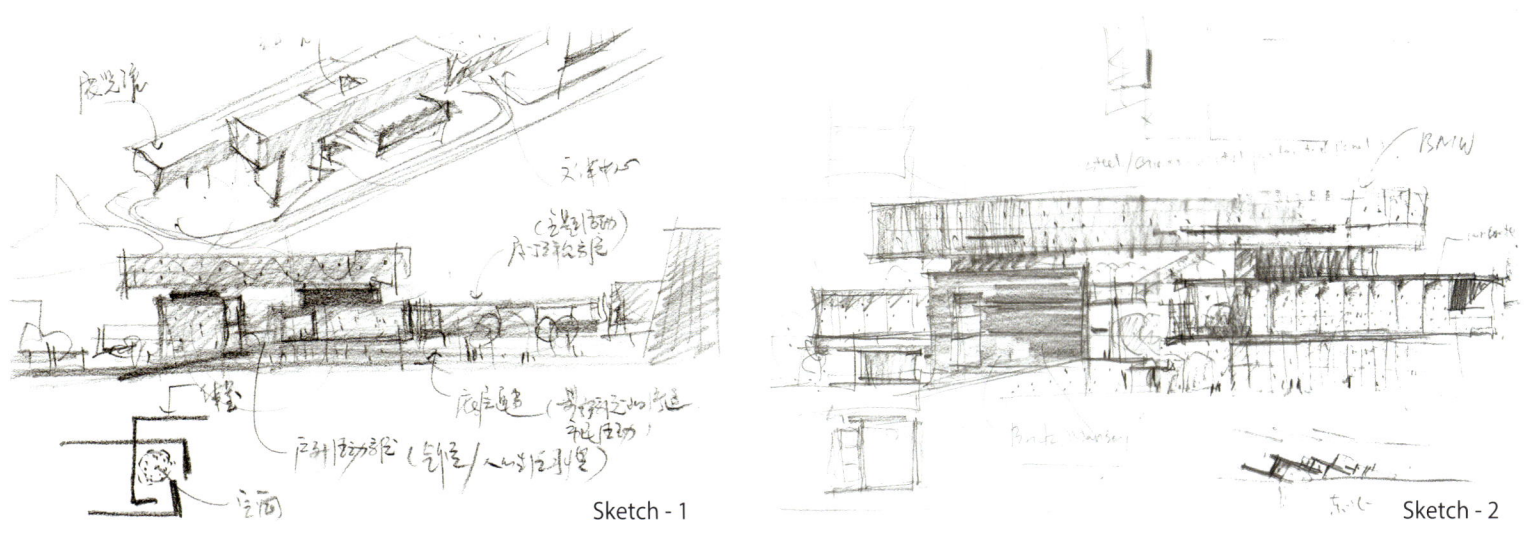

Sketch - 1   Sketch - 2

We hope that, outside of the operational hours of the Museum, the building complex and its public functions could also be open to the whole city for longer periods of time. Unlike the traditional interior circulation for exhibitions, we utilized the outdoor public platforms above the second floor to string together various exhibition spaces, with the trees interspersed along the way. The spaces under the overhangs provide ventilation as well as shelter from the sun and rain, specifically for the sub-tropical climate of Shenzhen. These transitional spaces, to the great extent, open up the sight bringing in the nature neighboring the site.

Material Type: **Concrete**
(Polymer exposed concrete)

In our imagination, the Art Museum would fully absorb the crowds along either side, as an interface that undertakes the daily life in the neighborhood. Furthermore, through our design, we hope to establish a new order of public spatial sceneries and to foster a linkage between the city and nature.

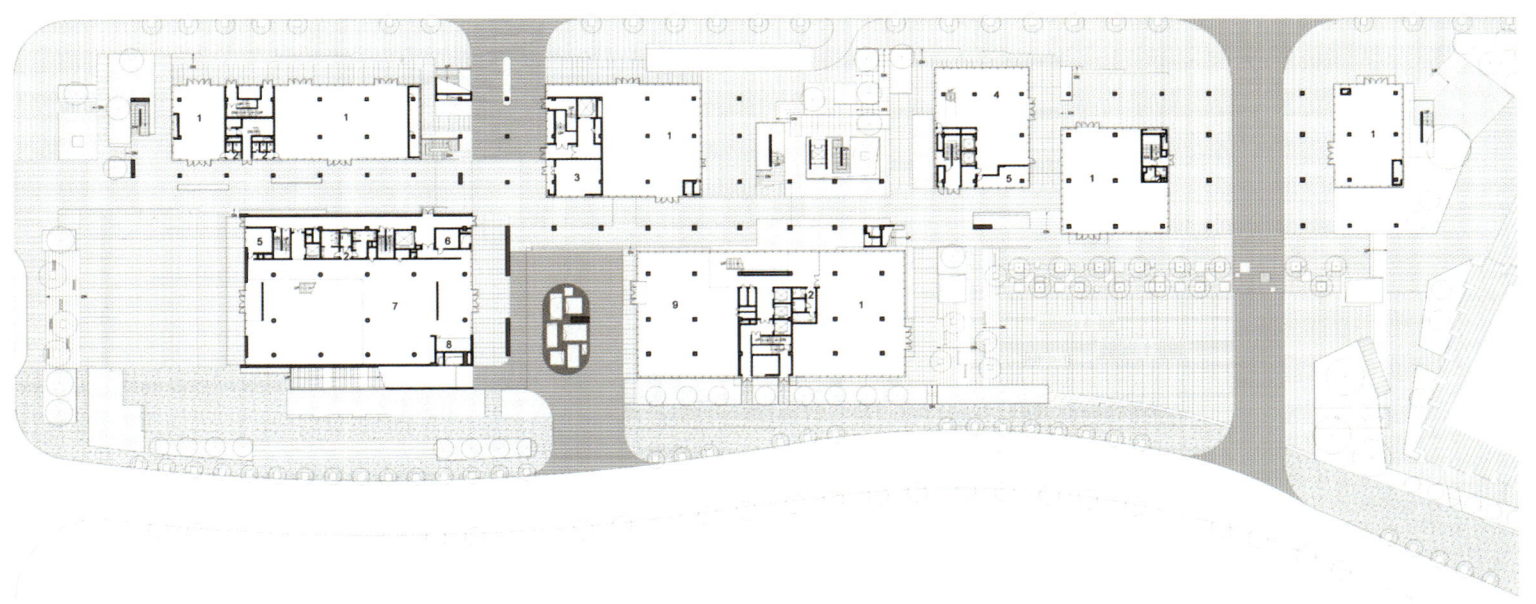

1 Retail Area	4 Lobby of Cultural and Creative Cener	6 Monitor Room	9 Lobby of Exhibition Hall
2 Toilet	5 Office	7 Lobby of Art Museum	
3 Fire Control Center		8 Baggage Deposit	

First Floor Plan

Wall Section 02

1. Silver Gray Metal Plate with Fluoro-Carbon Paint
2. Invisible Frame Curtain Wall with Ultra Clear Acid-Etched Glass
3. Silver Gray Steel Mullion with Fluoro-Carbon Paint
4. Cast-in-Place Fair-Faced Concrete
5. Balustrade with Ultra Clear Glass Panel
6. Bamboo Flooring
7. Silver Gray Metal Ceiling with Fluoro-Carbon Paint
8. Invisible Frame Curtain Wall with Ultra Clear Glass
9. Silver Gray Aluminum Alloy Mullion with Fluoro-Carbon Paint
10. Fabric Banner

Modeling

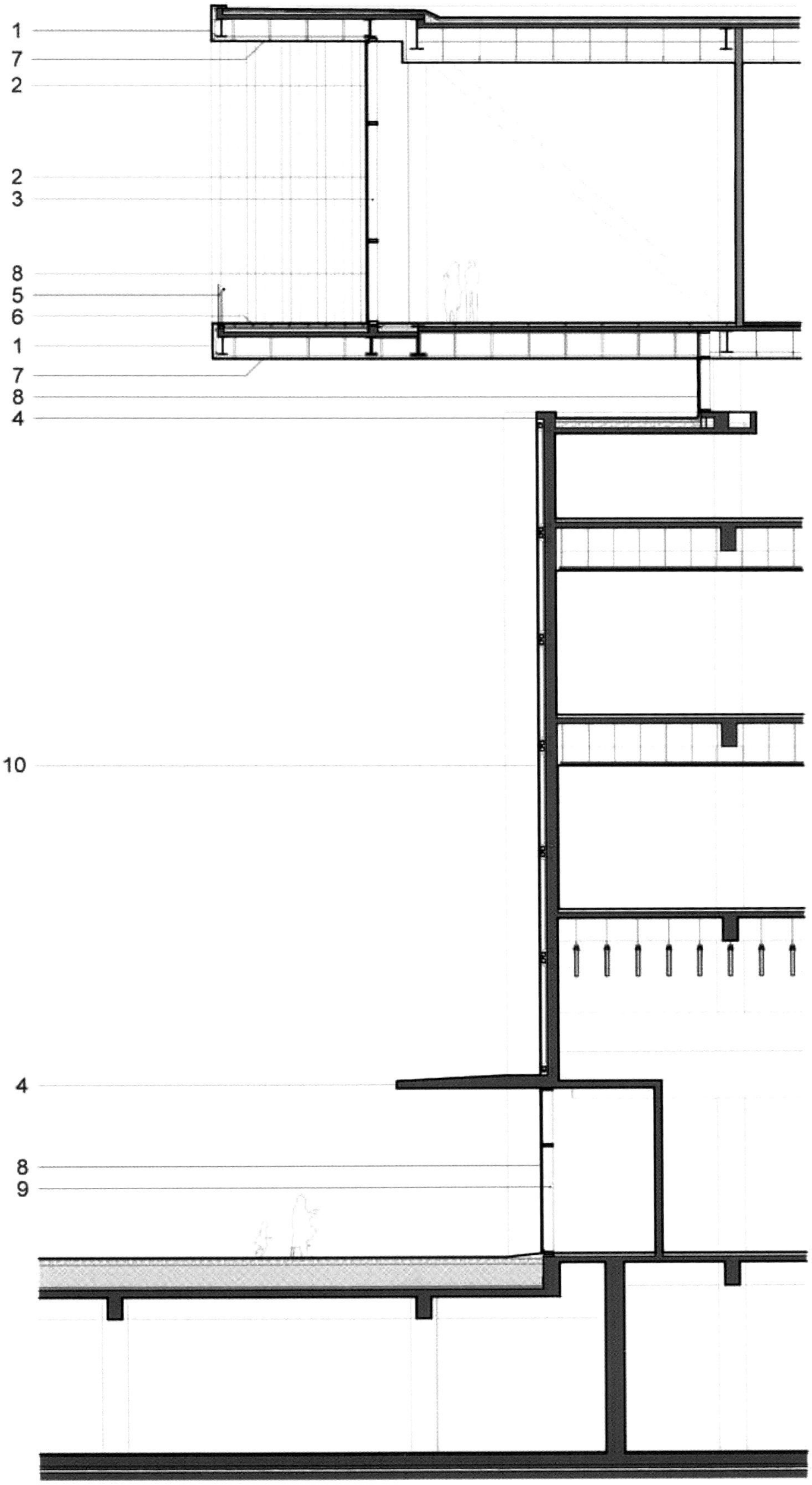

Detail Plan - A

Wall Section 01

1. Silver Gray Metal Plate with Fluoro-Carbon Paint
2. Invisible Frame Curtain Wall with Ultra Clear Acid-Etched Glass
3. Silver Gray Steel Mullion with Fluoro-Carbon Paint
4. Silver Gray Aluminum Alloy Lattice with Fluoro-Carbon Paint
5. Climbing Plants
6. Bamboo Planter
7. Cast-in-Place Fair-Faced Concrete
8. Balustrade with Ultra Clear Glass Panel
9. Bamboo Flooring
10. Tree with Planter Box
11. Invisible Frame Curtain Wall with Ultra Clear Glass
12. Silver Gray Aluminum Alloy Mullion with Fluoro-Carbon Paint

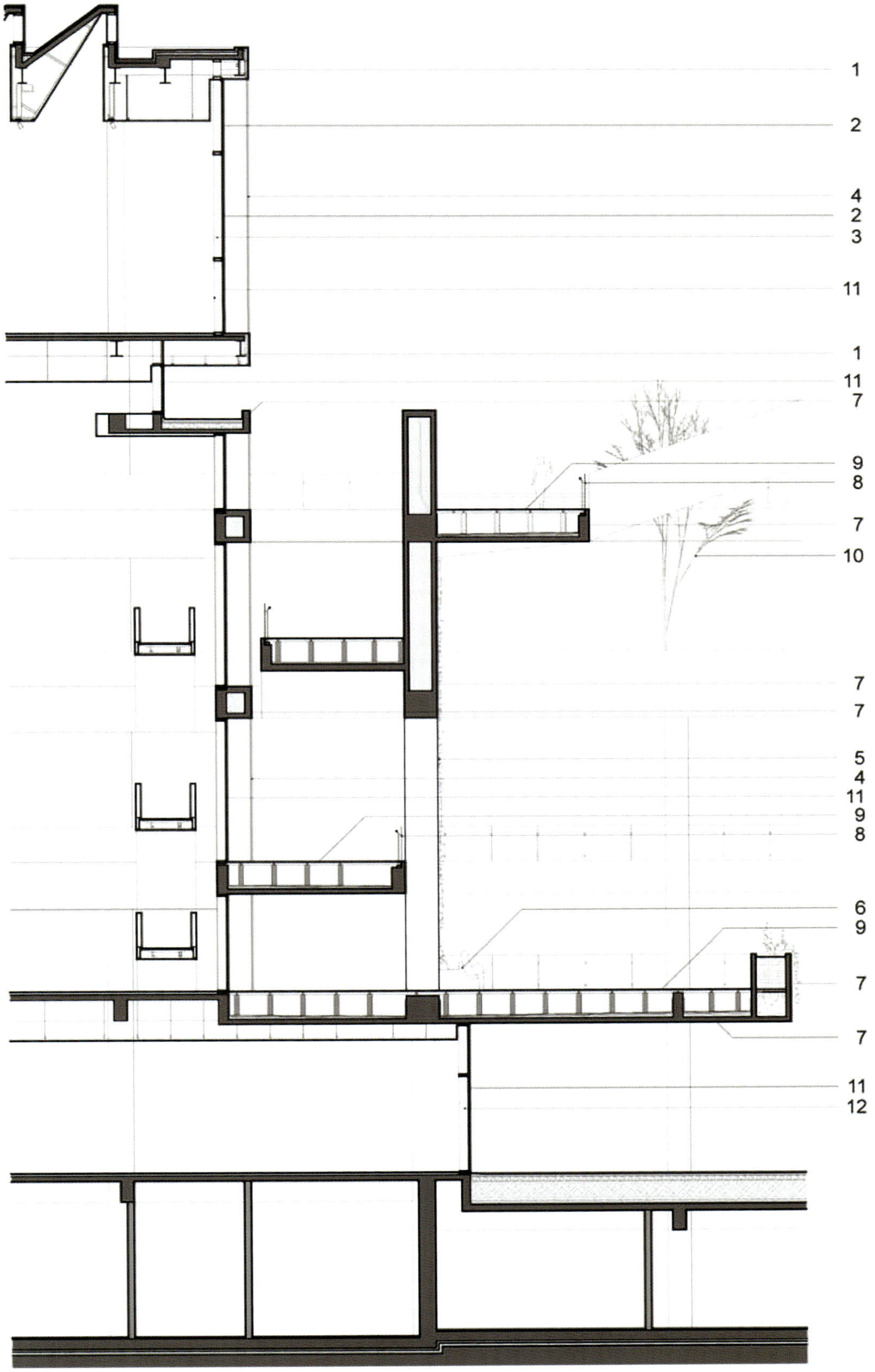

Detail Plan -B

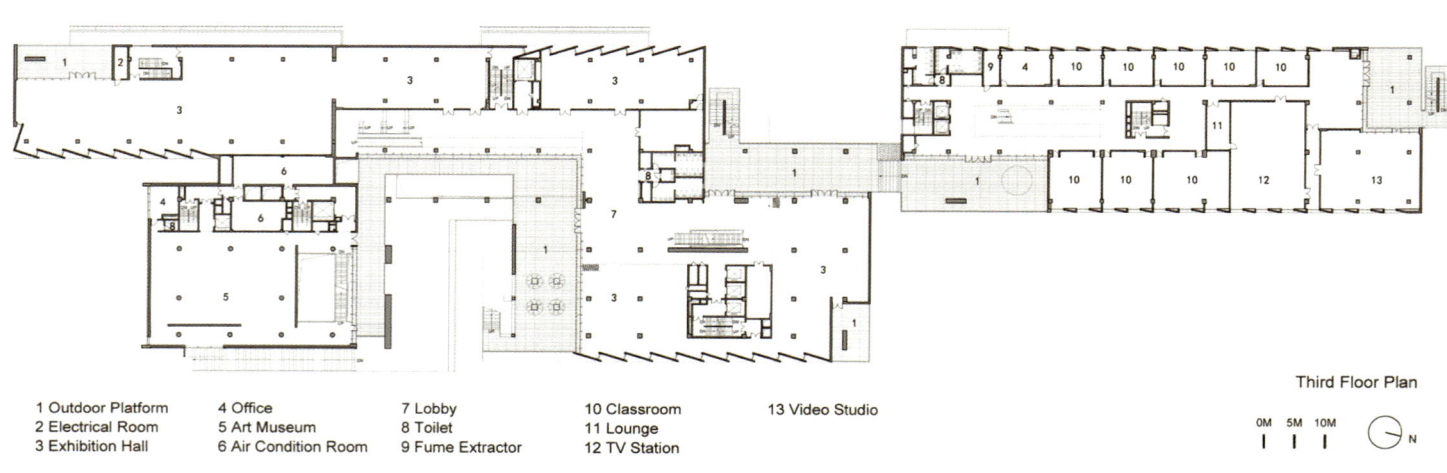

1 Outdoor Platform	4 Office	7 Lobby	10 Classroom	13 Video Studio		
2 Electrical Room	5 Art Museum	8 Toilet	11 Lounge			
3 Exhibition Hall	6 Air Condition Room	9 Fume Extractor	12 TV Station			

Third Floor Plan

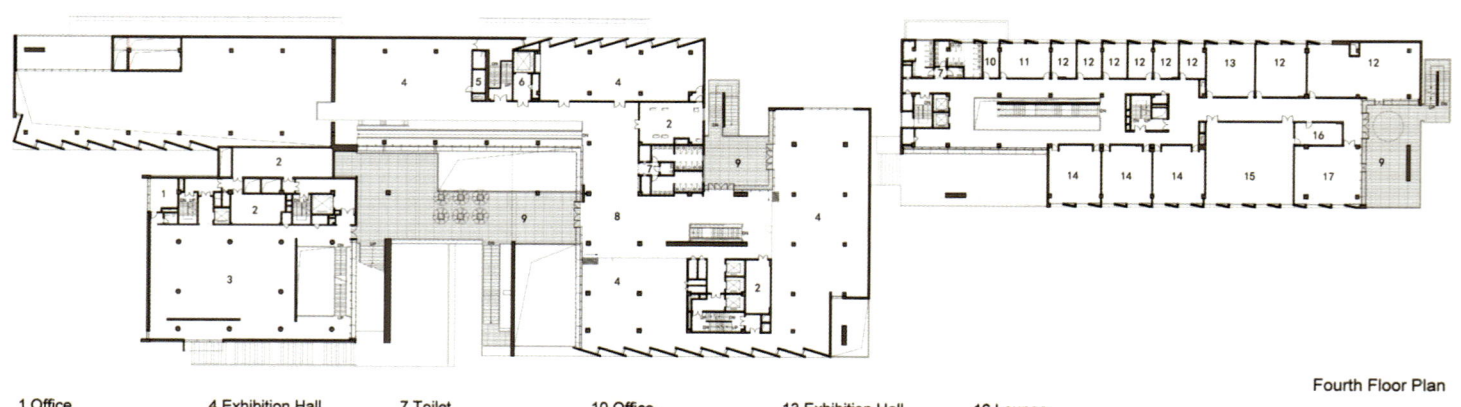

1 Office	4 Exhibition Hall	7 Toilet	10 Office	13 Exhibition Hall	16 Lounge
2 Air Condition Room	5 Electrical Room	8 Lobby	11 Document Room	14 Classroom	17 Chess Room
3 Art Museum	6 Freight Elevator	9 Balcony	12 Studio	15 Multi-purpose Hall	

Fourth Floor Plan

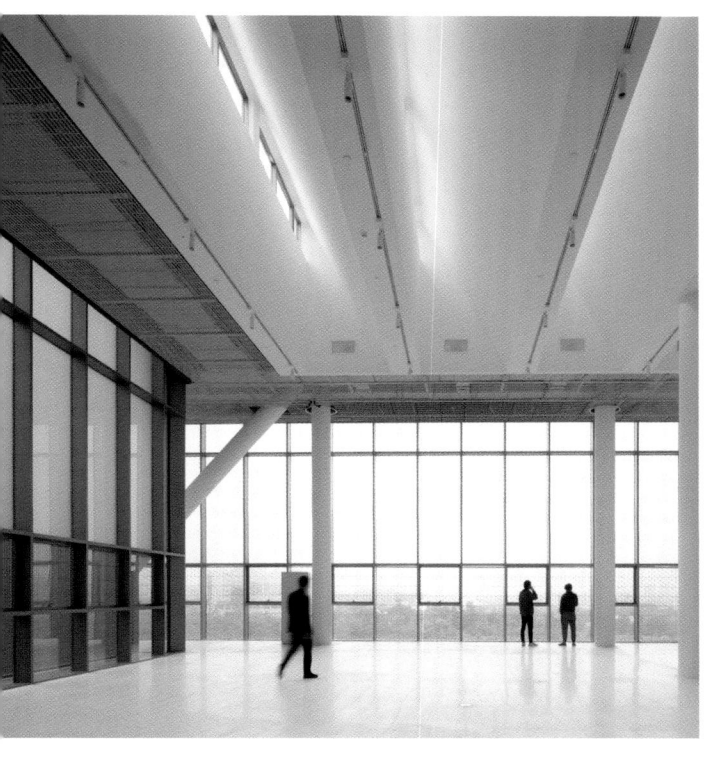

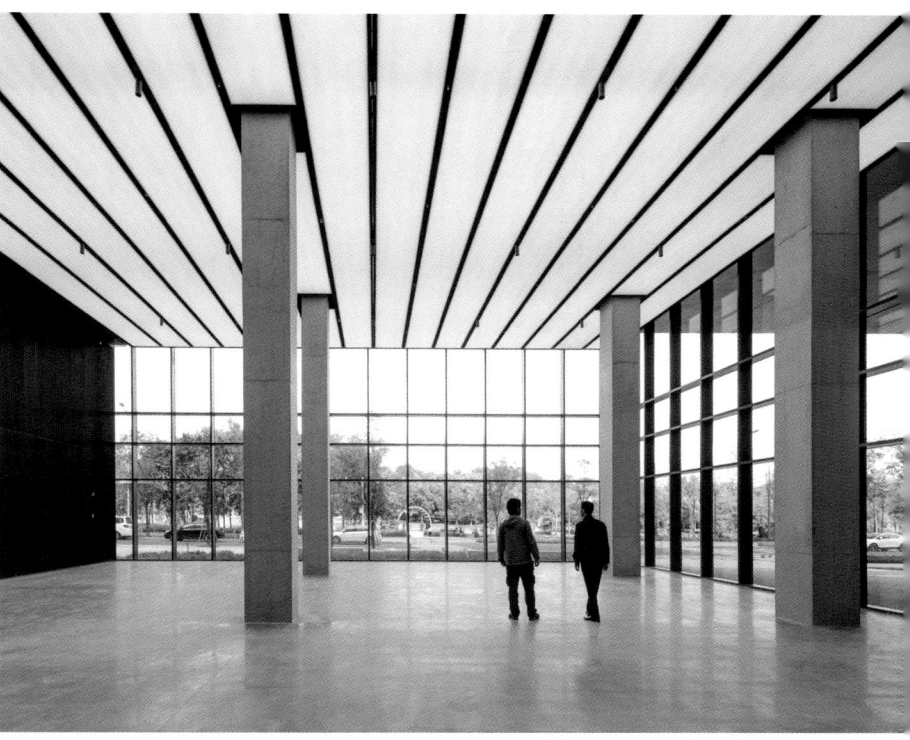

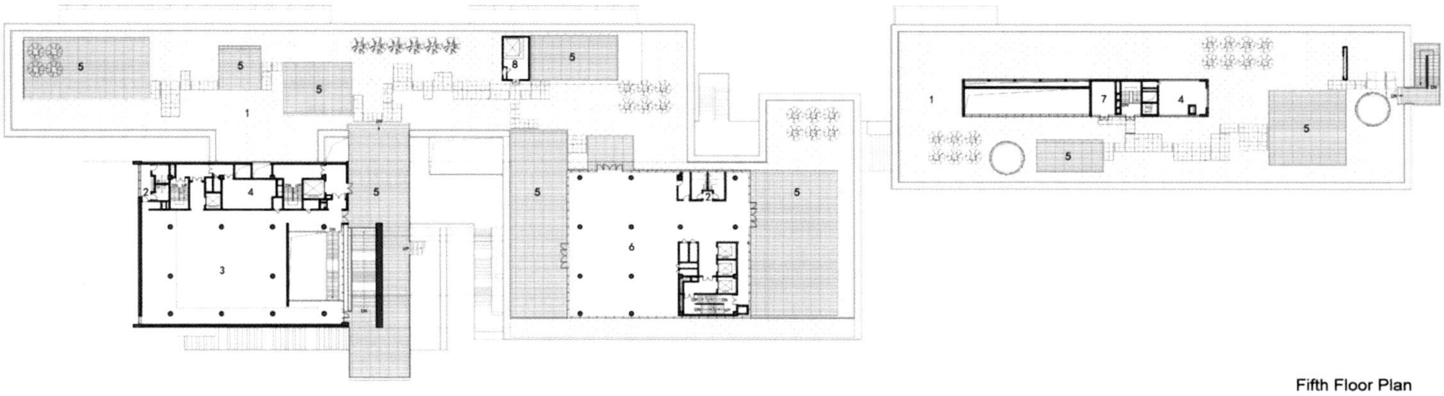

Fifth Floor Plan

1 Green Roof
2 Toilet
3 Art Museum
4 Air Condition Room
5 Outdoor Platform
6 Multi-purpose Hall
7 Air Condition Room
8 Elevator Machine Room

0M 5M 10M N

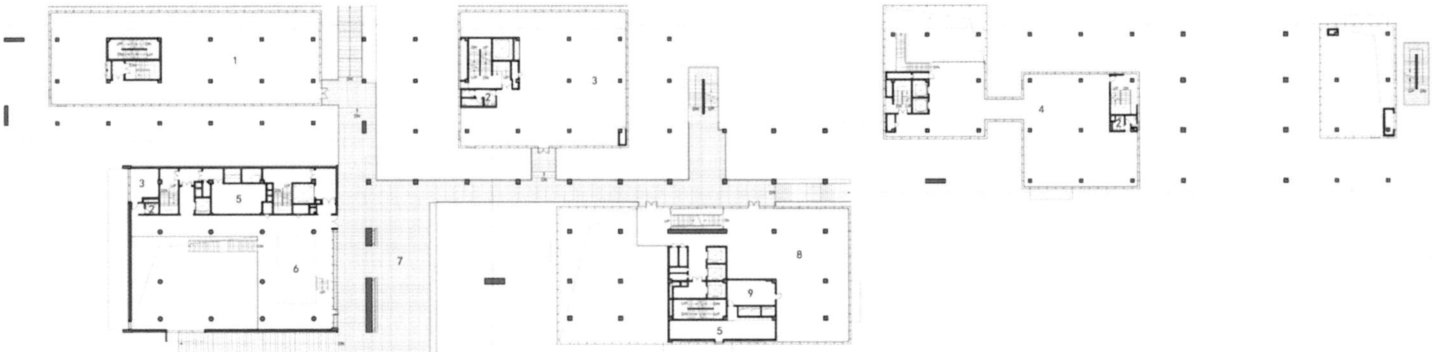

Second Floor Plan

1 Retail Area
2 Toilet
3 Office
4 Handcraft Studio
5 Air Condition Room
6 Art Museum
7 Outdoor Platform
8 Exhibition Hall
9 Guardhouse

0M 5M 10M N

Es Pou House in Formentera

Architects: Marià Castelló
Location: Spain
Area: 94 m²
Photographs: Clara Ott

Manufacturers: Firestone Building Products, Cerámica Mano Alzada, Cerámica la Andaluza, Cerámicas Ferrés, Corian, Diabla Outdoor, Marià Castelló, Architecture

Main Material: Concrete & Dryvit finish + Terracotta tiles + Ceramic lattices tiles (Mallorcan-style ceramic vaults and pressed terracotta tiles)

A rural plot where several pre-existent conditions the insertion of this small first residence in the territory. Among them, the network of centenarian dry stone walls stands out, as well as the organization of the crops. The intervention is located in the western area of the plot, parallel to a trace of more than a kilometer in length, oriented to the south and protected from the setting sun by a mass of vegetation, thus releasing the most fertile area to give continuity to the existing agricultural activity. The proposal is divided into three volumes, which order the program while providing it with a smaller grain and in accordance with the scale of the landscape. From south to north, the first volume houses a porch that offers solar protection, the second contains the more public program and the third two bedrooms.

Among them are transverse strips that physically separate the volumes, giving them ventilation and lighting, as well as providing them with services and connections. In front of the house, there is a cistern that makes it self-sufficient in terms of water supply while offering a solarium for the coldest months of the year.

From the inside and through the porch, deep perspectives are discovered towards the flat landscape of wheat and oat fields, where the soft and warm color of the earth and the muted greens of the almond and fig trees predominate. The light, color, and material from the outside enter the interior of the house thanks to ceramics and wood, two noble materials that are combined in a subtle and timeless way.

Modeling

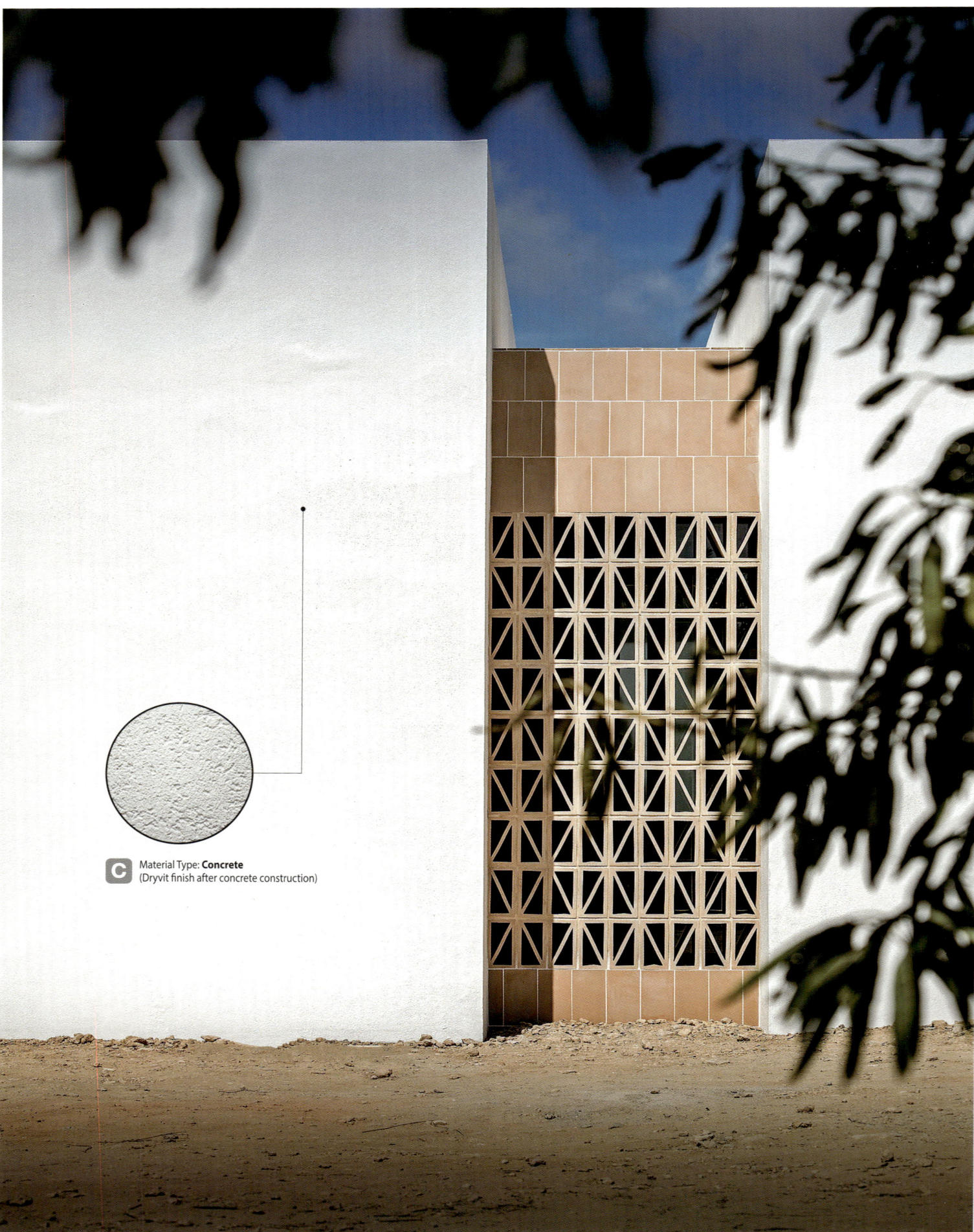

Material Type: Concrete
(Dryvit finish after concrete construction)

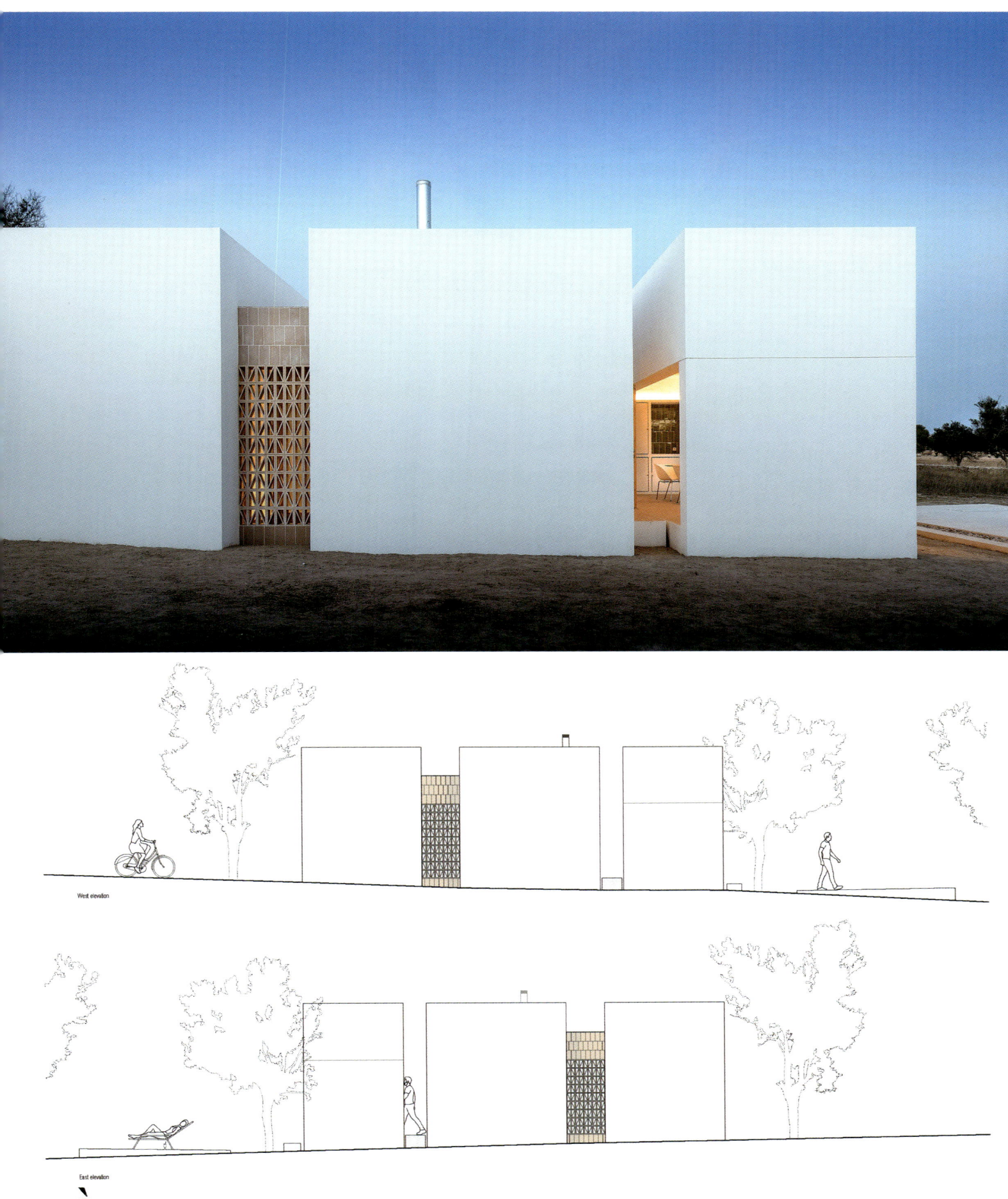

South elevation

North elevation

à **marià castelló** · architecture www.m-ar.net **Cross elevations** s. 1:100 0 1 2 4 m

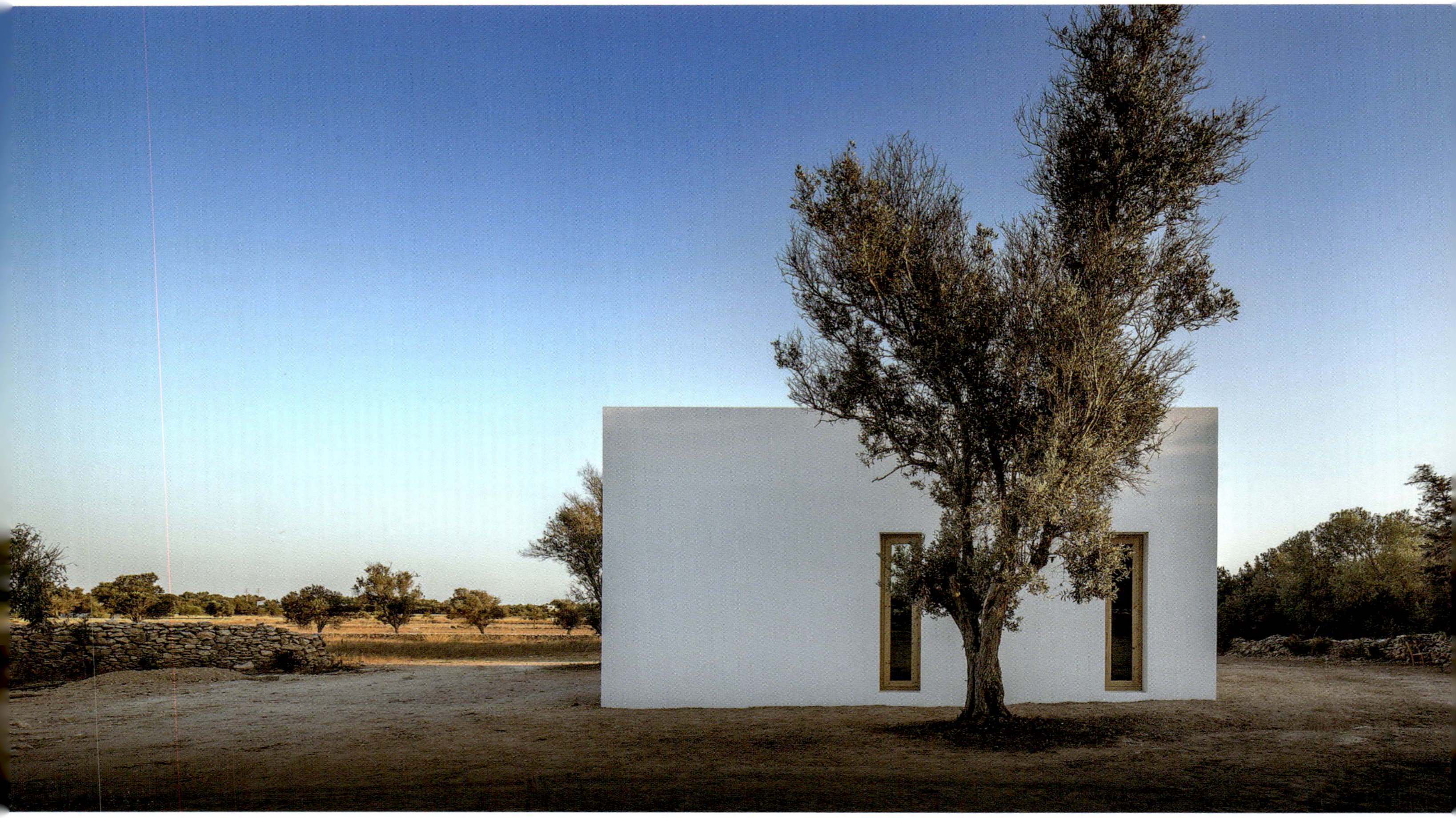

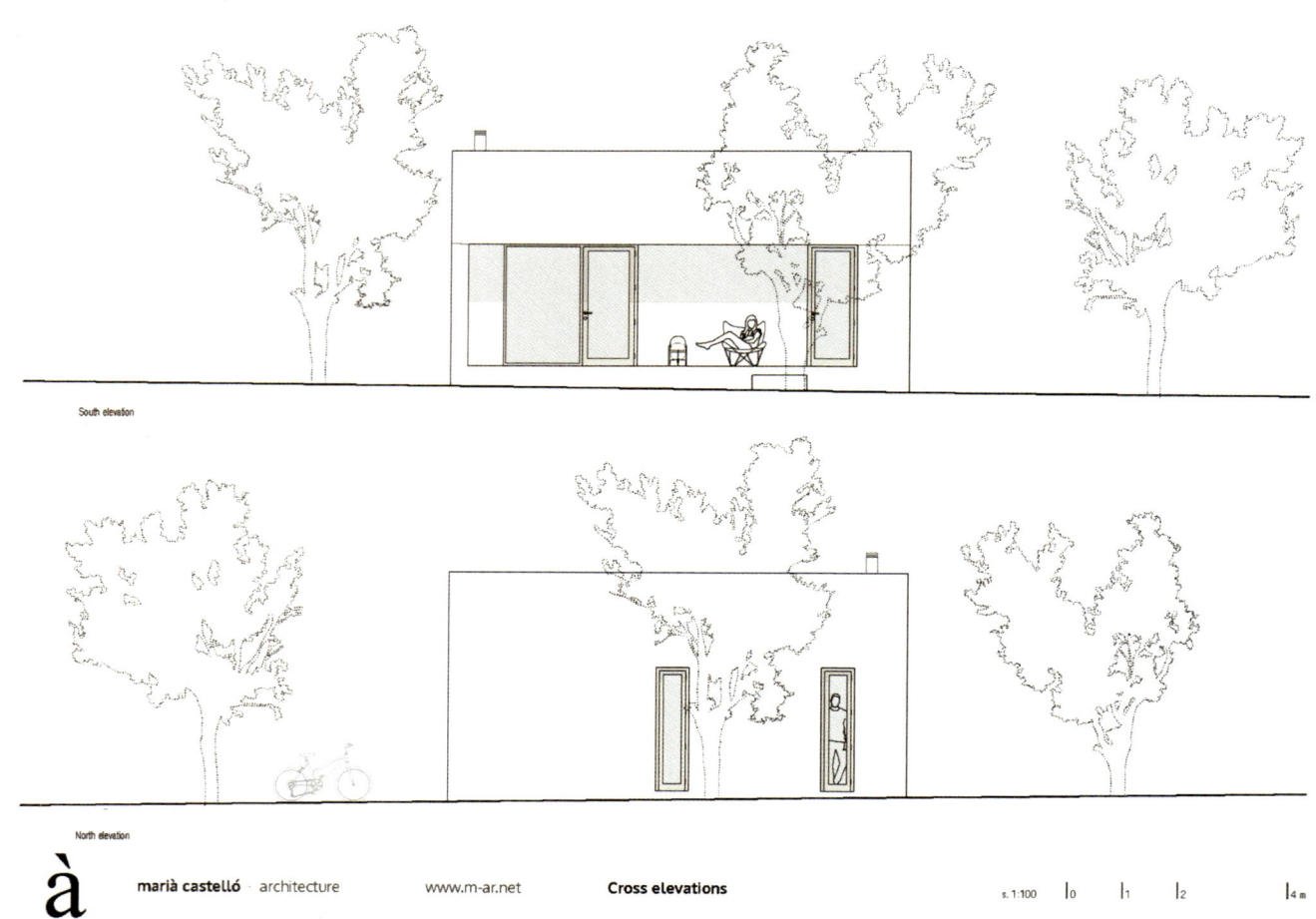

Location of the house along existing dry stone walls

à   marià castelló · architecture   www.m-ar.net   **Site plan**   s. 1:3000   0   30   60   150 m

Terracotta tiles

The warmth of the earth is transferred to the ceiling and pavements, resolved by means of Mallorcan-style ceramic vaults and pressed terracotta tiles. Likewise, the tiles are used to solve various other elements, such as façade cladding, roof finish, the headboard of the master bedroom, or pebble gravel, processing in situ the losses of the ceramic elements used. The freshness associated with the color of the vegetation predominates in the humid areas, where some vertical walls are covered with vitrified ceramic tiles of a diluted green color and identical dimensions to the rest of the pieces. The light filters inwards through its passage through ceramic lattices, generating, in turn, a constant evolution of lights and shadows.

S Material Type: **Stone & Tiles** (Terracotta tiles)

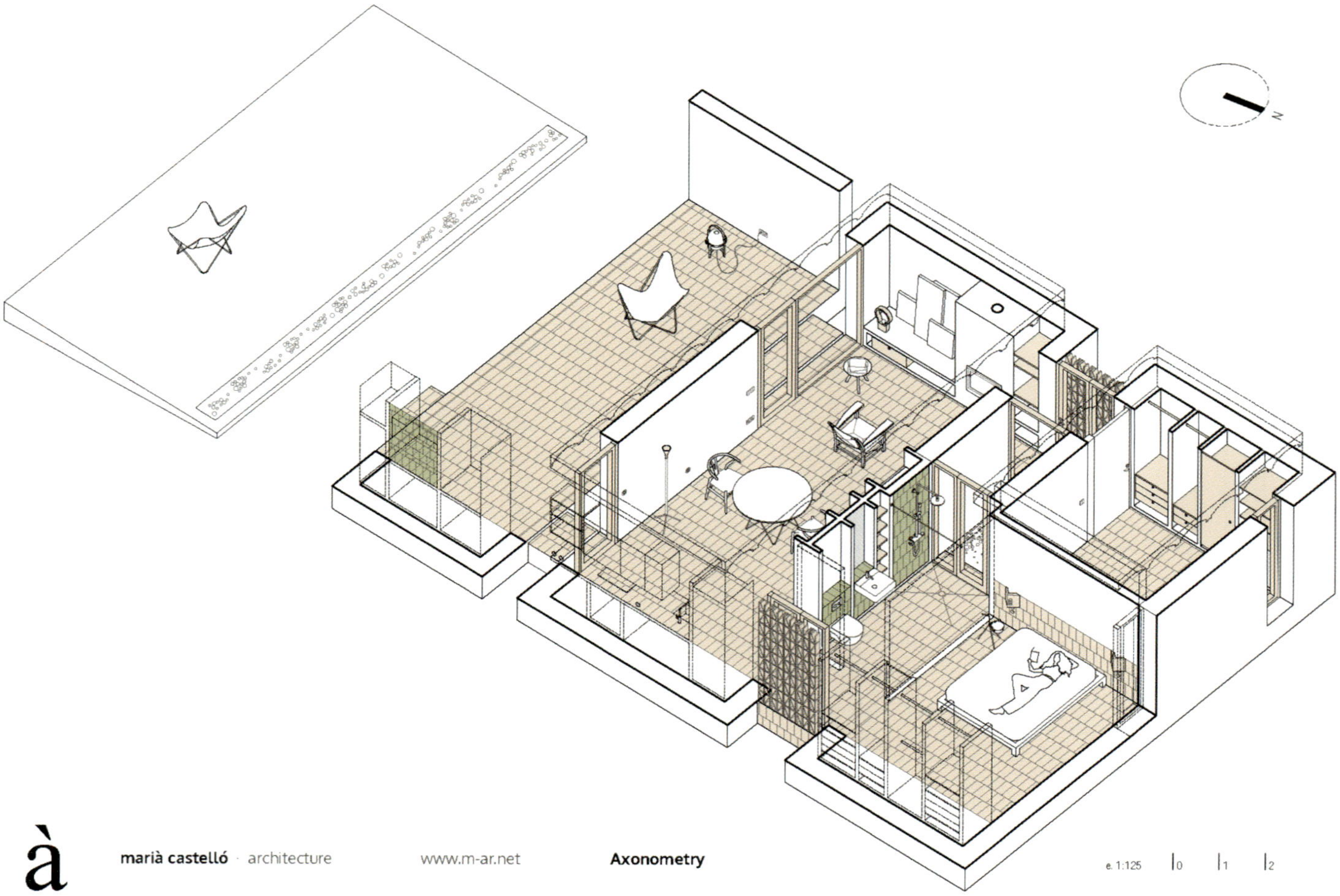

à **marià castelló** · architecture www.m-ar.net **Axonometry** e. 1:125 |0 |1 |2 |4 m

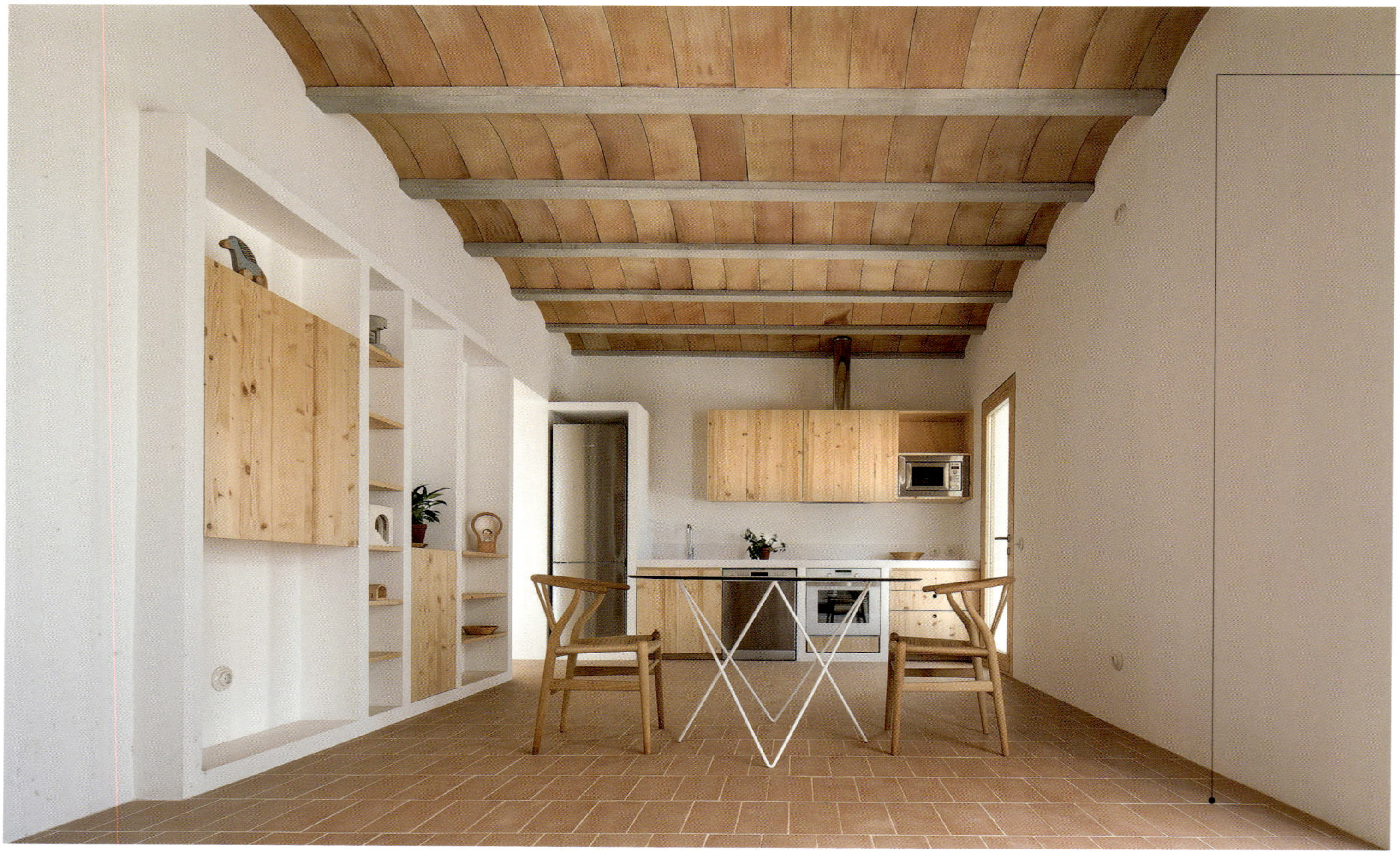

Material Type: **Stone & Tiles**
(Terracotta tiles)

à marià castelló · architecture   www.m-ar.net   Rooftop plan   s. 1:100

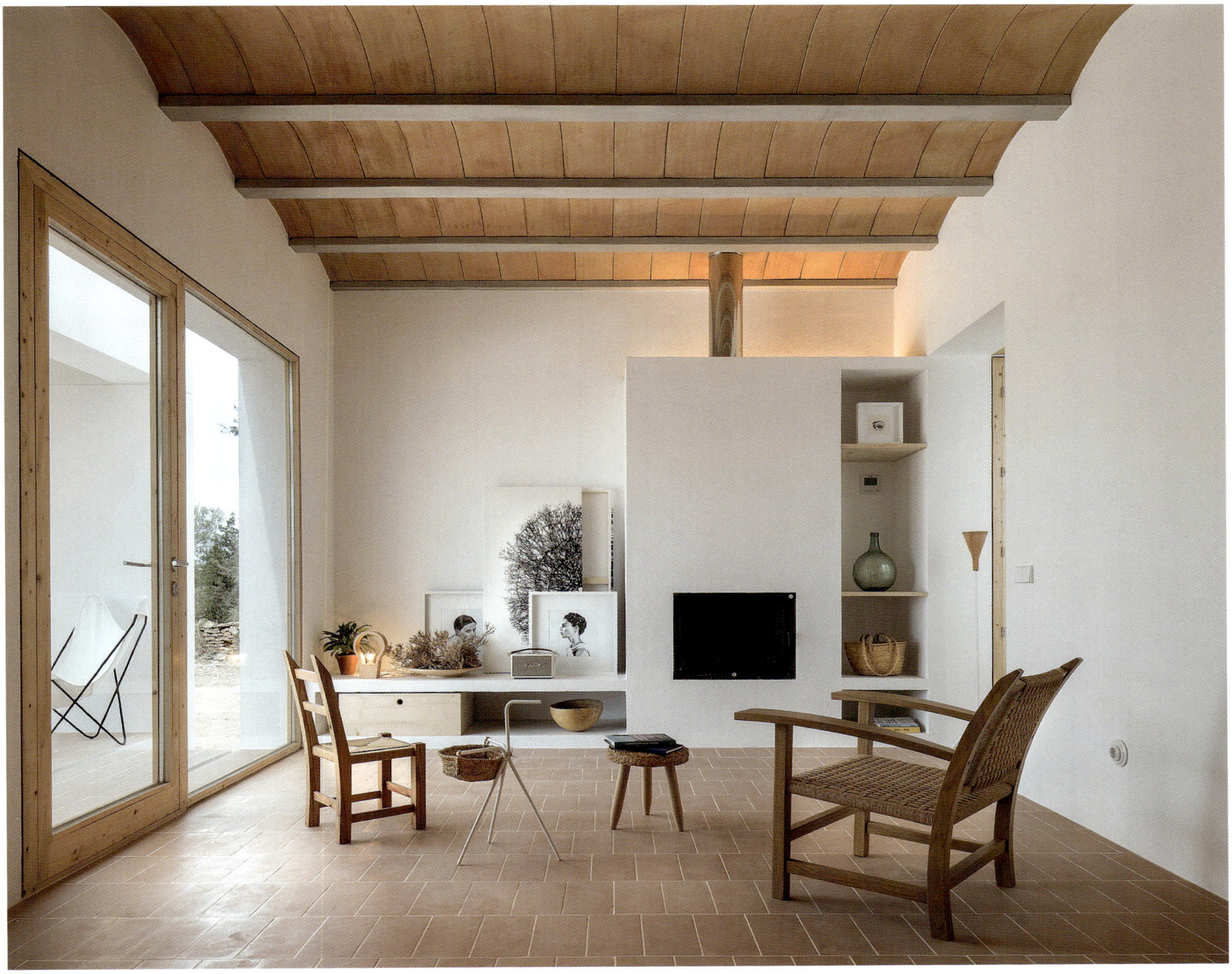

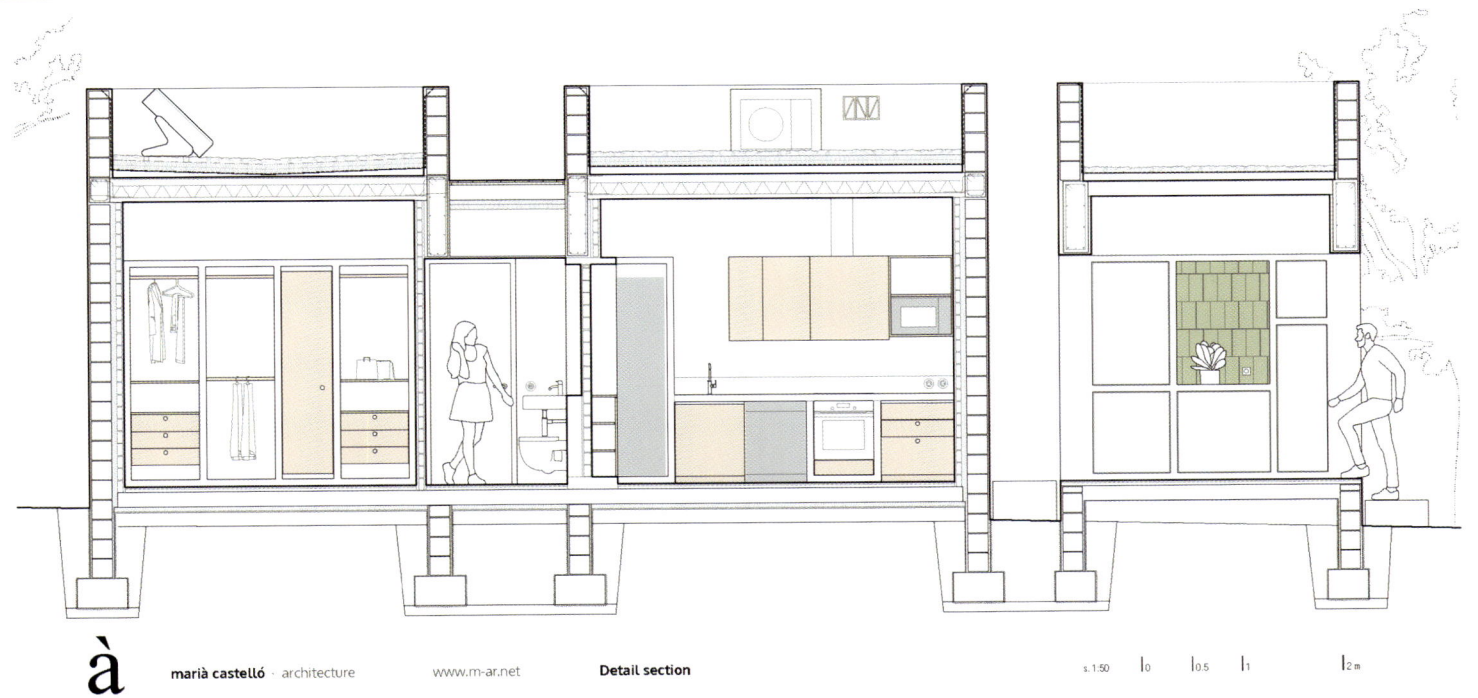

à **marià castelló** · architecture www.m-ar.net **Detail section**

s. 1:50 0 0.5 1 2 m

The coherence and material harmony have led to solving with white vitrified porcelain electrical mechanisms the integration of the installations in unique places such as the headboard of the master bedroom, as well as other more common such as lamp holders and toilets. A set of lights and special pieces made by hand with formwork made in our studio have also been specifically designed for this project, seeking their chromatic and dimensional integration in the context of the coatings.

Most of the furniture has been custom-designed integrated into the architecture itself, while icons such as the Torres Clavé armchair, from 1934, or the traditional chairs from Formentera pay homage to the Mediterranean artisan tradition. Other more contemporary pieces such as the table and coffee tables from the D12 collection designed by Marià Castelló and Lorena Ruzafa for the editor Diabla Outdoor, provide a slight material and chromatic counterpoint to the set.

à marià castelló · architecture   www.m-ar.net   Interior elevations

The Building of flats Sucre 812

Architects: Ana Smud + Alberto Smud
Location: Belgrano, Argentina
Area: 2165 m²
Photographs: Javier Agustín Rojas
Project Team: Ana Sol Smud, Alberto Smud, Pilar Esnagola, Sasha Molczadzki, Camila Jalife, Florencia Lopez Iriquin
Manufacturers: Alsa carpinterias, Listone Giordano, Pimux, WAGG
Main Material: Concrete + Steel + Stretched fabric sliding panels

The Casa Sucre building is located in the residential neighborhood of Belgrano, in the city of Buenos Aires, and has two fronts. From the beginning, the project was conceived from three central axes: the relationship between meeting spaces and rest spaces, the connection between experiences in covered spaces and traditionally open areas and, finally, the possible gradual dialogues between transparency and opacity generated on the building's facade, with a profound result inside. The building plan consists of two apartments per floor. In them, the meeting spaces (integrated living room and kitchen) were designed for the front, -while the back side was destined for the bedrooms. The design of the plant, not only in its distribution, but also in the dimensioning of each space, prioritizes shared experiences and encourages new ways of building intimacy. The living room of each unit occupies the entire width of the unit and its frames make it possible to merge with the balcony, building a new transition space between the open and the covered, between the outside and the inside.

Elevation   Section

321

Material Type: **Steel & Metal**
(Steel + Stretched fabric sliding panels)

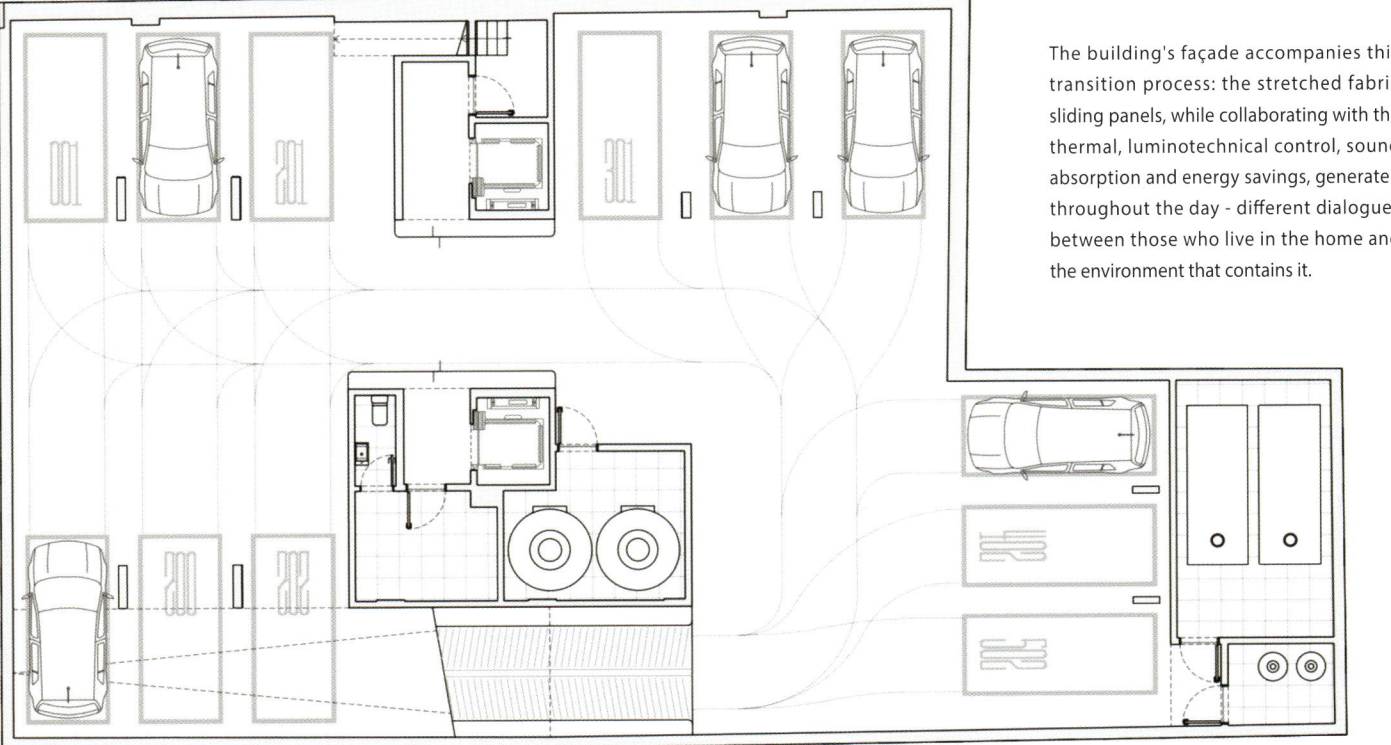

The building's façade accompanies this transition process: the stretched fabric sliding panels, while collaborating with the thermal, luminotechnical control, sound absorption and energy savings, generate - throughout the day - different dialogues between those who live in the home and the environment that contains it.

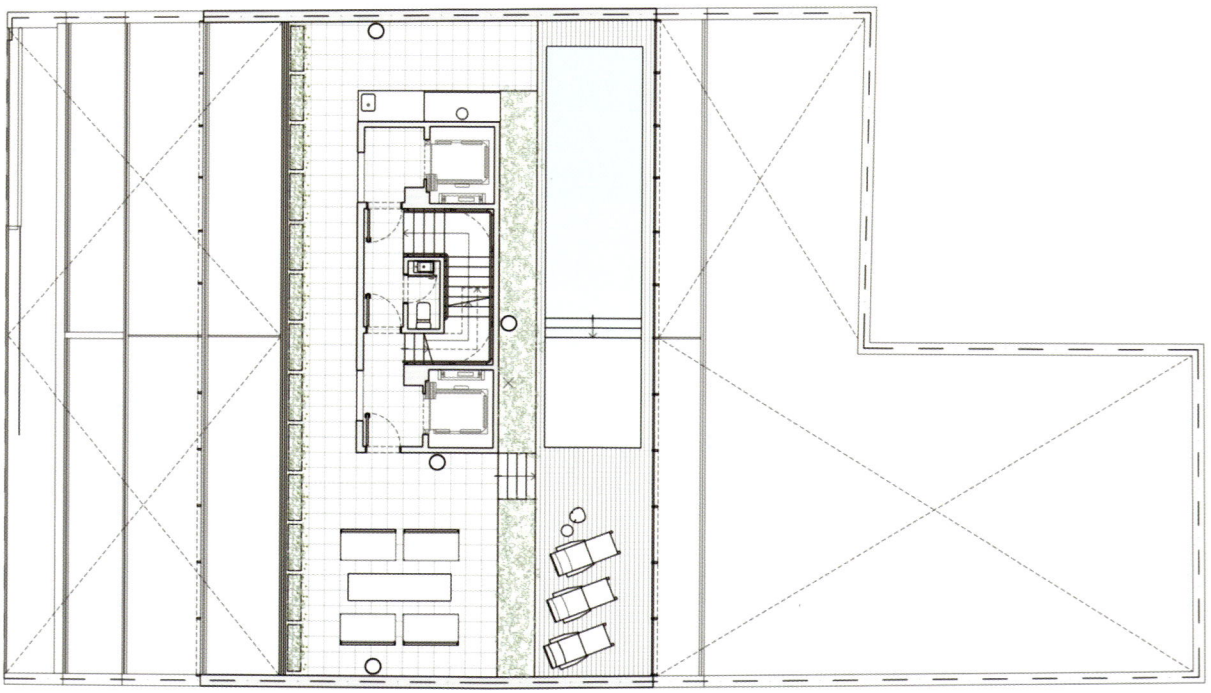

Floor Plan - A

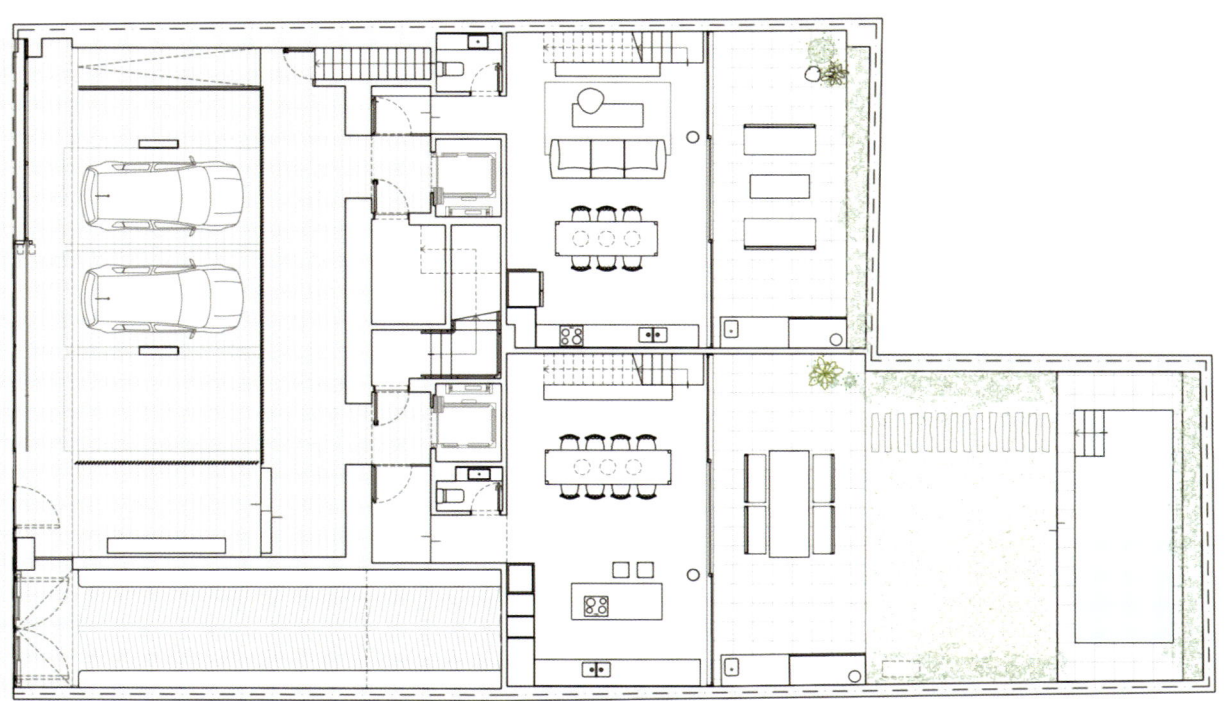

Floor Plan - B

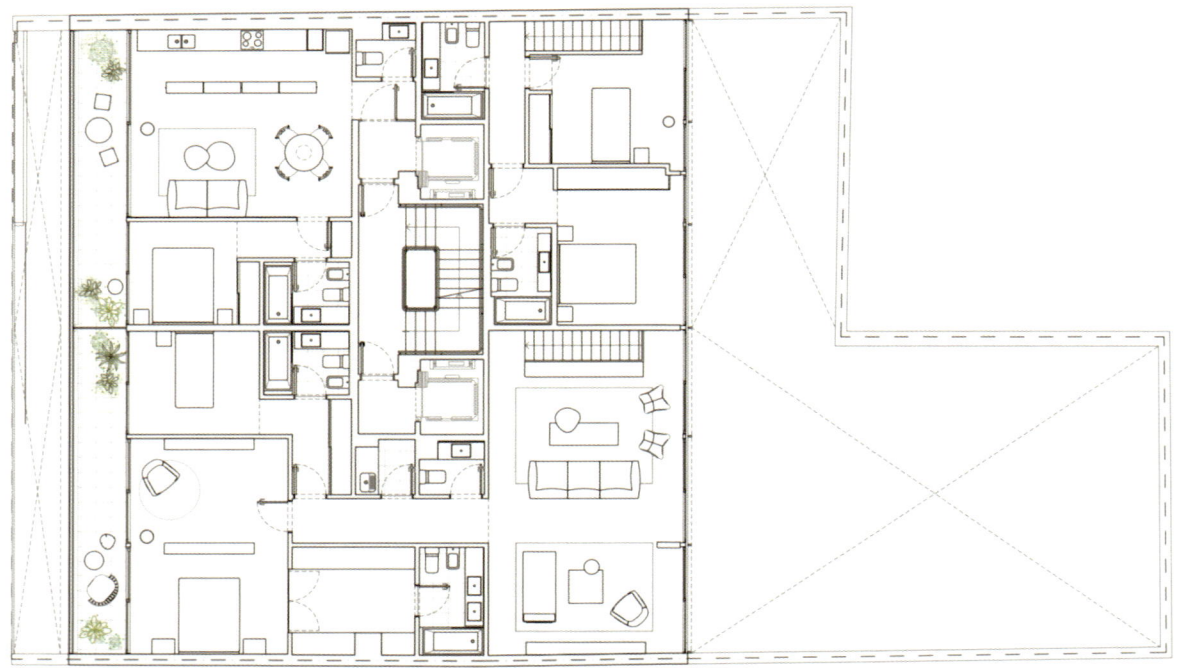

Floor Plan -C

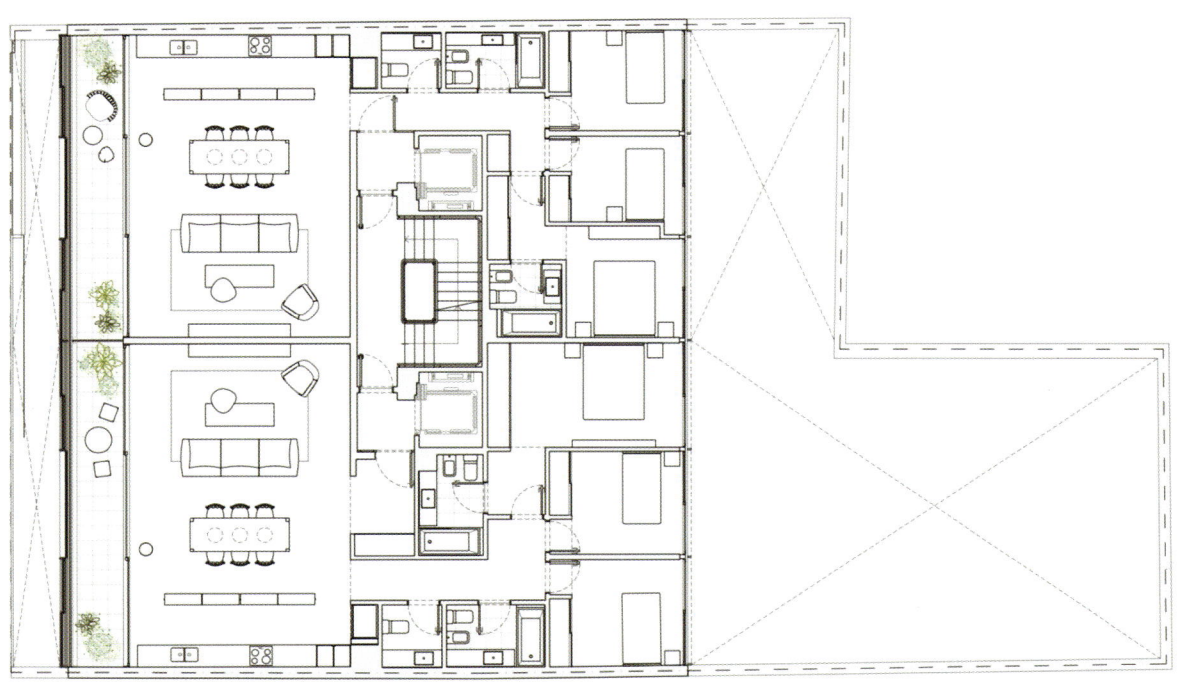

Floor Plan -D

327

Dong Phuong Y Dao Medical Center

Architects: Landmak Architecture
Location: Vietnam
Area: 2165 m²
Photographs: Trieu Chien

Manufacturers: Toto, Dulux pain, Lava Furniture, Tranphucable, Vietceramics
Main Material: Brick + Concrete + Glass

Vietnamese Oriental Medicine has developed throughout the country's history since ancient times, combining the basic theories of traditional Oriental Medicine, the experience in using tropical herbal medicine and healing from the ancient Vietnamese community. From the time of Hai Thuong Lan Ong and Zen master Tue Tinh who knew how to use herbal medicine in healing, an oriental medicine imbued with Vietnamese identity was created.

As Western medicine developed and showed rapid effect, our Traditional Medicine gradually lost its popularity. Thousands of great remedies, millions of precious medicinal plants are no longer used in healing. However, Western medicine has been increasingly revealing many shortcomings, among which is the inability to clear the root of diseases while Oriental medicine aims to regulate the human body to help eliminate the origin of diseases. Therefore, oriental medicine is still valuable and needs to be preserved and promoted properly.

Desiring to establish an oriental medicine center imbued with Vietnamese identity, contributing to the community, using Vietnamese traditional medicine to cure the elderly who cannot afford are the motivations for us to complete this project. The challenge is to renovate an old abandoned villa into an oriental medicine center with the full functions: Organize medical examination and treatment; Plant and preserve medicinal plants; Process medicinal plants; Organize workshops, educate and disseminate knowledge of oriental medicine. Sell and introduce oriental medicine products.

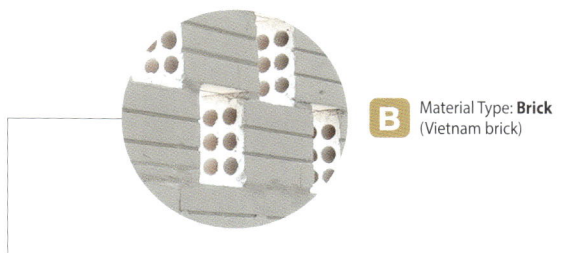

B Material Type: **Brick** (Vietnam brick)

Unlike any other project we have done before, all funding from the time of construction to operation were donated by individuals and organizations. The project was completed thanks to the contribution of everyone. In order to create a place that not only fully meets the requirements of simple comfort and medical functions but also helps patients relax, we decided to completely abandon the design of traditional hospitals which often have long narrow corridors between lines of treatment rooms. Instead, a central public space with stairs and a tea room from the first to the fourth floor helps connect the floors which have already been arranged from dynamically to statically.

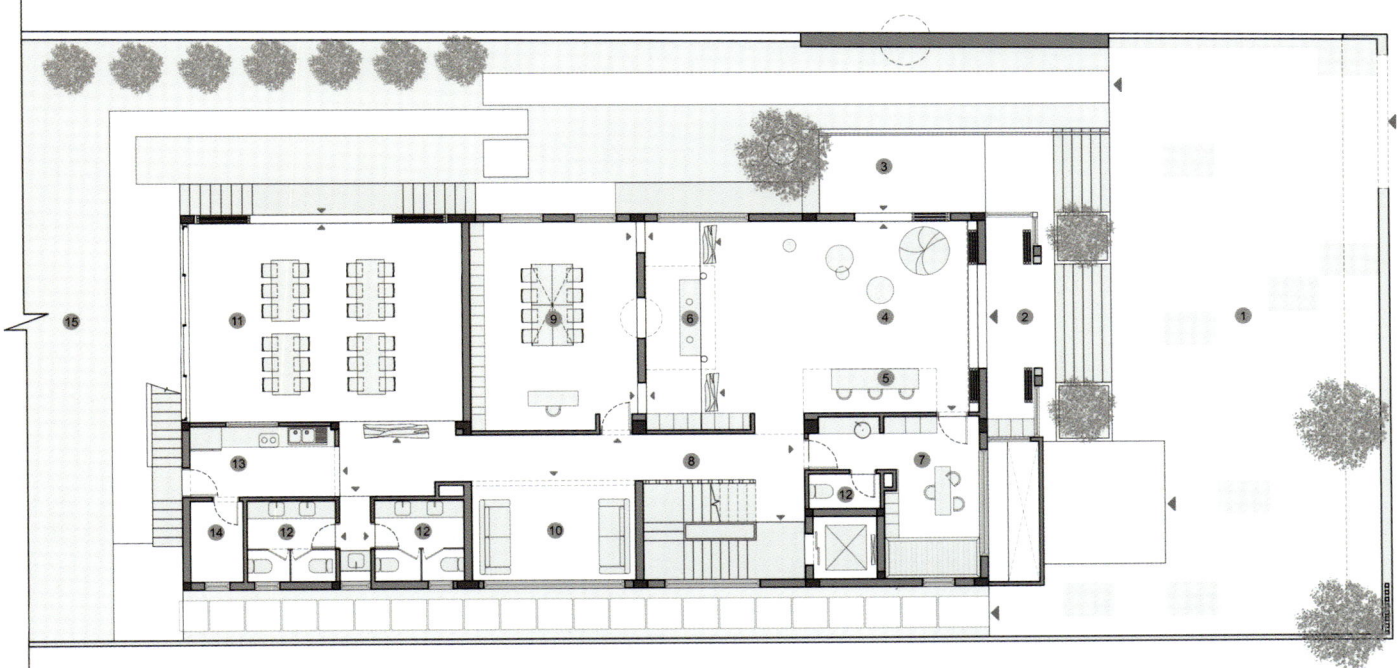

MẶT BẰNG TẦNG 1 / 1ST FLOOR PLAN

1. SÂN TRƯỚC / COURTYARD
2. LỐI VÀO CHÍNH / ENTRANCE
3. HIÊN NGHỈ / TERRACE
4. SẢNH ĐÓN TIẾP / ENTRANCE HALL
5. LỄ TÂN / RECEPTION
6. KHÔNG GIAN TRƯNG BÀY / SHOWROOM
7. PHÒNG KHÁM / CLINIC
8. SẢNH / HALLWAY
9. PHÒNG HỌP & ĐÀO TẠO / TRAINING ROOM
10. PHÒNG CHỜ / WAITING ROOM
11. PHÒNG ĂN / DINING ROOM
12. NHÀ VỆ SINH / RESTROOM
13. BẾP / KITCHEN
14. KHO / STORAGE
15. VƯỜN THUỐC / HERBAL GARDEN

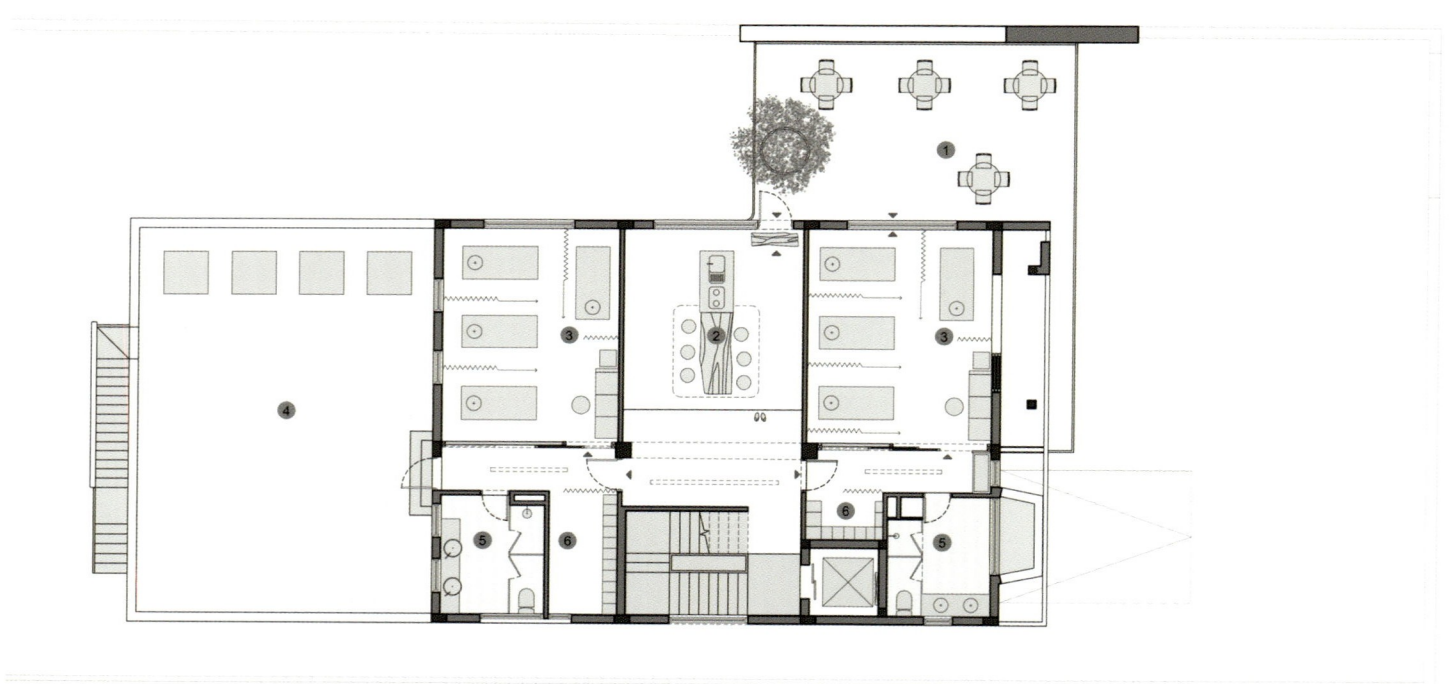

MẶT BẰNG TẦNG 2 / 2ND FLOOR PLAN

1. BAN CÔNG / BALCONY
2. THIỀN TRÀ / TEA ROOM
3. PHÒNG CHÂM CỨU / ACUPUNCTURE ROOM
4. SÂN PHƠI THUỐC / DRYING YARD
5. NHÀ VỆ SINH / RESTROOM
6. PHÒNG GỬI ĐỒ / LOCKER

MẶT BẰNG TẦNG 3 / 3RD FLOOR PLAN

1. BAN CÔNG / BALCONY
2. PHÒNG TRỊ LIỆU / THERAPY ROOM
3. PHÒNG CHÂM CỨU / ACUPUNCTURE ROOM
4. PHÒNG CHỜ / WAITING ROOM
5. NHÀ VỆ SINH / RESTROOM
6. PHÒNG GỬI ĐỒ / LOCKER

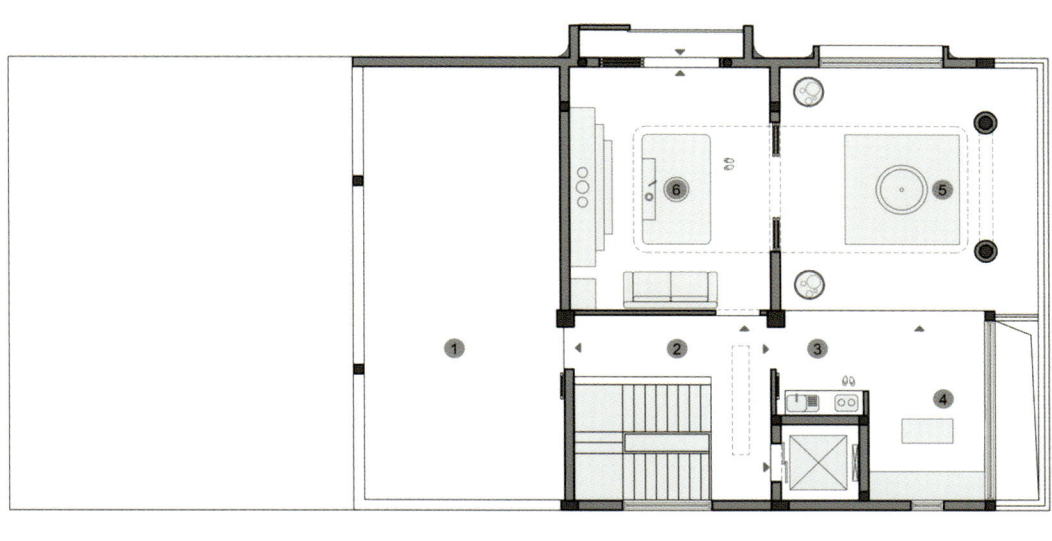

MẶT BẰNG TẦNG 4 / 4TH FLOOR PLAN

1. STUDIO
2. SẢNH / HALLWAY
3. BẾP / PANTRY
4. THIỀN TRÀ / TEA ROOM
5. KHÔNG GIAN TREO CHUÔNG / BELL SPACE
6. KHÔNG GIAN THỜ PHẬT / BUDDHA SPACE

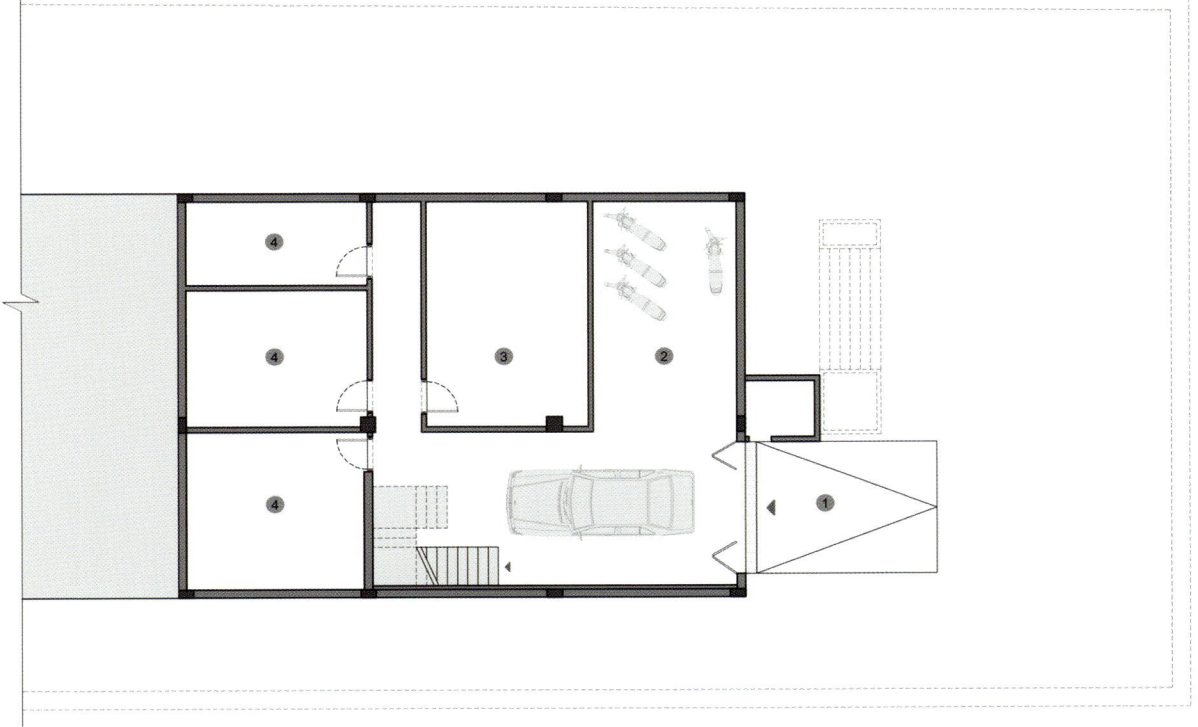

MẶT BẰNG TẦNG HẦM / LOWER GROUND FLOOR PLAN

1. DỐC XUỐNG HẦM / ENTRANCE RAMP
2. GARA ĐỖ XE MÁY / MOTORBIKE GARAGE
3. PHÒNG KỸ THUẬT / TECHNICAL ROOM
4. KHO CHỨA THUỐC / HERBAL STORAGE

The Biosphere

Architects: Chain10 Architecture & Interior Design
Location: Kaohsiung City, Taiwan
Area: 2165 m²
Photographs: KyleYu Photo Studio
Manufacturers: Toto, Dulux pain, Lava Furniture, Tranphucable, Vietceramics
Main Material: Concrete + Steel & Metal + Glass

The site is hampered by noise from the provincial highway, air pollution, and concerns about employees and guests transiting to the office. The building has a large number of suspended elements, deep balconies and sun visors. All of these components help to decrease the temperature inside the building. The windows along the front side were positioned in such a way that as the sun moves through the sky, a path of light can be seen moving through the office. This helps remind people in an office about the passage of time during the day. The corridor of the entrance is surrounded by a high degree of natural sloping embankments and densely planted trees. The exterior is dotted with stone chairs. This creates a natural forest farm that reduces the impact of the polluted air on the human body and achieves the purpose of reducing noise.

Abandoning The Traditional Can Lead to Luxury. The world is changing faster than any of us could expect, architects must transform or risk becoming obsolete along with their designs. This project continues the trend of trying to incorporate as many sustainable and eco-friendly principles as possible. The project is located in the south and close to the north-south provincial highway in Taiwan. The narrow plot is less than 30 meters wide so space constraints are something that had to be accounted for. The site is hampered by noise from the provincial highway, air pollution, and concerns about employees and guests transiting to the office. There are easier solutions to address some of these problems but sustainable choices have to be made in 2020.

The building was stacked back to the back, hoping to create a focal point around the openness of the building's entrance. The building also has a large number of suspended elements, deep balconies, and sun visors. All of these components help to decrease the temperature inside the building. The windowed area in the west was reduced in size to decrease the energy usage from air conditioning. The windows along the front side were positioned in such a way that as the sun moves through the sky, a path of light can be seen moving through the office. This helps remind people in an office about the passage of time during the day.

Climate change and global warming are an increasingly urgent issue, but the trend of people concentrating in urban areas is unavoidable. That means that architects must try to bring greenery into urban spaces as much as possible. This was done while accounting for the atmospheric conditions as well as the native plant species. The foliage tries to reverse the "hot island" effect that is common in dense urban areas. The facade of the building abandons the thinking of traditional architecture by removing all windows from the western side. Taiwan is in a subtropical environment. This means high humidity affects the maintenance of buildings and all the objects inside of it.

C Material Type: **Concrete** (a)
(Precast Concrete; makes the exterior 2500x more hydrophobic waterproof tiling + Painting finishing)

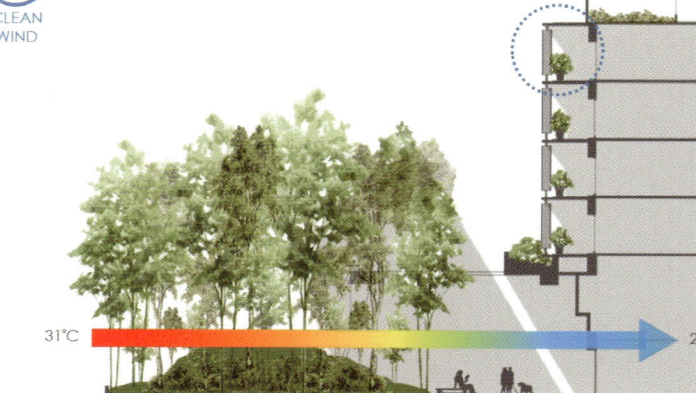

SECTIONAL VIEW OF THE BALCONY
The building has a deep balcony and sunshade to the south. The bonsai trees on the balcony gives the building good shade. Through the vents on the underside of the glass handrails, the seasonal ventilation enters the room, which greatly reduces the effect of the ambient temperature on the interior.

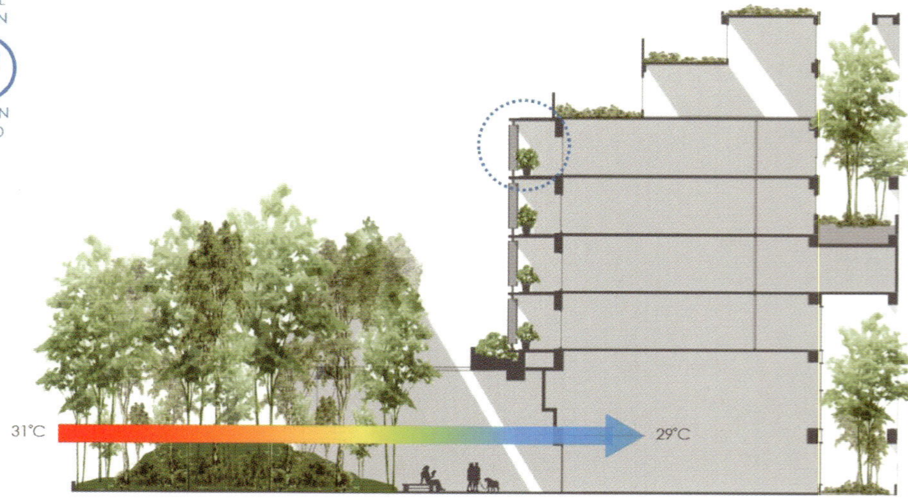

WIND

COOL DOWN

CLEAN WIND

DEEP BALCONY

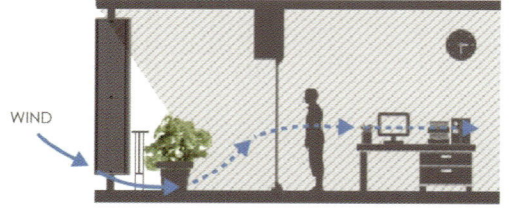

SUN VISORS

REDUCE AMBIENT TEMPERATURE

31°C 29°C

AUDITORY FACTORS

Many mounds were used to block the 70dB noise from the highway. In addition to the natural bird calls and insects, a light background music is also set up in the forest to cover the noise of the road with a slightly higher 2dB sound.

75Hz — NOISE POLLUTION

60Hz — MOUND

62Hz — NATURAL SOUND+MUSIC

WIND CONDITIONS

Summer in southern Taiwan is driven by a south by south-east wind. After being purified and cooled by the dense foliage, it can enter the large windows opening on the south side of the building, providing fresh air with high oxygen content for the interior and improving the workplace for the office staff. In these trying times, natural ventilation is essential to limit the spread of disease.

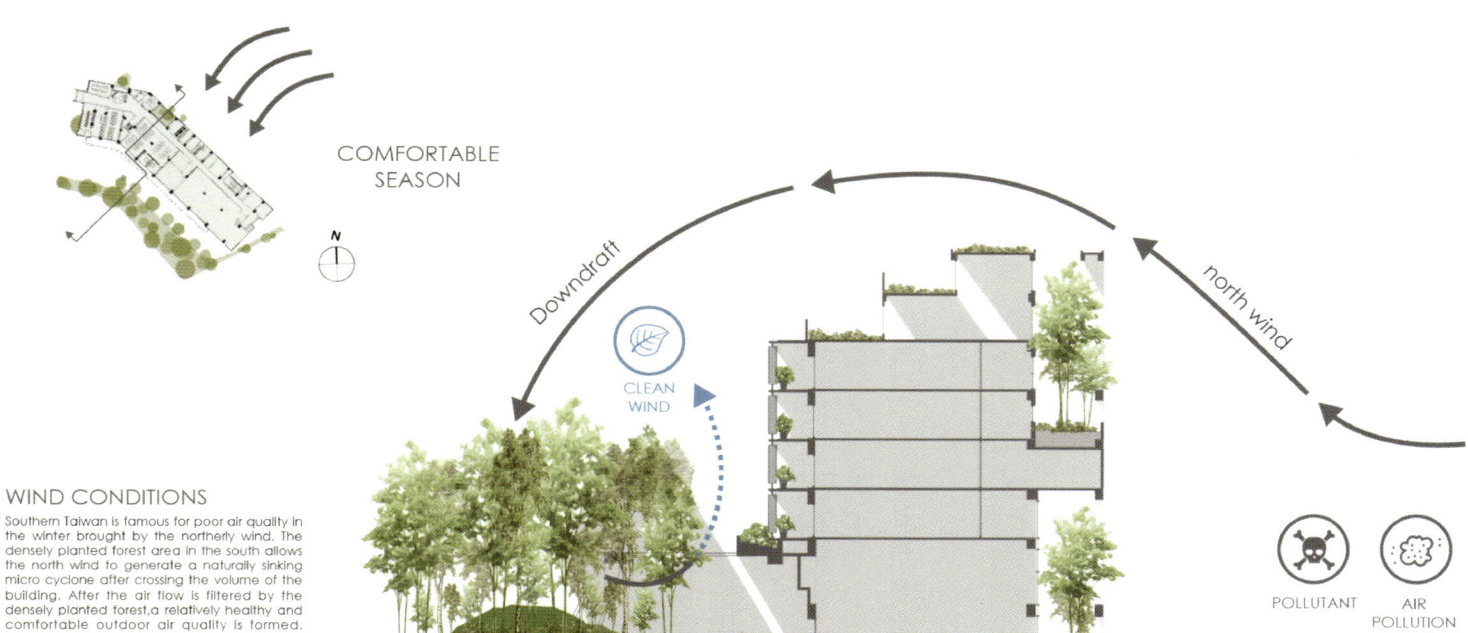

COMFORTABLE SEASON

WIND CONDITIONS

Southern Taiwan is famous for poor air quality in the winter brought by the northerly wind. The densely planted forest area in the south allows the north wind to generate a naturally sinking micro cyclone after crossing the volume of the building. After the air flow is filtered by the densely planted forest, a relatively healthy and comfortable outdoor air quality is formed.

The traditional concept of Asian structures was transformed by changing waterproof tiling into a waterproof coating and applying it to the exterior. This makes the exterior 2500x more hydrophobic while still allowing oxygen into and out of the building. This also solves the problem of unclean water coating the outside of a project while creating a living building that breathes. The color of the building and a large amount of greenery in the environment is intertwined with the greenery of the property to create a coordinated and friendly image of the entrance. The corridor of the entrance is surrounded by a high degree of natural sloping embankments and densely planted trees.

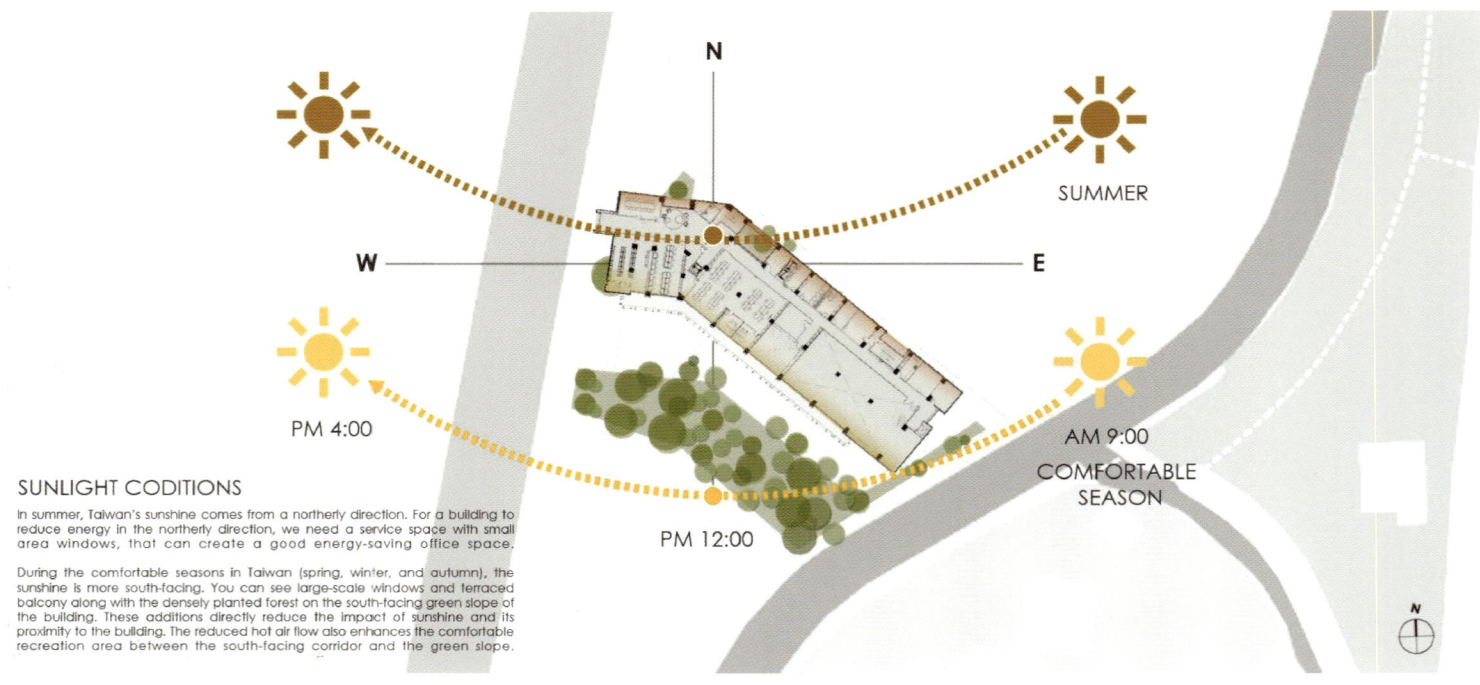

SUNLIGHT CODITIONS

In summer, Taiwan's sunshine comes from a northerly direction. For a building to reduce energy in the northerly direction, we need a service space with small area windows, that can create a good energy-saving office space.

During the comfortable seasons in Taiwan (spring, winter, and autumn), the sunshine is more south-facing. You can see large-scale windows and terraced balcony along with the densely planted forest on the south-facing green slope of the building. These additions directly reduce the impact of sunshine and its proximity to the building. The reduced hot air flow also enhances the comfortable recreation area between the south-facing corridor and the green slope.

The exterior is dotted with stone chairs. During times of high stress, the employees of the office can take a walk outside and feel like they have strolled into another biome. This is a critical component that more modern offices need, something to help employees deal with the stress of long working hours. By abandoning the traditional Asian model of architecture, a structure was created that is symbiotically linked to the environment. Sustainable while being a place for many generations of people to work.

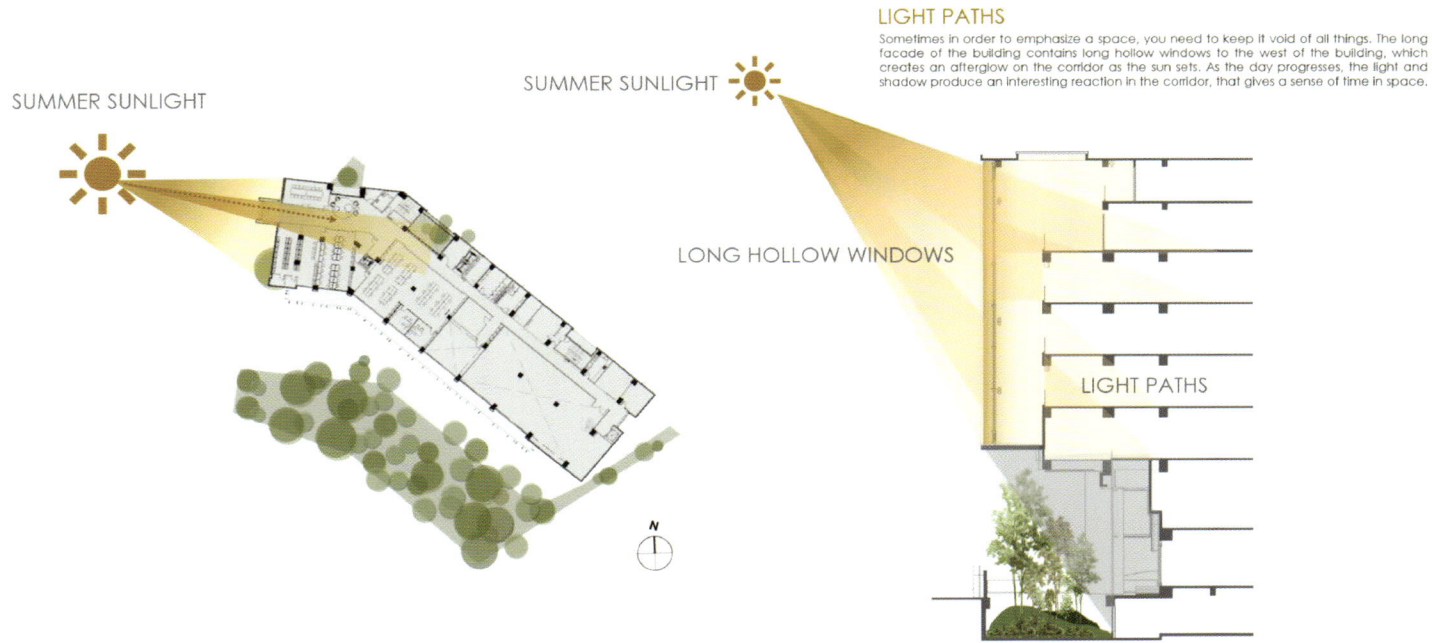

LIGHT PATHS
Sometimes in order to emphasize a space, you need to keep it void of all things. The long facade of the building contains long hollow windows to the west of the building, which creates an afterglow on the corridor as the sun sets. As the day progresses, the light and shadow produce an interesting reaction in the corridor, that gives a sense of time in space.

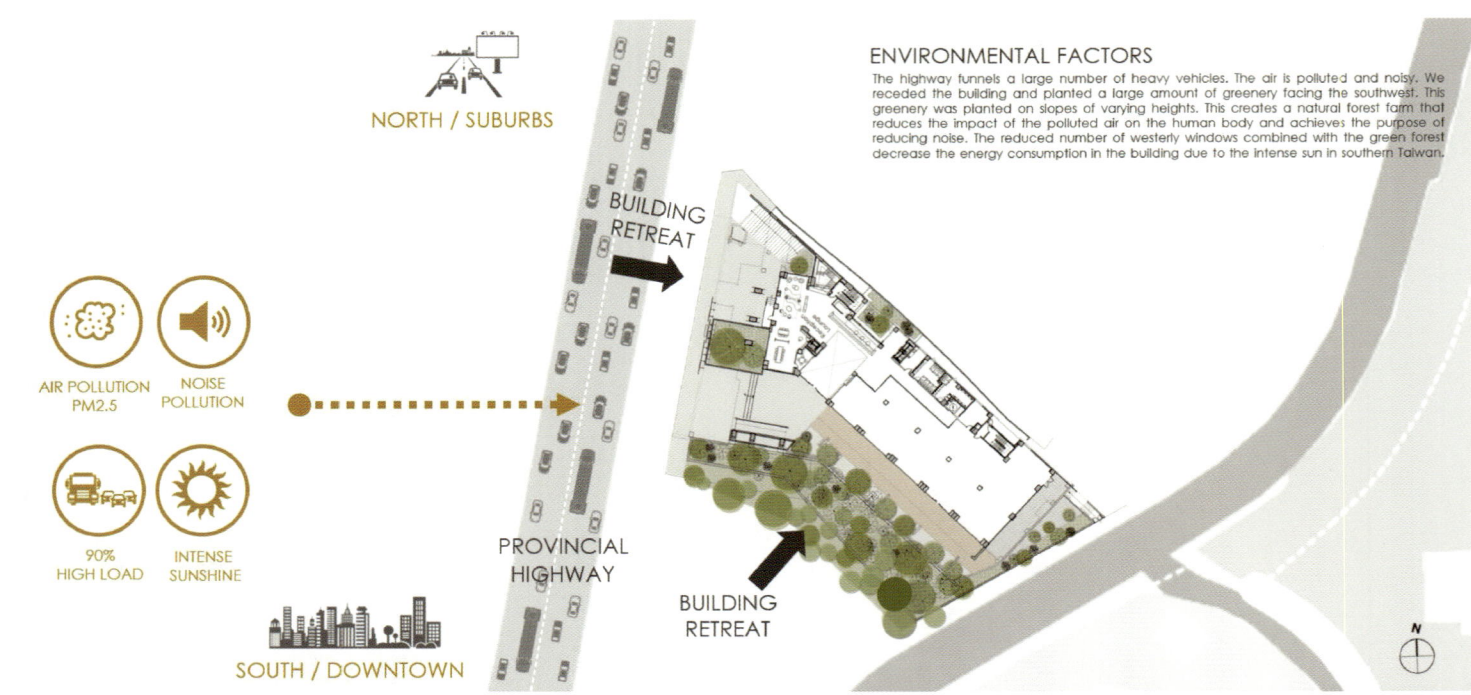

ENVIRONMENTAL FACTORS
The highway funnels a large number of heavy vehicles. The air is polluted and noisy. We receded the building and planted a large amount of greenery facing the southwest. This greenery was planted on slopes of varying heights. This creates a natural forest farm that reduces the impact of the polluted air on the human body and achieves the purpose of reducing noise. The reduced number of westerly windows combined with the green forest decrease the energy consumption in the building due to the intense sun in southern Taiwan.
NORTH / SUBURBS
BUILDING RETREAT
AIR POLLUTION PM2.5
NOISE POLLUTION
90% HIGH LOAD
INTENSE SUNSHINE
PROVINCIAL HIGHWAY
BUILDING RETREAT
SOUTH / DOWNTOWN
N

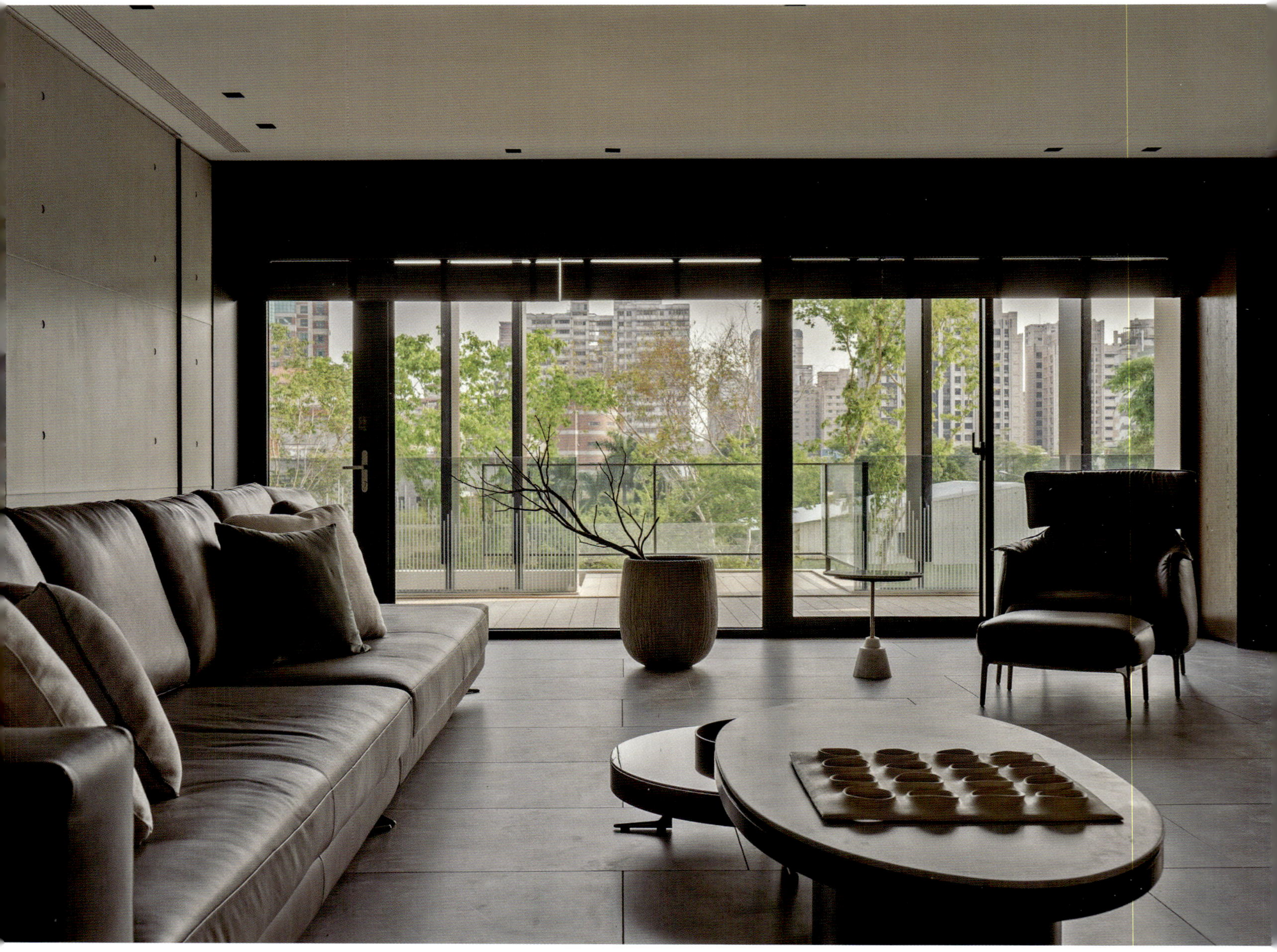

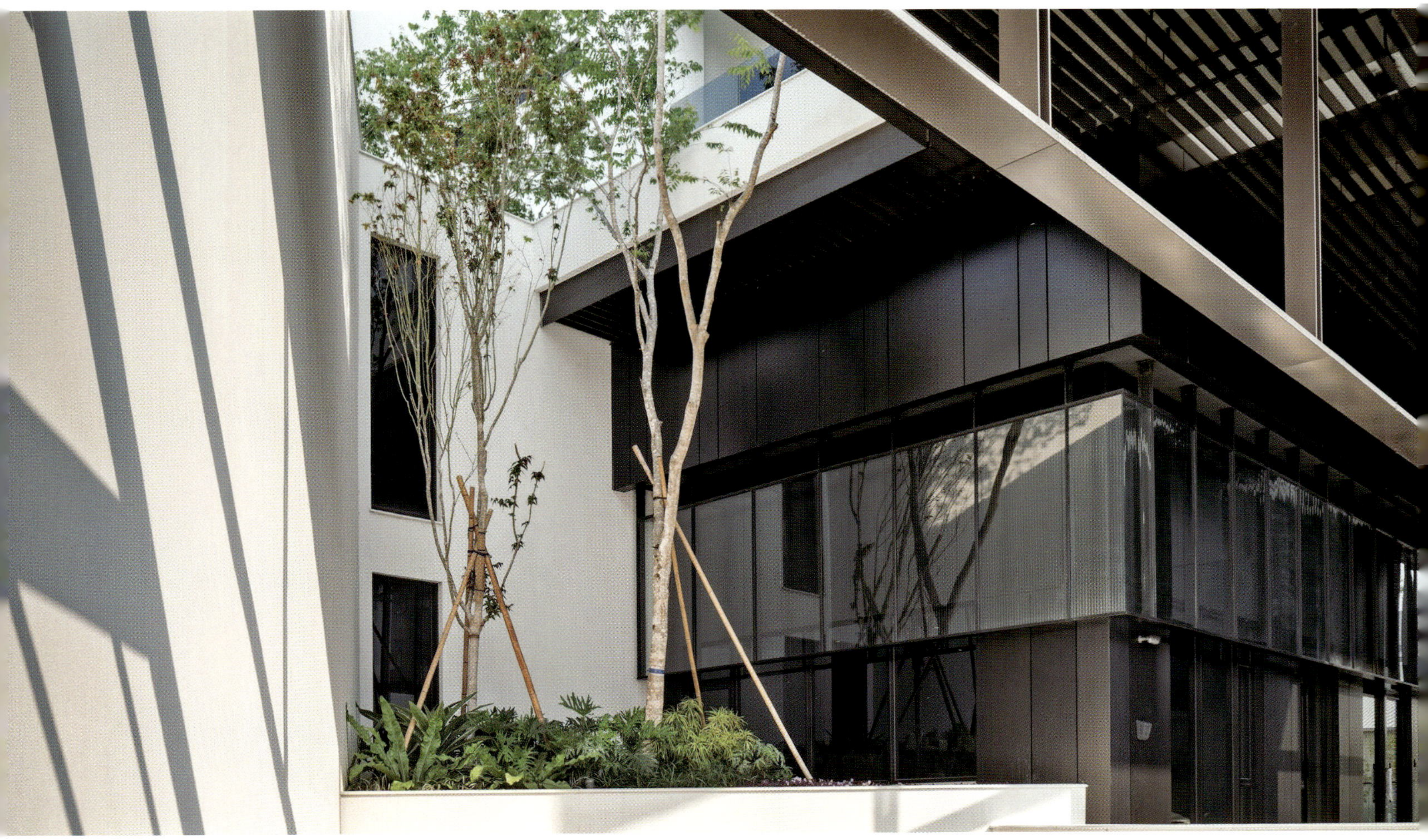

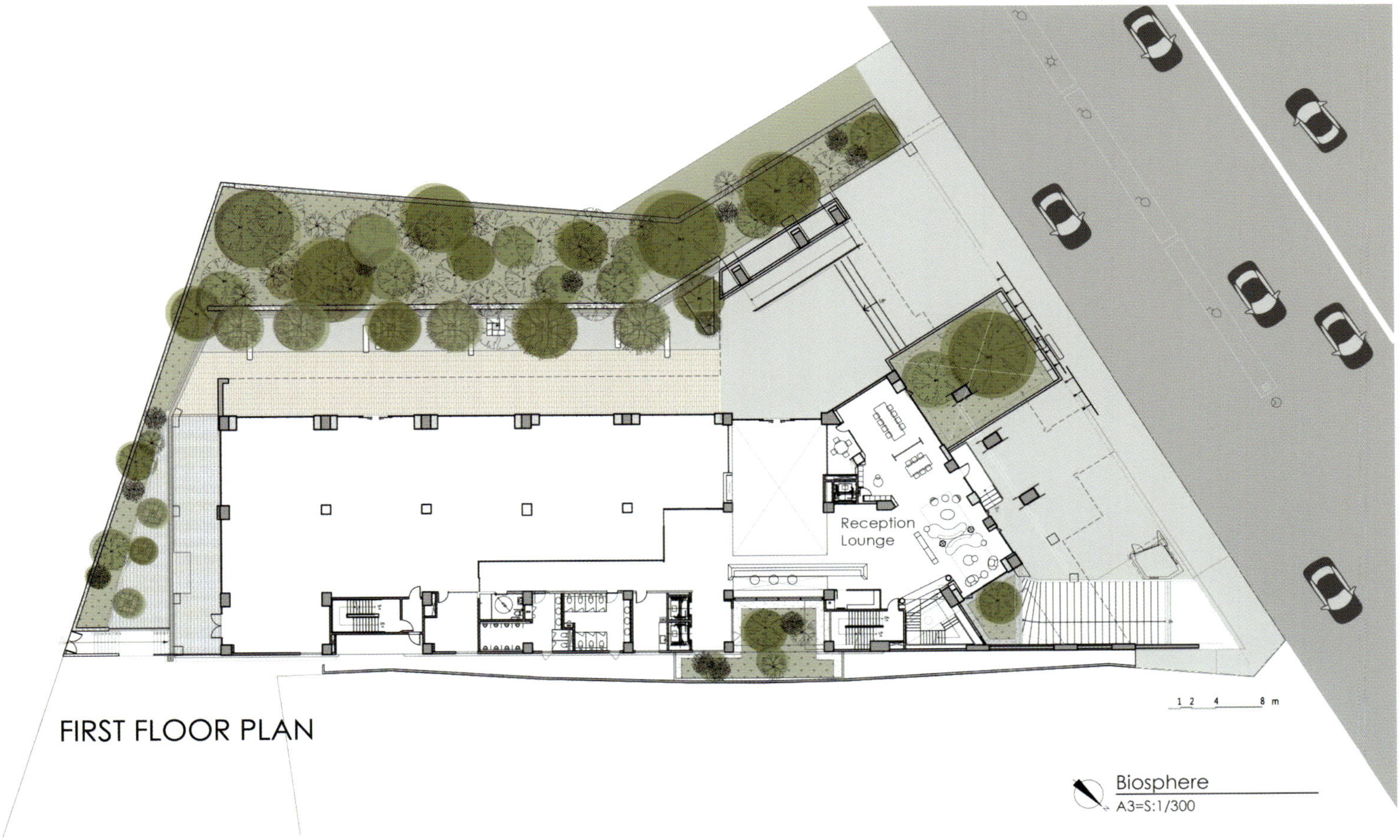
Reception
Lounge
1 2 4 8 m
Biosphere
A3=S:1/300

FIRST FLOOR PLAN

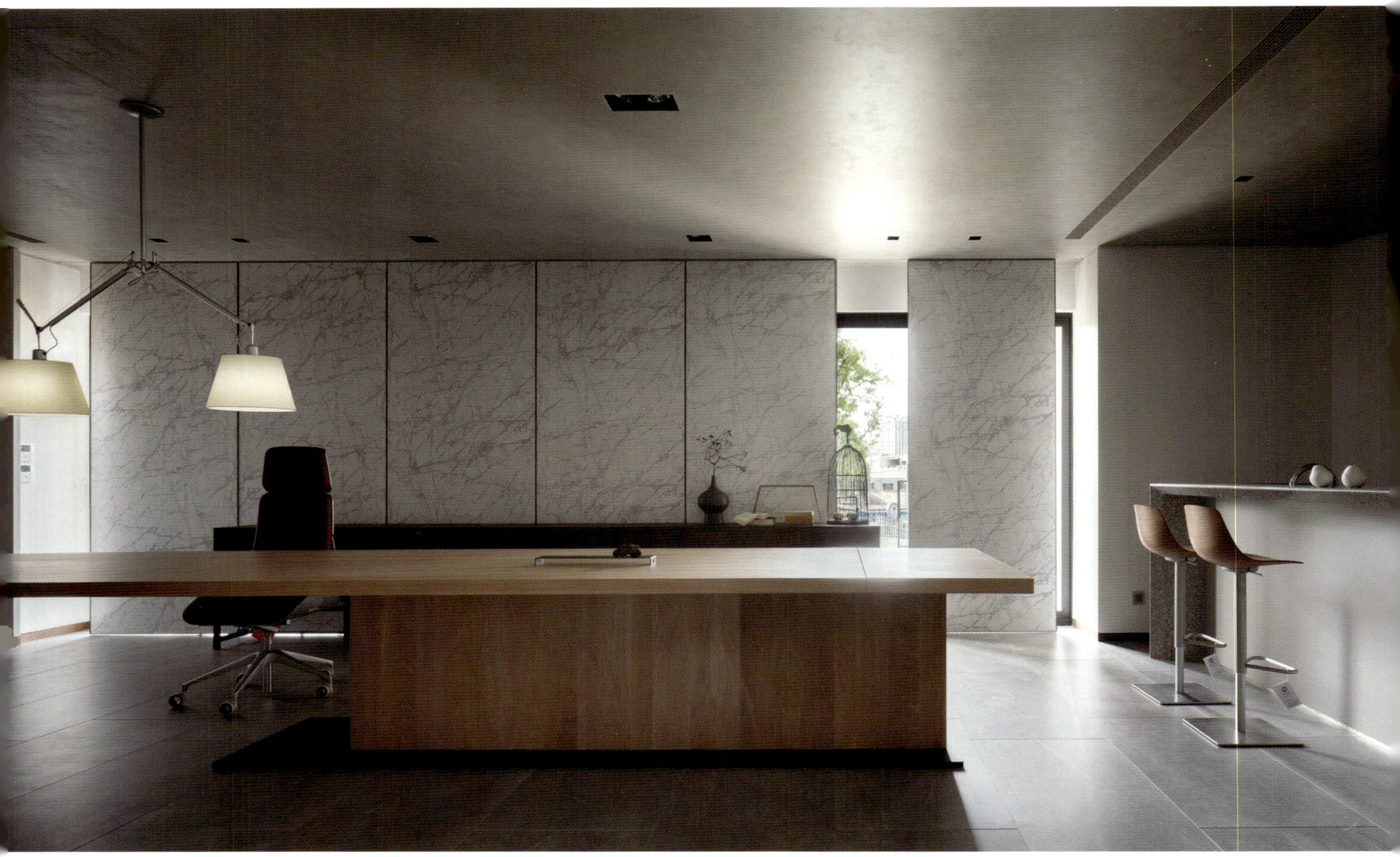

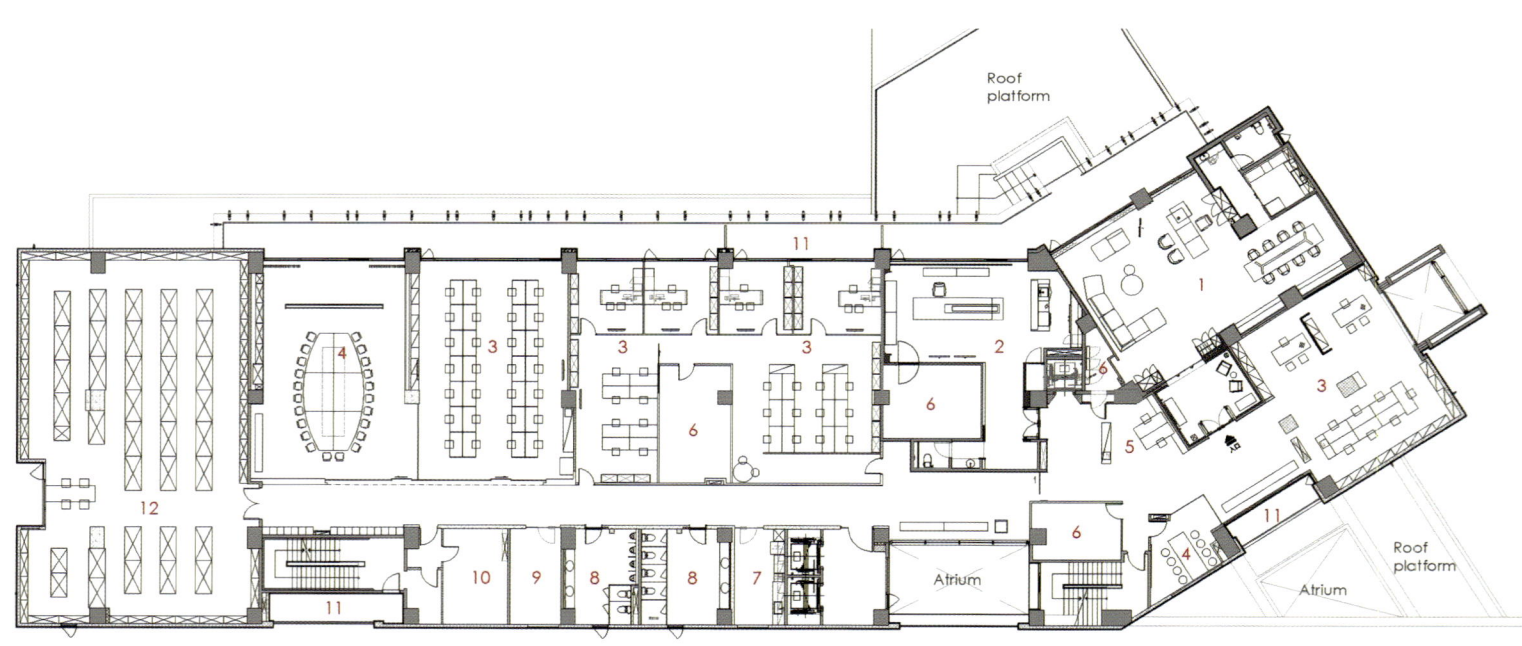

THIRD FLOOR PLAN

1. Chairman office
2. CEO's office
3. Office
4. Conference room
5. Secretary office
6. Storage
7. Pantry
8. Toilet
9. Information room
10. Server room
11. Balcony
12. Expansion space

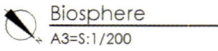

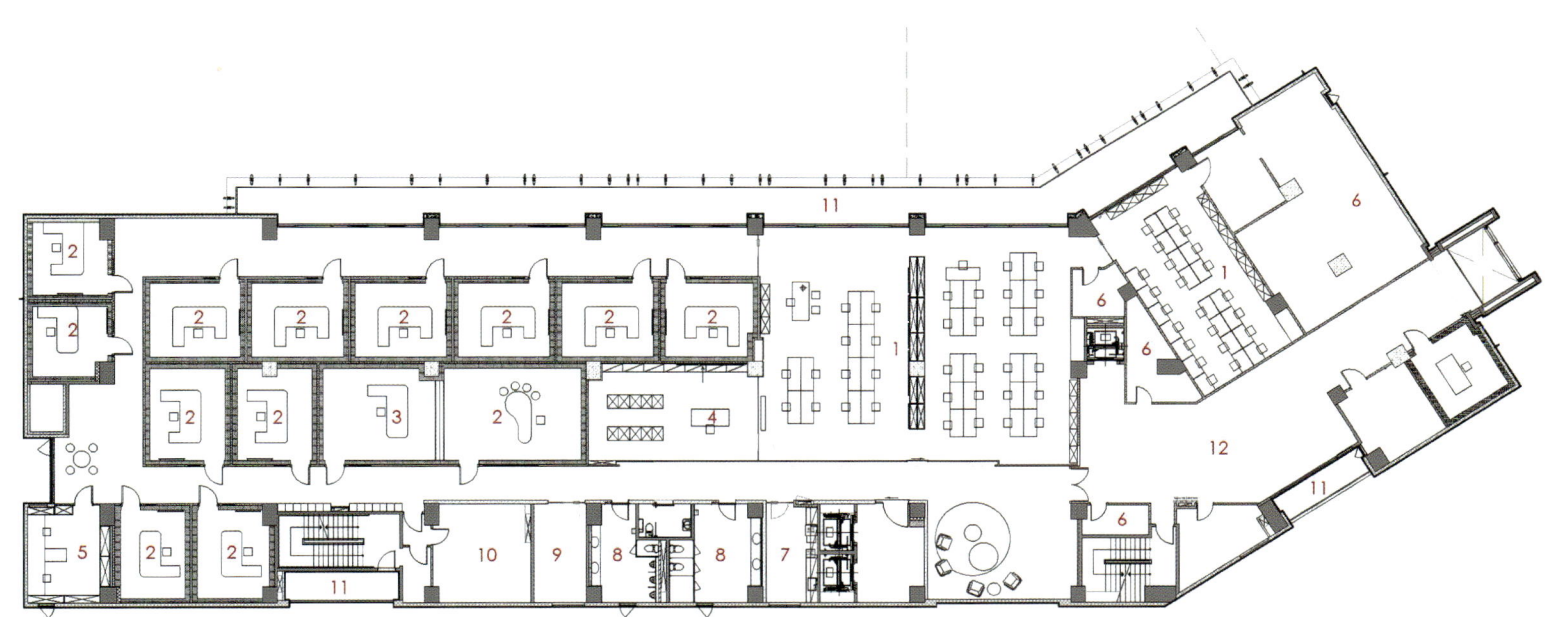

FORTH FLOOR PLAN
1. Office
2. Broadcasting studio
3. Sub control room
4. Master control room
5. Editing room
6. Storage
7. Pantry
8. Toilet
9. Information room
10. Server room
11. Balcony
12. Exhibition area

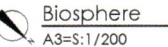
Biosphere
A3=S:1/200

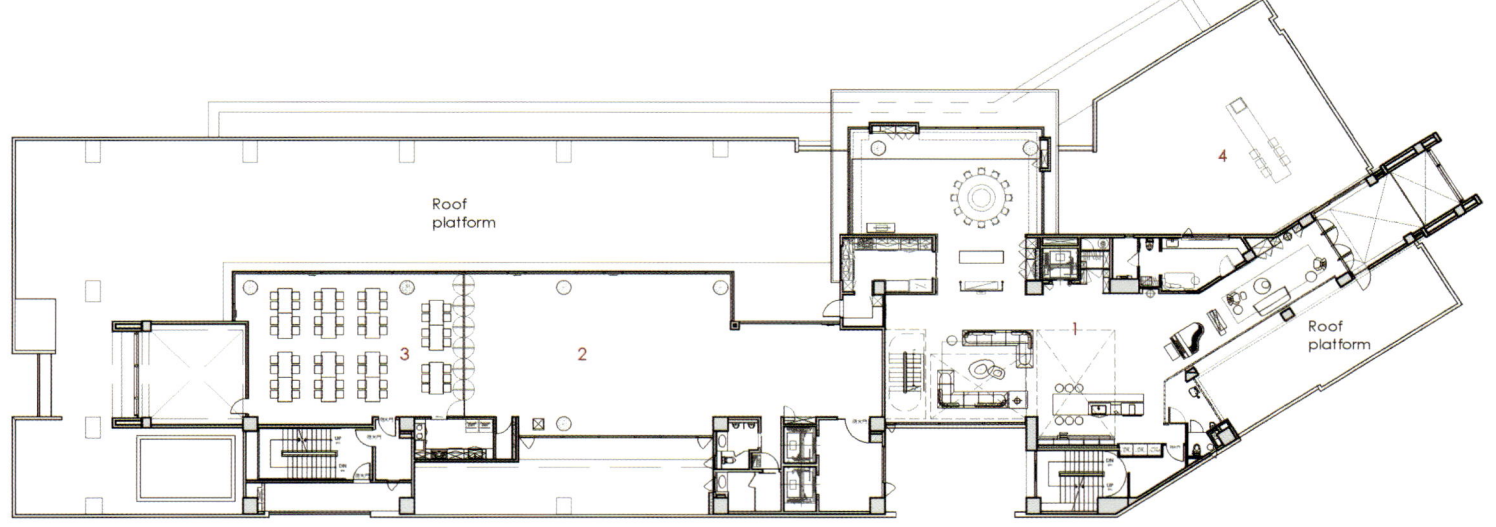

SEVENTH FLOOR PLAN

1. VIP reception lounge
2. Conference room
3. Staff restaurant
4. Terrace

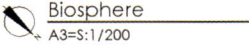

Biosphere
A3=S:1/200

Sala Samui Chaweng Beach Resort, Phase 02

Architects: onion Architecture
Location: Surat Thani, Thailand
Area: 16000m²
Photographs: Wworkspace

Manufacturers: Dulux, Hafele, Vista Inno
Main Material: Concrete + Glass + Wood + Steel & Metal

Sala Chaweng is a 144-room hotel in Samui Island's Chaweng Beach, one of the most beautiful beaches in the Gulf of Thailand. It has 2 phases, 52-room for the Beachfront and 82-room for the Roadside. These two phases are divided by a street. Sala Chaweng Beachfront has a single big swimming pool courtyard next to the sea, which is bright and active, whereas Sala Chaweng Roadside is composed of many smaller courtyards, without linear circulation, full of existing trees. The Roadside is shaded, it is more passive, private and quite than the Beachfront.

Each courtyard in the Roadside has a domestic characteristic because we think that it is easier for our guests to remember the spaces. These 5 courtyards are named the Outdoor Stairwell, the Outdoor Dining Room, the Outdoor Living Room, the Outdoor Play Room and the Outdoor Bedroom. Each of them is occupied by household furniture and elements such as single-flight stairs, 12-seat-round table, sofas, swings, beds and giant lamps. All of these domestic furniture are placed in-between existing trees and newly planted greenery.

Our guests can relax in beds, larger than king-size, shaded by big trees next to the public swimming pool, having a drink served from the pool bar and feel that, in this strange setting, they are indeed on holiday. We direct our attention to the sky. In every transitions between the courtyards, we open the vertical voids in different shapes, size and materials. Layers of bamboo rectangle voids are placed between the Outdoor Dining Room and the Outdoor Bedroom whereas oval concrete voids in white color are composed in the transition space between the Outdoor Dining Room and the Outdoor Living Room.

C Material Type: **Concrete** (a)
(Concrete + Painting finishing)

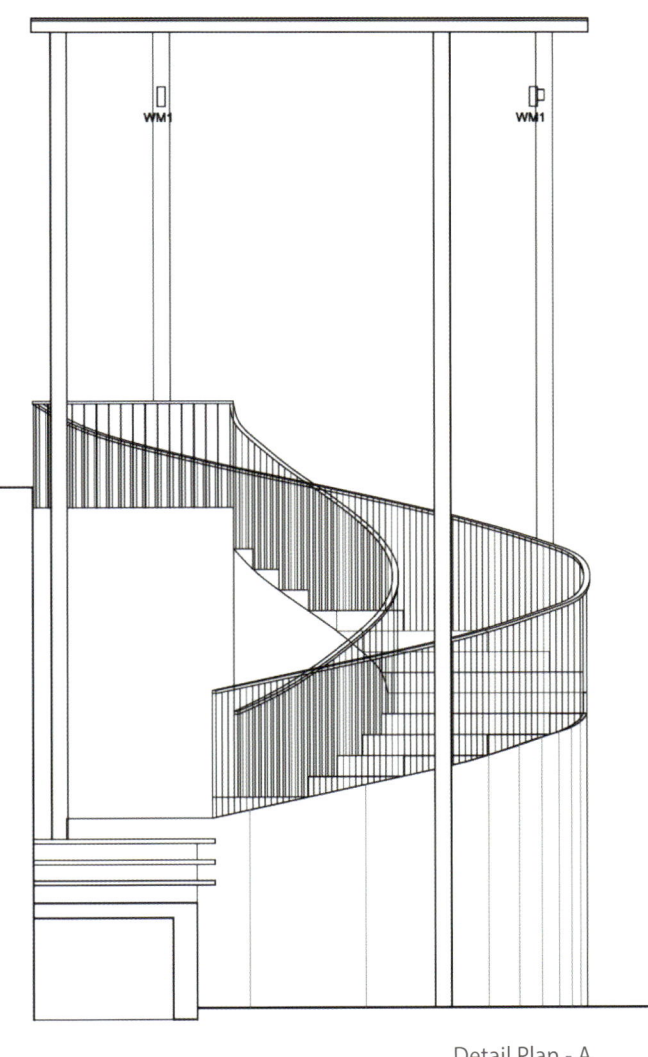

Detail Plan - A

We encourage our guests to look up at the sky and to feel the warmth of the sun. A dominant architectural element of Sala Chaweng Roadside are the white walls. They are long, massive, sometimes concave and sometimes convex. The effects of shadow are what we are interested in when we were drawing the walls. It is only in the spa area that we use colors, dark green and later on pink.

These coloured walls set a new spatial experience to our spa guests. As the sun moves, the light changes its direction, architecture becomes alive. Bathtub of the spa room is placed in the garden surrounded by the dark green walls. The clear glass door brings nature into the treatment room. There is no clear distinction between the interior and the exterior spaces in Sala Chaweng Roadside.

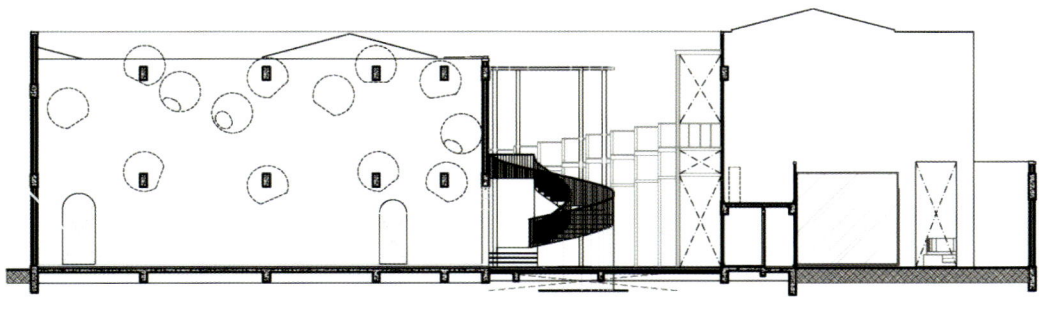

Section - A

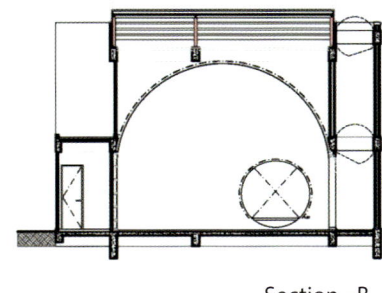

Section - B

There are 4 types of room in Sala Chaweng Roadside, namely Garden Two Bedroom Pool Suite 329 sq.m., Garden One Bedroom Duplex Pool Suite 181 sq.m., Garden Pool Villa 77 sq.m. and Garden Deluxe Balcony 45 sq.m. We also design a variety of objects composed in the room, such as the lamps, the beach bag, rattan shelves and knotted ropes for pulling the drawers. Each room tends to have a private backyard that our guests can relax on a huge circular outdoor bed next to their private swimming pool. The architecture of the interior in Sala Chaweng is the extension of the exterior and vice versa.v

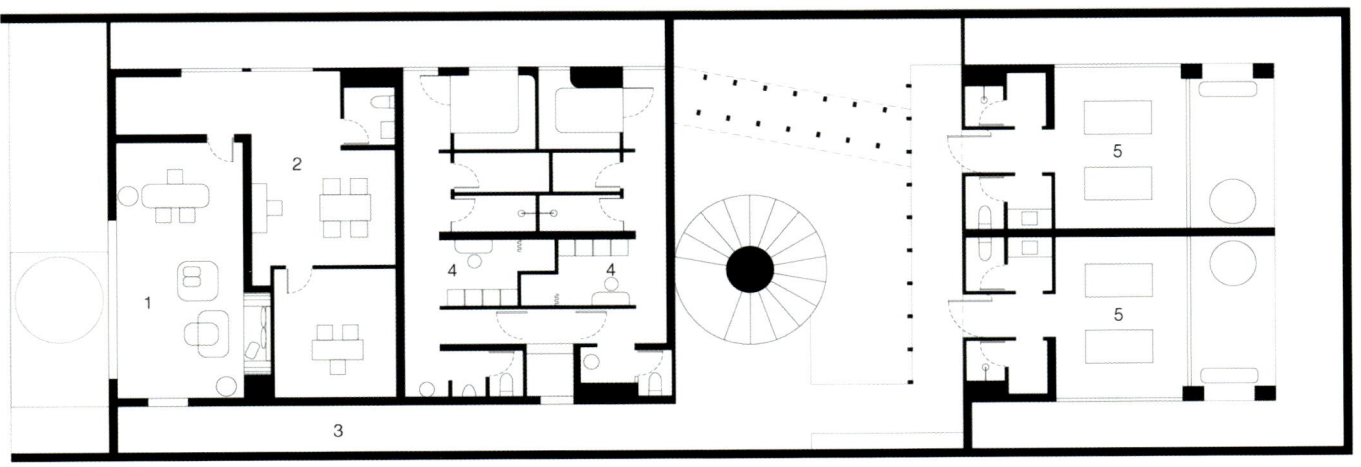

Floor Plan   1 LOBBY   2 OFFICE   3 CORRIDOR   4 WC   5 TREATMENT ROOM

367

Grand Palais Cinema

Architects: Antonio Virga Architecte
Location: Cahors, France
Area: 3653 m²
Photographs: Luc Boegly, Pierre Lasvenes

Manufacturers: Sculptform, Fastmount®, Parklex International S.L., Kalwall®
Main Material: Brick + Steel & Metal + Concrete + Wood + Glass

Located on the north side of the historic center of the town of Cahors and a few steps away from the banks of the Lot River, this cinema stands on a former site dedicated to the army (today renamed Place Bessières). The project offered the opportunity to recreate and reinterpret the symmetry of the preexisting army barracks by occupying the area of the east wing of this complex, destroyed by fire in 1943. Previously serving as a parking lot, the Place Bessières has been transformed into a broad and welcoming urban space dedicated to pedestrians and protected by an existing canopy of trees. The square is mostly paved in brick but benefits from a densely green area at the center called "the oasis".

The Museum of the Resistance, previously housed in a building on this site which was demolished to make way for this project, will be located on the building's top level with an entrance clearly separate from the cinema. The buildings and adjacent outdoor areas here are organized according to a rigorous, harmonious, and level layout, in keeping with the practices governing the 19th-century military and public facilities. To fulfill the aim of restoring their former scale, this group of spaces is treated with simplicity and unity of materials, furniture, and greenery.

The monolithic volume reveals awe-inspiring façades, but on the upper floors, it has been surrounded by a machrabiya composed of little perforations that lighten the façade and intrigue from a distance, attracting viewers toward its environment. The intricate alternation of solids and voids serves a functional purpose. The perforated skin enlivens the interior spaces during the day thanks to the penetrating light and the interplay of light and shadow whereas, at night, the façade creates a screen of tiny shimmering lights.

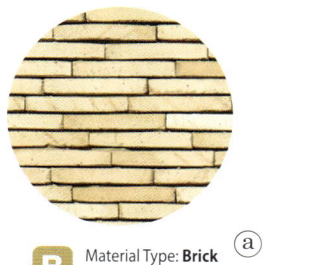

B Material Type: **Brick** (a)
(Monolithic volume brick)

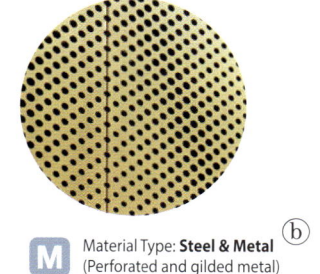

M Material Type: **Steel & Metal** (b)
(Perforated and gilded metal)

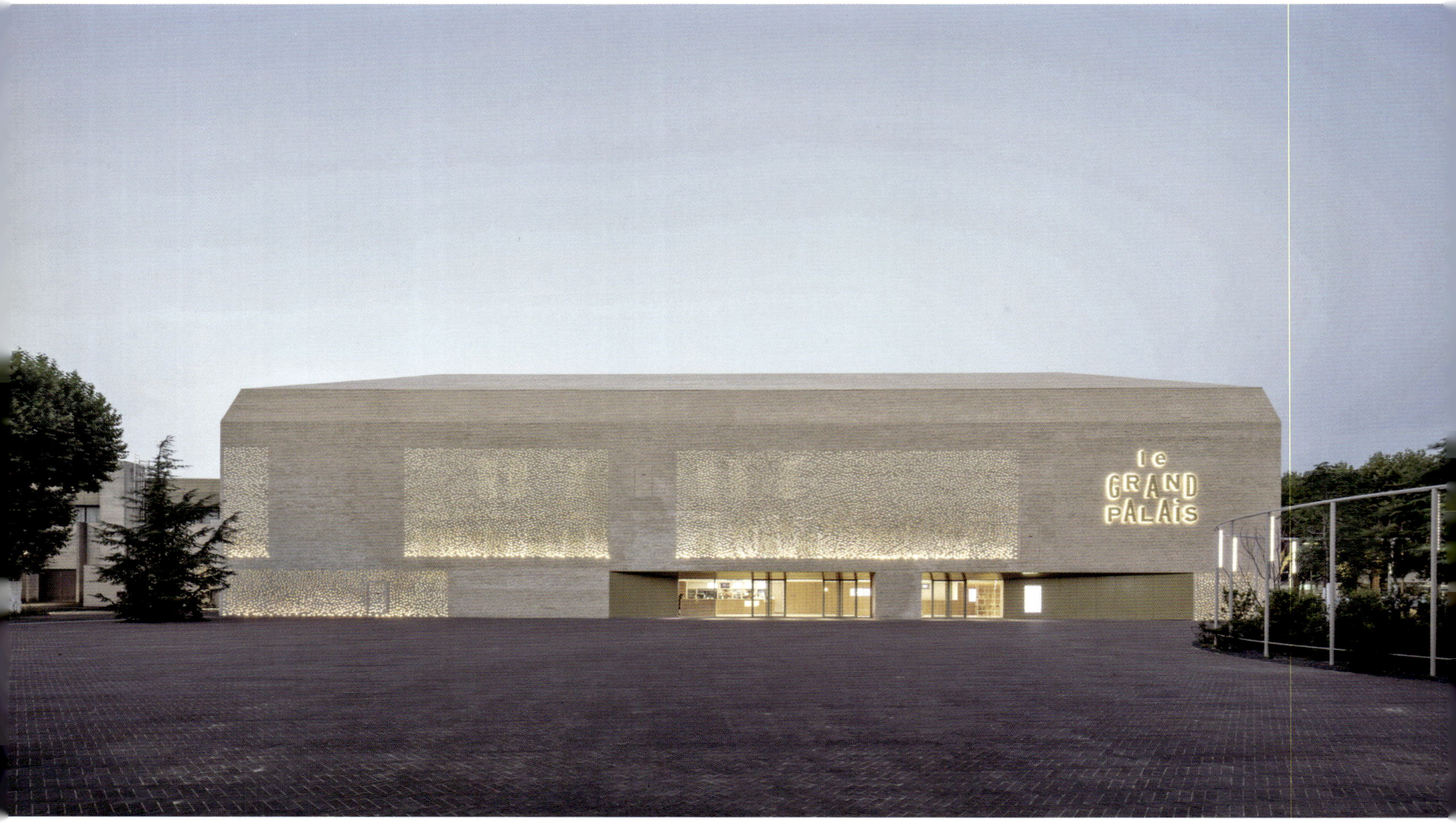
le
GRAND
PALAiS

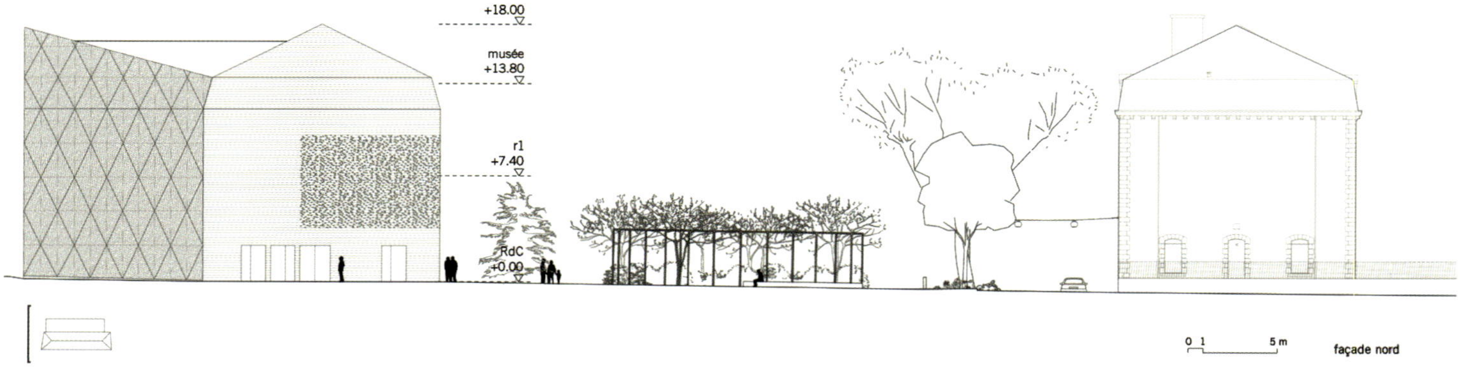
+18.00
musée
+13.80
r1
+7.40
RdC
+0.00
0 1 5 m
façade nord

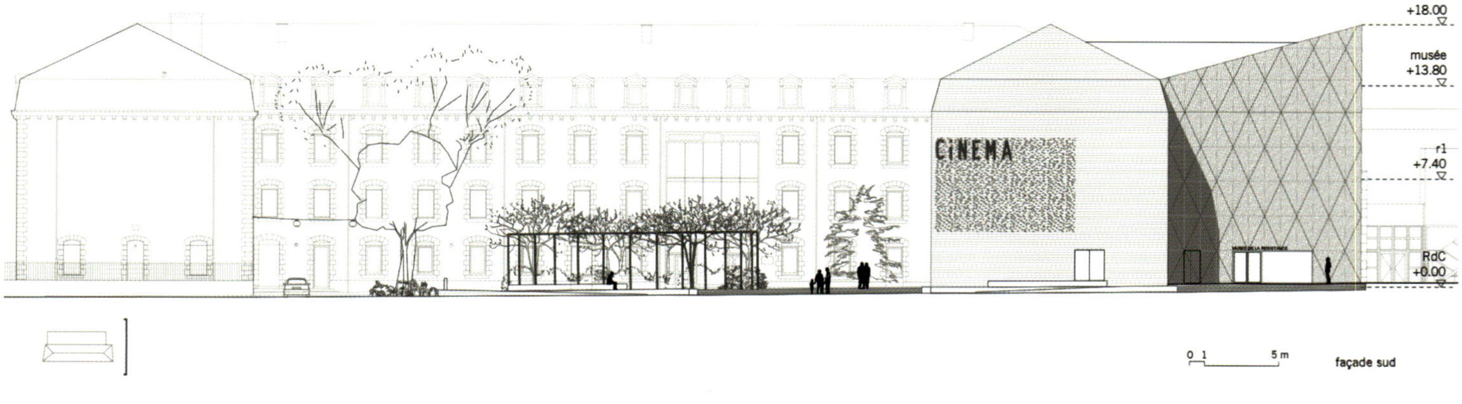
+18.00
musée
+13.80
CiNEMA
r1
+7.40
RdC
+0.00
0 1 5 m
façade sud

façade ouest

façade est

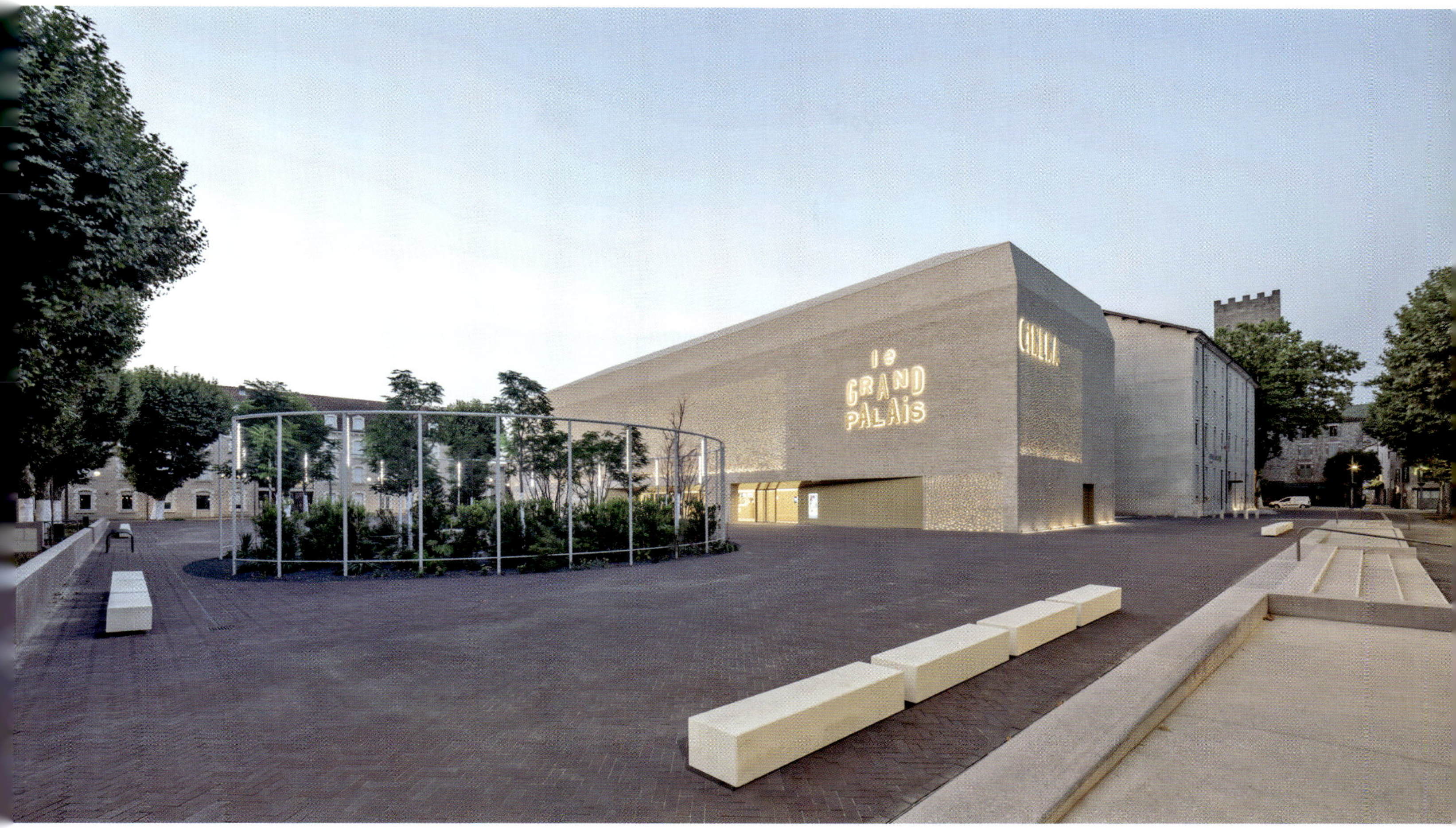

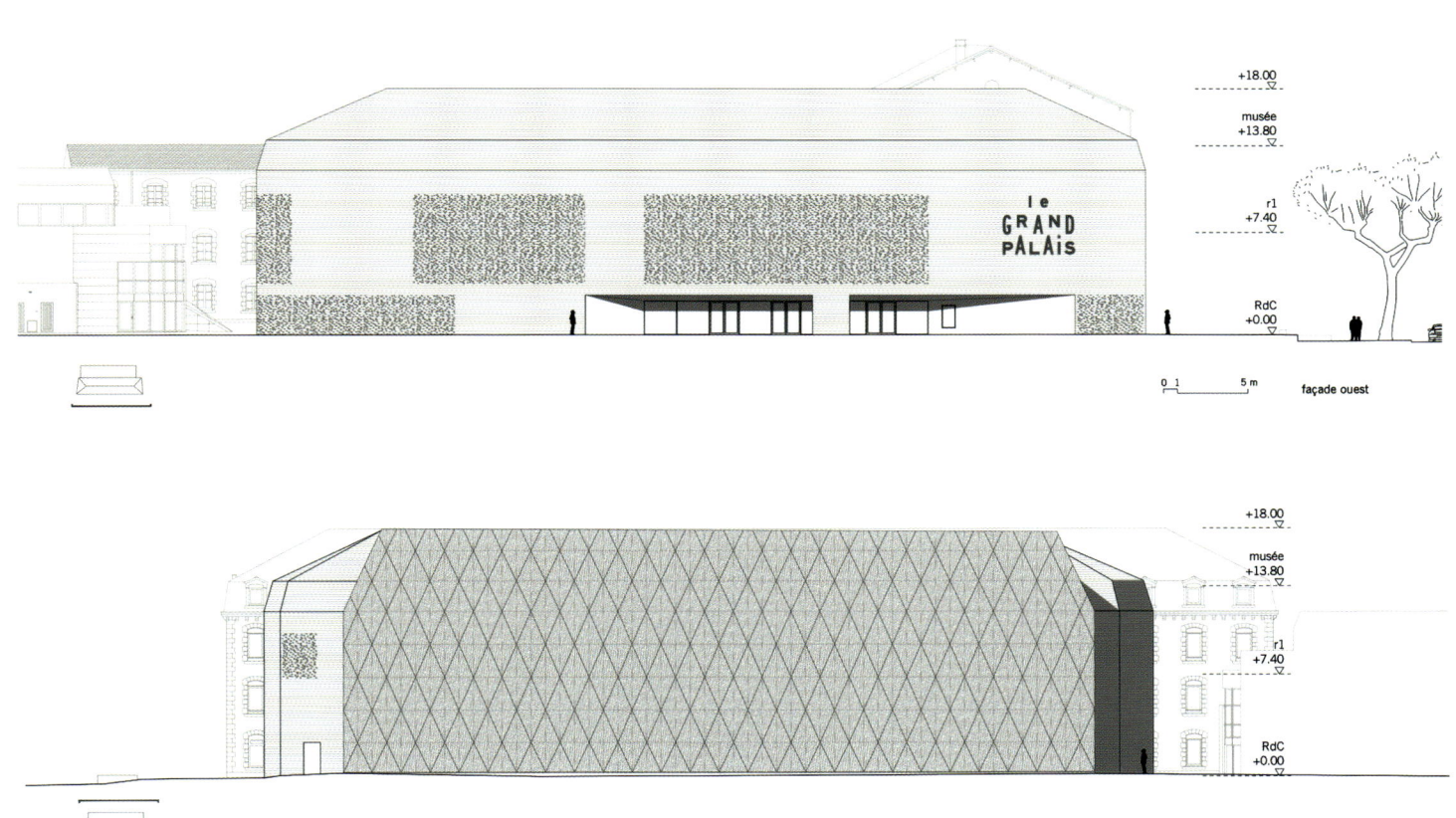

The architect's search for a powerful contemporary aesthetic seeks to carry the architecture of the cinema beyond the simple objective of recreating the morphology of the former barracks. The building is divided into two distinct, methodically created, and visually highlighted volumes: one built of brick and the other of perforated and gilded metal, each one playing a very precise role in relation to the public space. The brick volume mirrors the two buildings of the former barracks and is imagined as a contemporary and identifiable reinterpretation of these existing structures. It is the most striking and visible element on the square, owing to the direct reference to the town's history. Brick was chosen with the aim of enhancing the collective memory of the citizens of Cahors while avoiding any hint of pastiche.

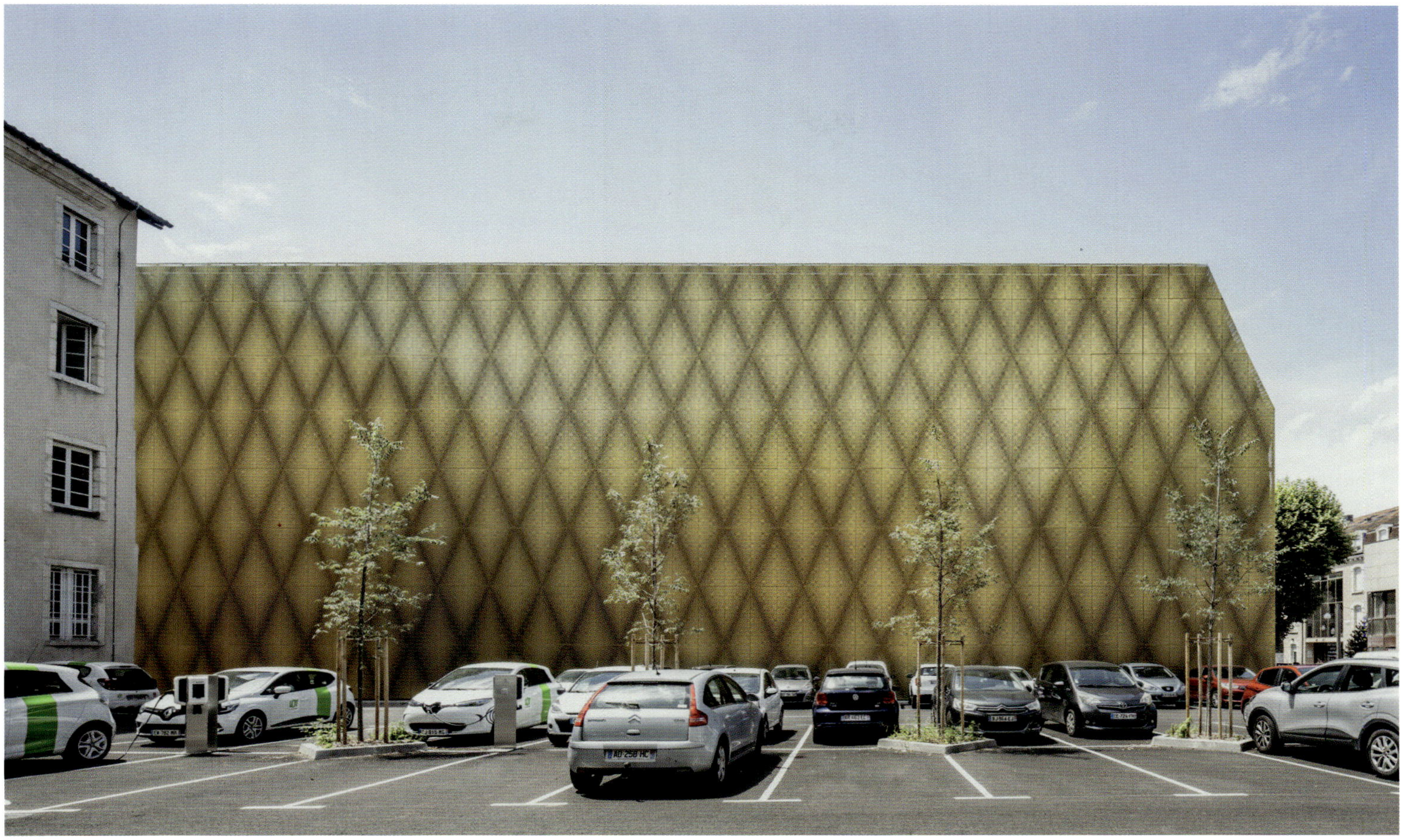

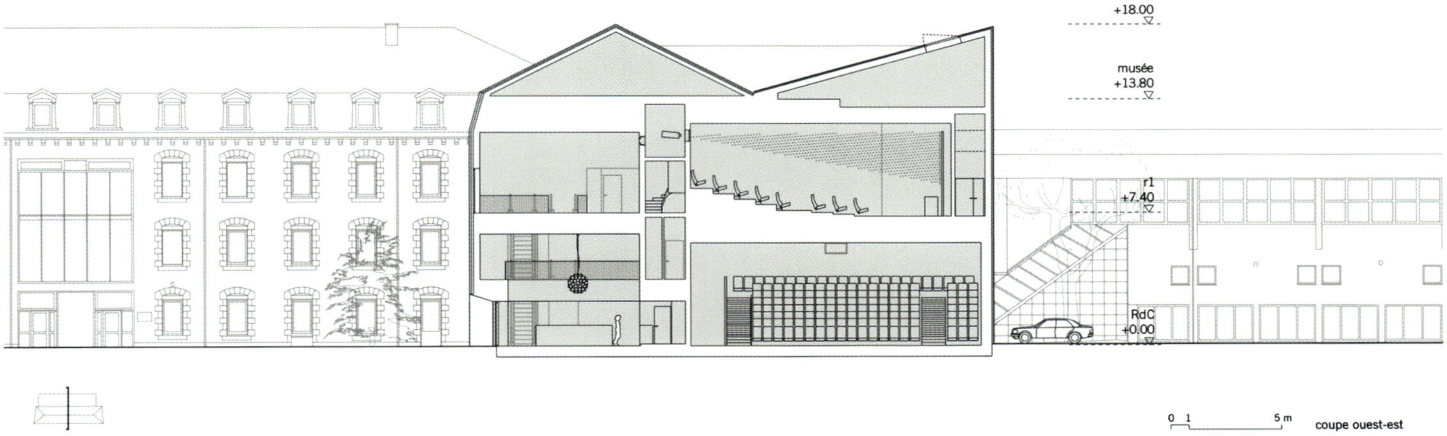

+18.00
musée
+13.80
r1
+7.40
RdC
+0.00
0 1 5 m
coupe ouest-est

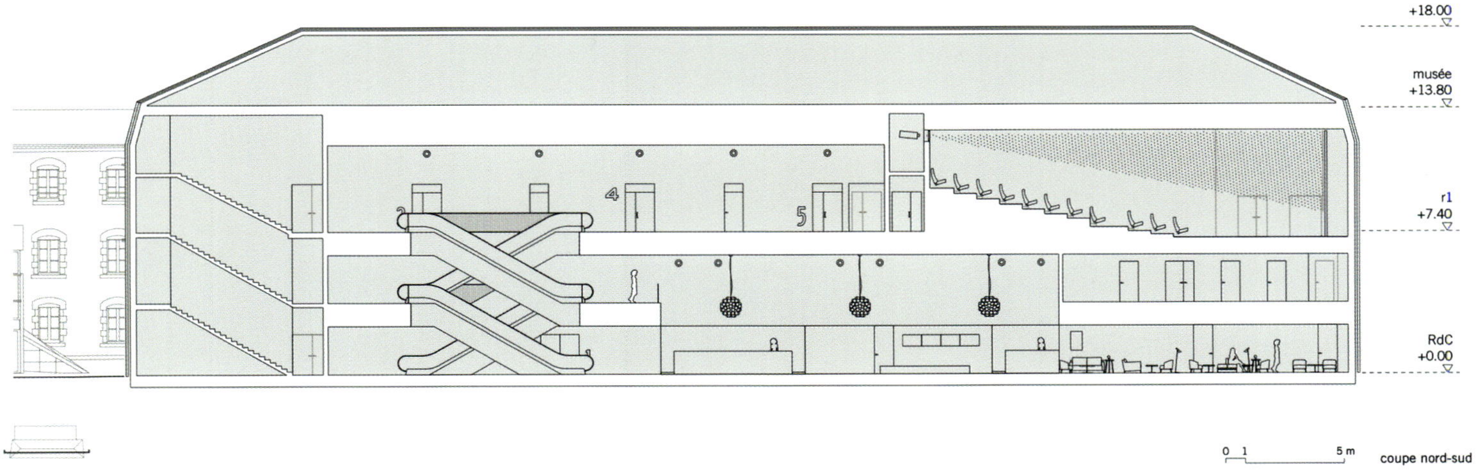

+18.00
musée
+13.80
r1
+7.40
RdC
+0.00
0 1 5 m
coupe nord-sud

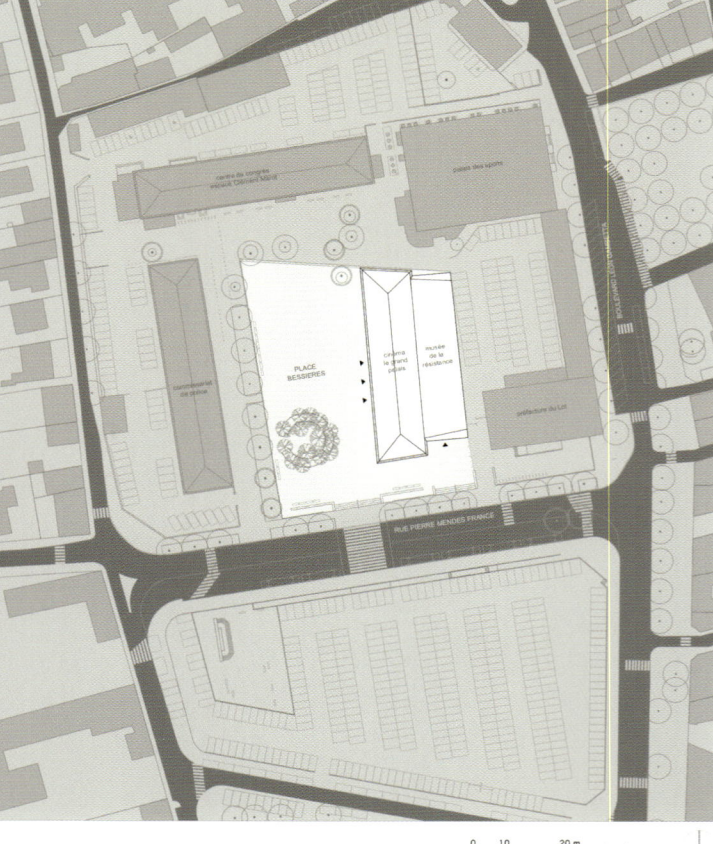

plan de masse

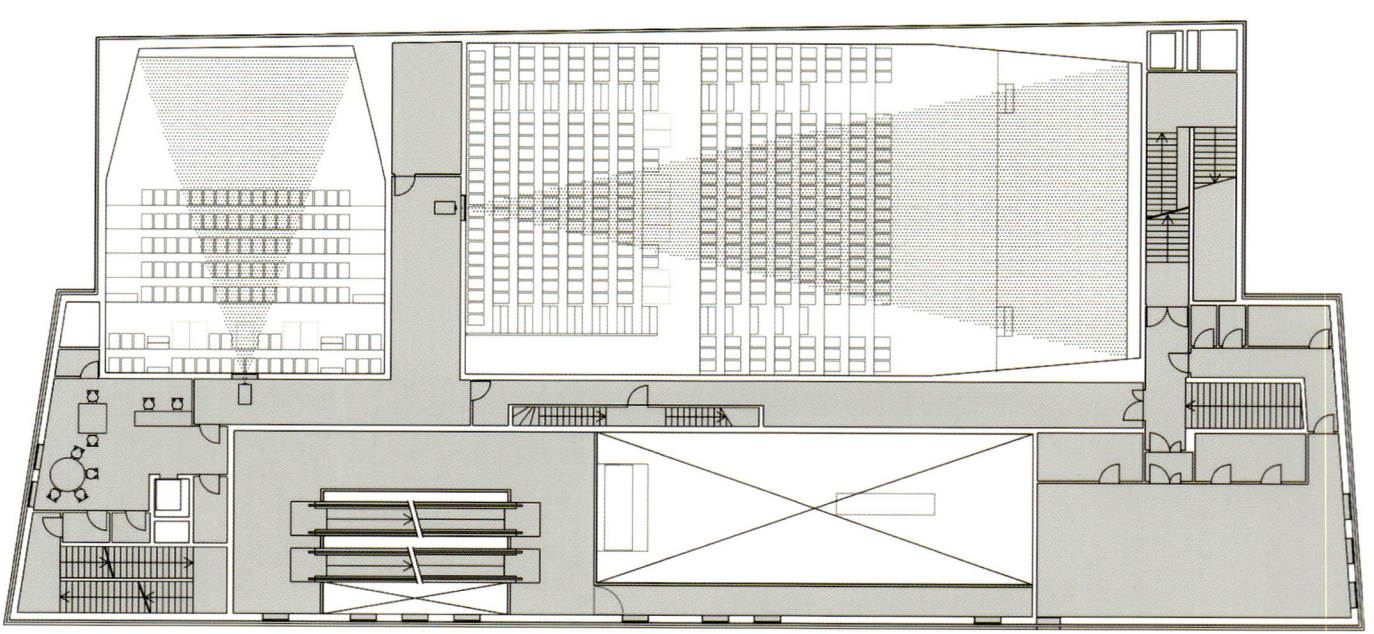

0 1   5 m   rdc haut

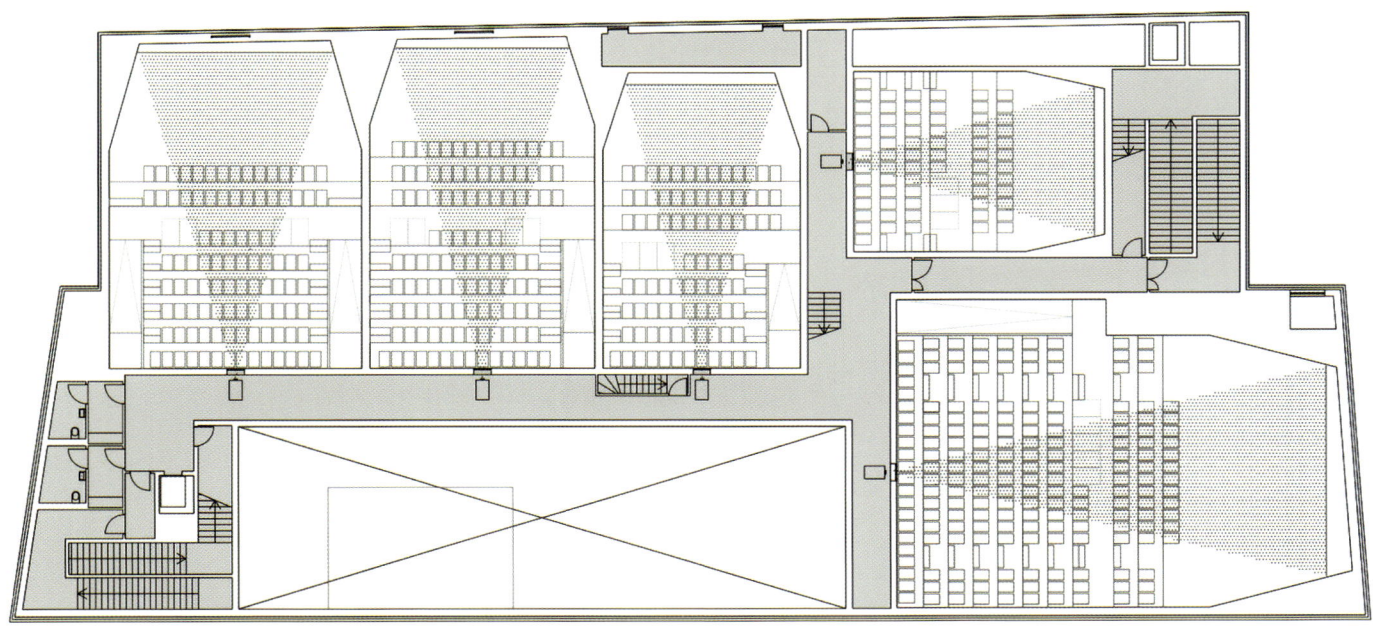

0 1 5 m r1 haut

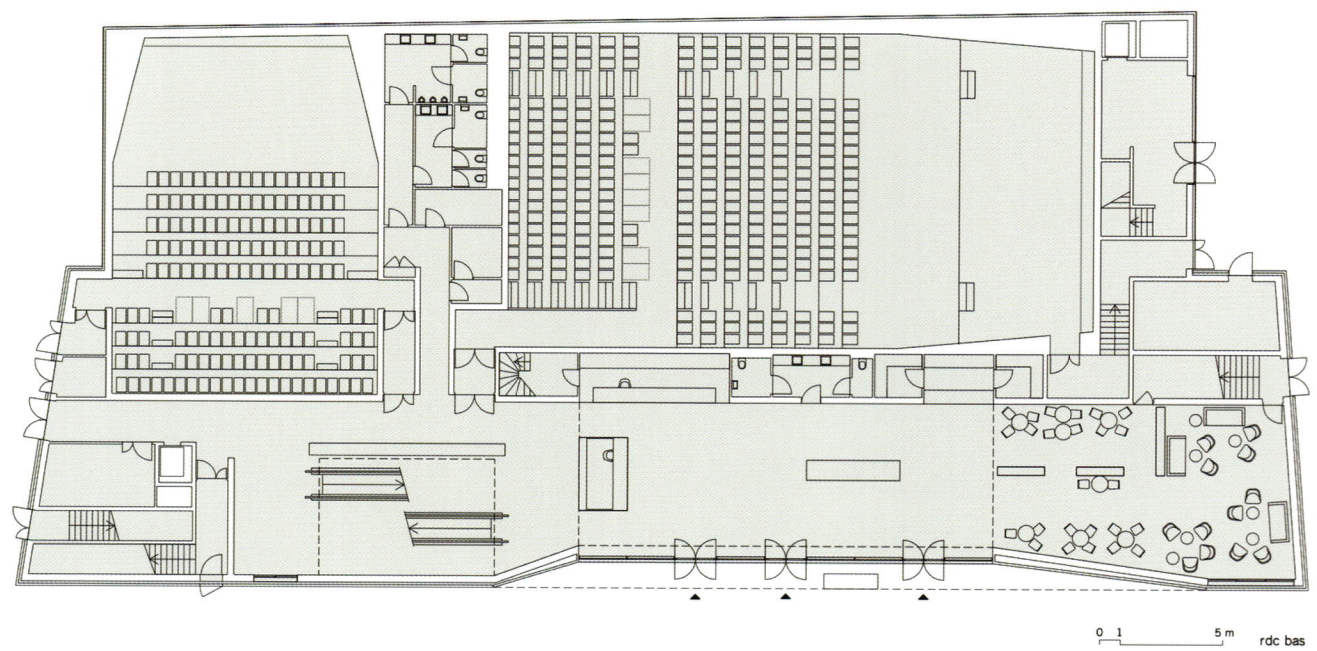

rdc bas

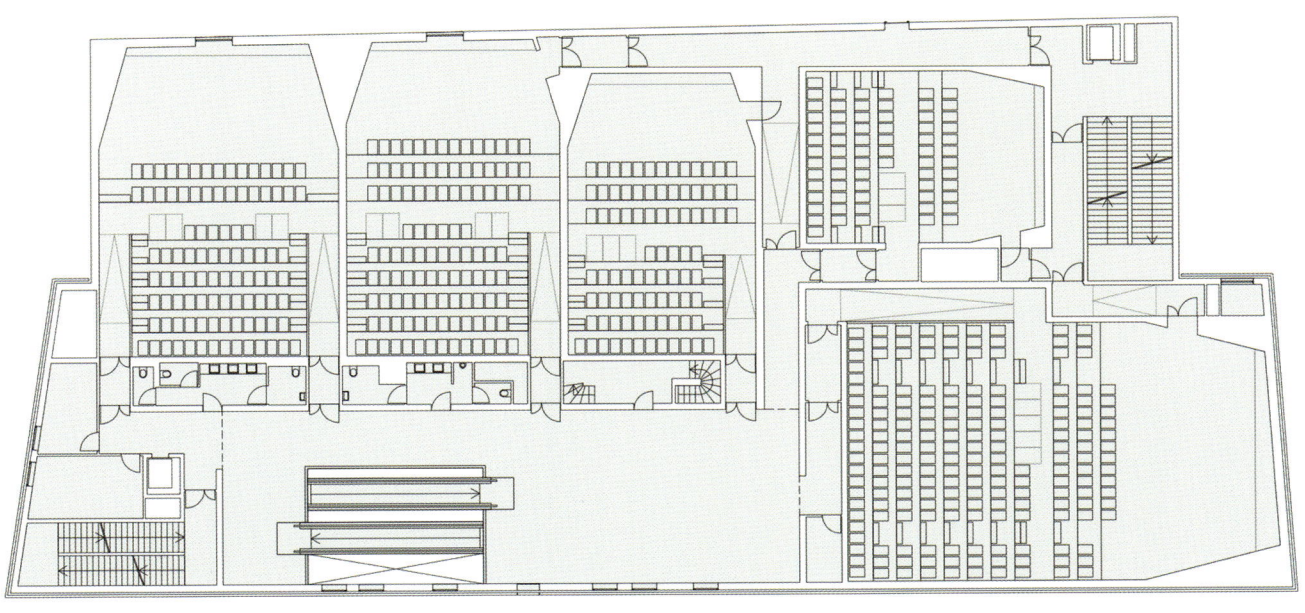

Cima Morelos Complex

Architects: Minuz Workshop + Miguel Montor Architecture Workshop
Location: Leon, Mexico
Area: 468 m²
Photographs: Onnis Luque
Manufacturers: Novaceramic, URREA
Main Material: Brick + Concrete + Glass

Located in the northern part of the city of León, this project is located on one of the geographically highest and fastest growing real estate parts of the country. From the beginning we seek to respond to create a semi-vertical housing pilot, thinking of giving greater benefits to its inhabitants, taking advantage of the property and generating play areas and gardens, working on the idea of a small neighborhood endowed with a contemporary air and a desire strong to convey the architectural materiality. This idea arises from observing that, not only in this region but in the entire country, there are few worthy and honest proposals for multi-family housing. At least when it comes to small real estate developments, there are always spatial and experience sacrifices, not to mention the lack of areas where you can have a richer neighborhood life.

The project seeks to give its inhabitants a host of routes where coexistence and encounter are encouraged, we seek at all times to have paths full of surprises sheltered by specific acoustics and lighting through a material palette and a highly studied volumetry. Thus, the project consists of 16 apartments, distributed in 4 blocks of 4 levels that are woven through their circulations, both horizontal and vertical, generating transits with finishes to their natural environment. With this we achieved that each apartment was completely separated from its neighbor in all directions, achieving better acoustic intimacy. One of the main premises of the development was to respect a traditional tectonics, this with two purposes, the first that was very practical and economical when building, -

and the second most important for us was that it be anchored to the environment and that in some way it did not scream , in addition to looking for materials that through aging were giving personality to the whole, and have a lower maintenance cost. This is how we decided to work masonry walls based on a ceramic hollow block, by means of which we could hide the vertical reinforcements, freeing the work from falsework. Likewise, a joist and vault system was selected for the slabs. The decision was made to make the construction system apparent at all times, since honesty and clarity in the design is really important to us, making it clear how the elements behave.

We place wooden panels on the paths, sheltering these areas and generating less reverberation in addition to giving warmth when walking. Likewise, within the departments in larger-scale areas such as the dining room, the same material was chosen in their panels, with this we managed to give it better acoustics and an ideal climate. The finishing of the walls was worked through an earth bath, a mixture of earth from the same site was made with some sealants, in order to generate a waterproofing system with natural pigmentation. This allowed us to give it an ideal color scheme and play with the environment.

B Material Type: **Brick** (a)
(Ceramic hollow block)

Gravel, crushed and some precast make up most of the exterior soil treatments allowing to leave the greatest amount of permeable soil, in addition to a very simple and endemic landscape project. The exercise resulted in an example of timelessness, which we look for in all our projects.

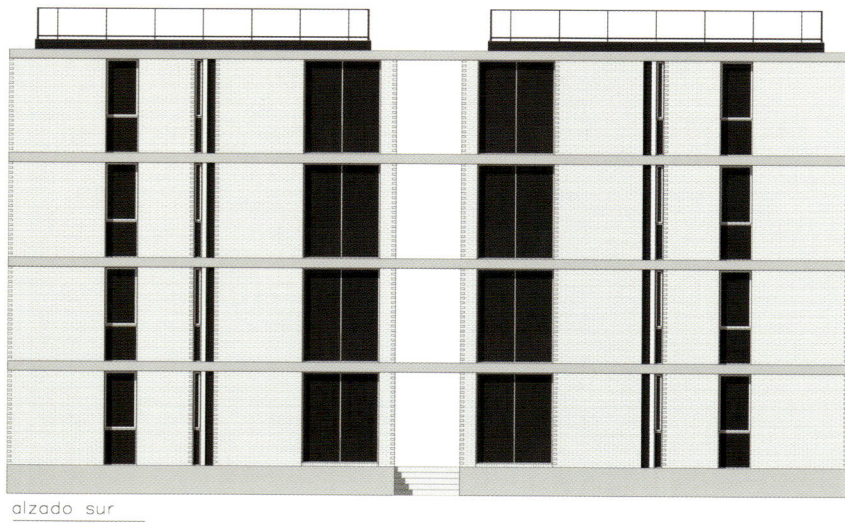

alzado sur
e s c. 1 : 5 0
proyecto   arquitectónico

alzado poniente
e s c. 1 : 5 0
proyecto   arquitectónico

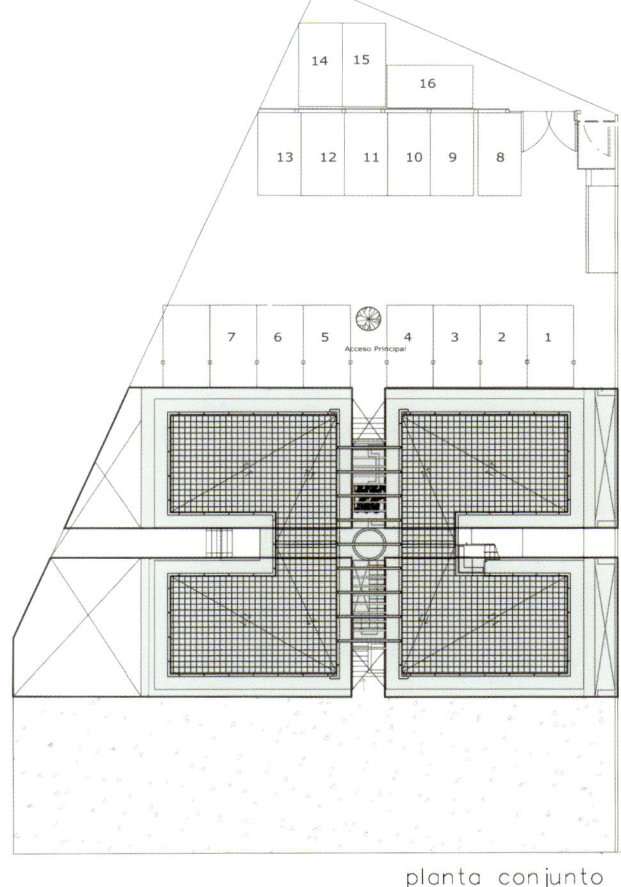

planta conjunto
esc. 1:100
proyecto arquitectónico

383

PLUGIN
Publishing company

Architecture Material Sourcebook

Copyright by Plugin Publishing Company

All rights reserved. No part of this book may be reproduced in any form without written permission of the copyright owners. All images in this book have been reproduced with the knowledge and prior consent of the artists concerned and no responsibility is accepted by producer publisher or printer for any infringement of copyright or otherwise, Arising from the contents of the publication.

- Publication: Plugin Publishing Company
- Publisher: Sukam, Im
- Writer: Sukjwa, Park
- Printing: Abcds Co., Ltd.
- Address: 502 / 9, Hugok-ro, Ilsan 3(Sam)-dong, Ilsanseo-gu, Goyang-si, Gyeonggi-do, South Korea.
- Tel(Fax): +82-31-907-1772 (+82-31-907-1773)
- E-mail: 2twodesign@naver.com

₩97,000WON